大数据与人工智能技术丛书

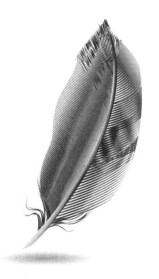

大数据技术与应用
微课视频版

◎ 肖政宏 李俊杰 谢志明 编著

清华大学出版社
北京

内 容 简 介

本书共分12章,分为基础篇、核心篇和应用篇。基础篇包括大数据概论、大数据集群系统基础、Hadoop分布式系统、HDFS分布式文件系统、MapReduce分布式计算、HBase分布式数据库应用;核心篇包括YARN资源分配、Spark集群计算、Spark机器学习、Hive数据仓库应用、ZooKeeper协调服务;应用篇包括医药大数据案例分析。

本书可以作为高等院校大数据技术相关课程本科生教材,也可以作为大数据技术基础相关课程研究生教材,还可以作为从事大数据相关工作的工程技术人员的参考用书。

本书封面贴有清华大学出版社防伪标签,无标签者不得销售。
版权所有,侵权必究。举报: 010-62782989, beiqinquan@tup.tsinghua.edu.cn。

图书在版编目(CIP)数据

大数据技术与应用:微课视频版/肖政宏,李俊杰,谢志明编著. —北京:清华大学出版社,2020.5(2025.1重印)
(大数据与人工智能技术丛书)
ISBN 978-7-302-53843-1

Ⅰ. ①大… Ⅱ. ①肖… ②李… ③谢… Ⅲ. ①数据处理 Ⅳ. ①TP274

中国版本图书馆CIP数据核字(2019)第209027号

策划编辑:魏江江
责任编辑:王冰飞
封面设计:刘　键
责任校对:梁　毅
责任印制:沈　露

出版发行:清华大学出版社
　　　　网　址:https://www.tup.com.cn,https://www.wqxuetang.com
　　　　地　址:北京清华大学学研大厦A座　　邮　编:100084
　　　　社 总 机:010-83470000　　　　　　　邮　购:010-62786544
　　　　投稿与读者服务:010-62776969, c-service@tup.tsinghua.edu.cn
　　　　质量反馈:010-62772015, zhiliang@tup.tsinghua.edu.cn
　　　　课件下载:https://www.tup.com.cn,010-83470236
印 装 者:涿州市般润文化传播有限公司
经　　销:全国新华书店
开　　本:185mm×260mm　　印　张:21.5　　字　数:522千字
版　　次:2020年6月第1版　　　　　　　印　次:2025年1月第7次印刷
印　　数:10501～11300
定　　价:49.80元

产品编号:082527-01

前　言

编写目的

随着云计算、大数据、物联网、人工智能等信息技术的迅猛发展,大数据在电子商务、媒体营销、旅游、物流交通、农业、工业、企业服务、娱乐、汽车、物联网、生命科技、金融科技、房产、教育及政府等诸多行业得到了广泛的应用,大数据的相关课程也逐渐成为各个高等学校数据科学与大数据技术等专业的核心课程以及计算机相关硕士专业的必修课程。

平台支撑

大数据分布式系统的学习开发需要有实验平台,而一般大数据的实验平台的建设需要较多经费支持,同时一些基于这些平台的学习、训练也不是很方便。本书基于普通的PC,充分利用Linux操作系统、VMware虚拟软件的特点,通过虚拟多台计算机组建分布式计算机系统,搭建Hadoop大数据分析平台,非常适合读者从底层学懂弄通大数据的搭建过程,以及分布式文件系统、分布式计算框架、分布式数据库、Spark内存计算、分布式机器学习及大数据的分析系统的开发和应用。

本书内容

本书共分12章,分为基础篇、核心篇和应用篇。

基础篇包括第1~6章。第1章大数据概论,涉及的内容有大数据定义,大数据分析过程、技术与工具,以及大数据的应用;第2章大数据集群系统基础,讲解Linux操作系统、虚拟化技术和大数据集群的搭建;第3章Hadoop分布式系统,讲解Hadoop的原理和运行机制,以及Hadoop系统的配置与安装;第4章HDFS分布式文件系统,主要讲解大数据文件系统的读写过程和HDFS的操作;第5章MapReduce分布式计算,讲解MapReduce的架构、原理与机制,以及MapReduce应用案例;第6章HBase分布式数据库应用,主要讲解HBase的架构、部署和应用。

核心篇包括第7~11章。第7章YARN资源分配,讲解YARN架构、流程及操作应用;第8章Spark集群计算,主要讲解Spark架构、RDD、部署和应用;第9章Spark机器学习,讲解机器学习库和相关应用;第10章Hive数据仓库应用,讲解Hive的组成、安装、配置和应用;第11章ZooKeeper协调服务,讲解分布式应用程序的定义、部署、命令和应用。

应用篇包括第12章。第12章医药大数据案例分析,主要内容包括大数据系统的需求分析、架构设计、关键技术、存储设计、数据分析与数据展示。

本书特点

本书的作者团队具有多年大数据教材的编写经验,同时具有丰富的高校教学和大数据的培训经验,也具有实际的大数据项目开发经验。作者团队在"云计算大数据与智能制造论坛"、国家级、省级职业教育培训、研究生课程教学,以及本科生课程教学等方面进行了多次大数据技术的教学实践,本书也是这些系列教学的成果之一。

本书的主要特点如下。

1. 组织结构高效合理

作为大数据技术方面的教材,其内容全面、逐步递进,完整地呈现了一个大数据分析系统所涉及的各类技术。

2. 适合"线上线下"混合式教学模式

本书的每章首先描述的是基本原理,然后是操作实践,最后是课后作业,方便教师在原理部分讲授时利用一些公共的教学平台,把教学资源在讲课之前发送给学生,课堂讲授时突出重点、难点及实现部分,课后布置作业。

3. 基于项目的案例教学

为方便读者对大数据的相关系统进行开发学习,第12章通过项目概述、功能需求分析、软件关键技术、效果展示、系统架构设计、数据存储设计、数据分析及数据展示来呈现一个实际的大数据分析系统,让读者真正实现边学习、边训练、边实践。

读者对象

本书可以作为高等院校数据科学与大数据技术相关课程本科生教材,也可以作为大数据技术基础相关课程研究生教材,还可以作为从事大数据相关工作的工程技术人员的参考用书。

资源下载

本书提供教学大纲、教学课件、电子教案、习题答案、全部实例的源代码和教学进度表,扫描封底的课件二维码可以下载。本书还提供600分钟的视频讲解,扫描书中相应位置的二维码可以在线观看、学习。

致谢

本书由肖政宏、李俊杰、谢志明编写,编写过程中得到广东技术师范大学、汕尾职业技术学院、汕尾市创新工业设计研究院、广州市乐商软件科技有限公司、广州五舟科技股份有限公司、北京普开数据技术有限公司的大力协助,感谢梅阳阳、闫艺婷、吴进、周健烨、黄镇生、曾静、徐胜东等的全力支持。

编者关于大数据分析技术的研究及本书写作还得到了广东省省级科技计划项目——

基于医药电商大数据的服务系统研发(No：2016A010101029)、广州市科技计划项目——大数据分析平台的关键技术研究及应用示范(No：201607010152)的资助,在此表示感谢。

大数据技术发展很快,涉及的内容也较多,加上编者的水平有限,在内容的安排、表述方面难免有不当之处,希望广大读者在阅读本书的过程中能够批评指正。

编 者

2020 年 1 月

目 录

配套资源下载

基 础 篇

第 1 章　大数据概论 3
　1.1　大数据概述 3
　　1.1.1　大数据的定义 4
　　1.1.2　大数据的特征 4
　1.2　大数据的分析过程 7
　　1.2.1　大数据的采集 7
　　1.2.2　大数据的存储方式 8
　　1.2.3　大数据分析技术 9
　　1.2.4　大数据的展示及应用 9
　1.3　大数据的价值、挑战与风险 10
　　1.3.1　商业价值 10
　　1.3.2　社会生活价值 12
　　1.3.3　大数据的挑战与风险 12
　1.4　大数据的应用 13
　1.5　大数据的处理流程 17
　1.6　大数据成为人工智能产业的燃料 18
　1.7　大数据技术的发展前景 18
　小结 18
　习题 19

第 2 章　大数据集群系统基础 20
　2.1　大数据集群系统概述 20
　　2.1.1　集群的分类 21
　　2.1.2　集群的目的 21
　2.2　Linux 操作系统 22
　　2.2.1　Linux 操作系统简介 22
　　2.2.2　Linux 操作系统的特性 22

2.2.3　Linux 安装与基础操作 …………………………………… 22
　　2.2.4　Linux 常用命令 ………………………………………… 23
2.3　虚拟化技术 ……………………………………………………… 26
　　2.3.1　虚拟化技术简介 ………………………………………… 26
　　2.3.2　虚拟技术的原理 ………………………………………… 27
　　2.3.3　常见的虚拟化软件 ……………………………………… 28
　　2.3.4　虚拟化技术的优势和劣势 ……………………………… 29
2.4　CentOS 大数据集群系统的组成 ……………………………… 30
2.5　大数据集群技术的架构 ………………………………………… 31
2.6　操作实践：大数据集群的部署 ………………………………… 32
　　2.6.1　集群规划 ………………………………………………… 32
　　2.6.2　网络配置 ………………………………………………… 32
　　2.6.3　安全配置 ………………………………………………… 33
　　2.6.4　时间同步 ………………………………………………… 35
　　2.6.5　SSH 登录 ………………………………………………… 37
小结 …………………………………………………………………… 39
习题 …………………………………………………………………… 39

第 3 章　Hadoop 分布式系统 …………………………………………… 40

3.1　Hadoop 概述 …………………………………………………… 40
　　3.1.1　Hadoop 简介 …………………………………………… 40
　　3.1.2　Hadoop 的发展历程 …………………………………… 41
　　3.1.3　Hadoop 原理及运行机制 ……………………………… 42
3.2　Hadoop 相关技术及生态系统 ………………………………… 44
3.3　操作实践：Hadoop 安装与配置 ……………………………… 46
　　3.3.1　安装 JDK ………………………………………………… 46
　　3.3.2　安装 Hadoop …………………………………………… 49
　　3.3.3　配置 Hadoop …………………………………………… 49
　　3.3.4　格式化 …………………………………………………… 54
　　3.3.5　运行 Hadoop …………………………………………… 54
小结 …………………………………………………………………… 56
习题 …………………………………………………………………… 56

第 4 章　HDFS 分布式文件系统 ………………………………………… 57

4.1　HDFS …………………………………………………………… 57
　　4.1.1　设计前提和设计目标 …………………………………… 58
　　4.1.2　Namenode 和 Datanode ……………………………… 58
　　4.1.3　文件系统的名字空间 …………………………………… 59
　　4.1.4　数据复制 ………………………………………………… 61
　　4.1.5　HDFS 读流程 …………………………………………… 64

  4.1.6 HDFS 写流程 ………………………………………………………… 65
4.2 HDFS 操作实践 …………………………………………………………………… 66
  4.2.1 HDFS Shell ………………………………………………………… 66
  4.2.2 HDFS Java API …………………………………………………… 68
  4.2.3 Eclipse 开发环境 ………………………………………………… 68
  4.2.4 综合实例 …………………………………………………………… 71
小结 …………………………………………………………………………………………… 78
习题 …………………………………………………………………………………………… 78

第 5 章　MapReduce 分布式计算 ………………………………………………………… 79

5.1 MapReduce 简介 ………………………………………………………………… 79
  5.1.1 MapReduce 架构 ………………………………………………… 79
  5.1.2 MapReduce 的原理 ……………………………………………… 81
  5.1.3 MapReduce 的工作机制 ………………………………………… 83
5.2 MapReduce 操作实践 …………………………………………………………… 86
  5.2.1 MapReduce WordCount 编程实例 …………………………… 86
  5.2.2 MapReduce 倒排索引编程实例 ………………………………… 89
小结 …………………………………………………………………………………………… 92
习题 …………………………………………………………………………………………… 92

第 6 章　HBase 分布式数据库应用 ……………………………………………………… 93

6.1 HBase 简介 ………………………………………………………………………… 93
  6.1.1 HBase 架构 ………………………………………………………… 93
  6.1.2 HBase 的存储 ……………………………………………………… 98
6.2 HBase 集群部署 …………………………………………………………………… 99
  6.2.1 HBase 参数配置 ………………………………………………… 100
  6.2.2 HBase 运行与测试 ……………………………………………… 101
6.3 HBase Shell 操作命令 ………………………………………………………… 103
  6.3.1 general 操作 ……………………………………………………… 105
  6.3.2 namespace 操作 ………………………………………………… 105
  6.3.3 DDL 操作 ………………………………………………………… 106
  6.3.4 DML 操作 ………………………………………………………… 108
  6.3.5 授权 ………………………………………………………………… 111
6.4 HBase 过滤器 …………………………………………………………………… 112
6.5 HBase 编程 ……………………………………………………………………… 116
  6.5.1 HBase 表操作编程 ……………………………………………… 117
  6.5.2 HBase 过滤查询编程 …………………………………………… 121
小结 ………………………………………………………………………………………… 122
习题 ………………………………………………………………………………………… 123

核 心 篇

第 7 章 YARN 资源分配 ……………………………………………………………… 127
- 7.1 统一资源管理和调度平台引例 ………………………………………………… 127
  - 7.1.1 背景 ……………………………………………………………………… 128
  - 7.1.2 特点 ……………………………………………………………………… 128
  - 7.1.3 典型的统一资源调度平台 ……………………………………………… 129
- 7.2 YARN 简介 ……………………………………………………………………… 130
  - 7.2.1 YARN 架构 ……………………………………………………………… 130
  - 7.2.2 YARN 的工作流程 ……………………………………………………… 132
  - 7.2.3 YARN 的优势 …………………………………………………………… 134
- 7.3 操作实践：YARN Shell 实例 …………………………………………………… 134
- 小结 …………………………………………………………………………………… 136
- 习题 …………………………………………………………………………………… 136

第 8 章 Spark 集群计算 ……………………………………………………………… 137
- 8.1 Spark 简介 ……………………………………………………………………… 137
  - 8.1.1 Spark 生态系统 ………………………………………………………… 138
  - 8.1.2 Spark 架构 ……………………………………………………………… 139
- 8.2 Spark RDD ……………………………………………………………………… 140
  - 8.2.1 RDD 的依赖关系 ……………………………………………………… 140
  - 8.2.2 作业调度 ………………………………………………………………… 141
  - 8.2.3 内存管理 ………………………………………………………………… 142
  - 8.2.4 检查点支持 ……………………………………………………………… 142
- 8.3 Spark 集群部署及应用案例 …………………………………………………… 143
  - 8.3.1 Spark 参数配置 ………………………………………………………… 143
  - 8.3.2 Spark 集群运行 ………………………………………………………… 144
  - 8.3.3 Spark 交互 ……………………………………………………………… 144
  - 8.3.4 Spark 算子 ……………………………………………………………… 146
  - 8.3.5 Spark 算法实例 1：词频统计 ………………………………………… 167
  - 8.3.6 Spark 算法实例 2：相关系数 ………………………………………… 170
- 小结 …………………………………………………………………………………… 173
- 习题 …………………………………………………………………………………… 173

第 9 章 Spark 机器学习 ……………………………………………………………… 174
- 9.1 机器学习概述 …………………………………………………………………… 174
  - 9.1.1 机器学习的发展史 ……………………………………………………… 175
  - 9.1.2 机器学习步骤 …………………………………………………………… 175
- 9.2 Spark MLlib 概述 ……………………………………………………………… 176
  - 9.2.1 数据类型 ………………………………………………………………… 176

  9.2.2 基本统计——基于 DataFrame 的 API ·· 184
  9.2.3 基本统计——基于 RDD 的 API ··· 188
 9.3 Spark 实例 ·· 196
  9.3.1 聚类问题 ·· 196
  9.3.2 随机森林 ·· 199
小结 ··· 204
习题 ··· 204

第 10 章　Hive 数据仓库应用 ·· 205

 10.1 Hive 简介 ·· 205
  10.1.1 Hive 组成模块 ·· 205
  10.1.2 Hive 执行流程 ·· 206
  10.1.3 MetaStore 存储模式 ··· 207
 10.2 Hive 安装与配置 ·· 209
  10.2.1 Hive 参数配置 ·· 209
  10.2.2 Hive 运行与测试 ··· 212
  10.2.3 Hive Beeline ·· 213
 10.3 数据类型和文件格式 ·· 214
  10.3.1 数据类型 ·· 215
  10.3.2 文件格式 ·· 216
 10.4 Hive 数据定义与数据操作 ·· 218
  10.4.1 基本概念 ·· 218
  10.4.2 数据库管理 ··· 219
  10.4.3 表的管理 ·· 220
  10.4.4 外部表的管理 ·· 222
  10.4.5 分区管理 ·· 224
  10.4.6 数据操作 ·· 226
  10.4.7 桶的操作 ·· 228
  10.4.8 索引 ··· 229
 10.5 Hive 数据查询 ·· 230
  10.5.1 简单查询 ·· 231
  10.5.2 复杂查询 ·· 232
  10.5.3 JOIN 连接查询 ··· 235
  10.5.4 其他语句 ·· 238
 10.6 Hive 编程 ·· 241
  10.6.1 JDBC 函数 ··· 241
  10.6.2 完整实例 ·· 243
小结 ··· 244
习题 ··· 244

第 11 章 ZooKeeper 协调服务 ... 245
11.1 ZooKeeper 简介 ... 245
11.1.1 ZAB 协议 ... 246
11.1.2 ZooKeeper 数据模型 ... 248
11.1.3 会话 ... 250
11.1.4 事件监听器 ... 251
11.1.5 访问权限 ... 252
11.2 ZooKeeper 集群部署 ... 253
11.3 ZooKeeper 基本命令 ... 254
11.4 ZooKeeper 应用 ... 258
11.4.1 Hadoop ... 258
11.4.2 Spark ... 264
11.4.3 Hive ... 267
11.5 ZooKeeper 编程 ... 269
11.5.1 ZooKeeper 读/写操作 ... 271
11.5.2 集群状态监控 ... 272
小结 ... 277
习题 ... 277

应 用 篇

第 12 章 医药大数据案例分析 ... 281
12.1 项目概述 ... 281
12.2 功能需求 ... 282
12.3 软件关键技术 ... 282
12.4 效果展示 ... 282
12.5 系统构架设计 ... 283
12.5.1 系统组成 ... 283
12.5.2 系统协作方式 ... 283
12.5.3 系统网络拓扑 ... 283
12.5.4 系统建设方案 ... 284
12.6 数据存储设计 ... 286
12.7 数据分析 ... 287
12.8 数据展示 ... 313
小结 ... 328
习题 ... 329

参考文献 ... 330

基 础 篇

第 1 章

大数据概论

1.1 大数据概述

2011年5月全球知名咨询机构麦肯锡全球研究所发布的《大数据：下一个创新、竞争和生产力的前沿》研究报告中首次提及大数据（big data）概念，麦肯锡称："数据已经渗透到当今每个行业和业务职能领域，成为重要的生产因素。人们对于海量数据的挖掘和运用，预示着新一波生产率增长和消费者盈余浪潮的到来。"

随着云计算、大数据、物联网、人工智能等信息技术的迅猛发展，这些技术与人类世界政治、经济、军事、科研、生活等方面不断交叉融合，催生了超越以往任何年代的巨量数据。*Nature* 于 2008 年出版了大数据专刊 *Big Data*，专门讨论了巨量数据对互联网、经济、环境以及生物等各方面的影响与挑战。*Science* 于 2011 年出版了如何应对数据洪流（data deluge）的专刊 *Dealing with Data*。2014 年 3 月，"大数据"首次出现在全国人大会议的《政府工作报告》中。2014 年 5 月，美国白宫发布了 2014 年全球"大数据"白皮书的研究报告《大数据：抓住机遇、守护价值》，鼓励使用大数据以推动社会进步。大数据已成为国家战略布局的重要组成部分，与人工智能技术相结合，将给各行各业带来根本性变革。

然而，大数据也是一把双刃剑。一方面，海量数据的迅速增长为社会发展提供了更多宝贵的数据资源。网络和数据库中所记载的各种巨量数据，是现实生产劳动的真实反映。人们可以利用这些数据去分析问题、解决问题，并且促成新的理论和技术。伴随着数据处理能力的提升，运算与存储成本的井喷，以及越来越多的设备中嵌入各种传感技术，使得数据的收集、存储与分析正处于一个近乎无限上升的趋势。另一方面，大数据前所未有的运算能力也给人们带来挑战，不可控的持续爆炸式增长的大数据正向人们的数据中心基础设施和数据处理及分析的各个环节发起严峻的挑战，也给人们的法律、伦理及社会规范发起了挑战，考验人们能否在大数据的世界中保护隐私和其他价值观。大数据时代的战略意义不仅在于掌握庞大的数据信息,还在于发现和理解信息内容及信息与信息之间的关系。

1.1.1 大数据的定义

自 2012 年以来,"大数据"一词越来越引起人们的关注。但是,到目前为止,在学术研究领域和产业界中,大数据并没有一个标准的定义。

通常来说,大数据是指数据量超过一定大小,无法用常规的软件工具在规定的时间内进行抓取、管理和处理的数据集合。

1.1.2 大数据的特征

从以上对大数据定义的各种宽泛的描述来看,大数据的定义并不可以简而概之。要想更为深入地了解大数据的深层含义,可从大数据的主要特征出发。大数据的主要特征可用"5V+1C"来进行概括,即数据量大(Volume)、数据类型多(Variety)、价值密度低(Value)、数据时效性强(Velocity)、准确性高(Veracity)、复杂性高(Complexity),如图 1-1 所示。

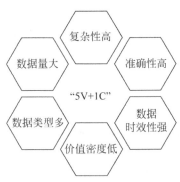

图 1-1 大数据特征图

1. 数据量大

首先来看一组公式:1024MB=1GB,1024GB=1TB,1024TB=1PB,1024PB=1EB,1024EB=1ZB。大数据的起始计量单位最少是 PB 级以上的。根据国际数据公司(IDC)的《数据宇宙》报告显示,2008 年全球数据量为 0.5EB,2010 年为 1.2ZB;谷歌公司高级副总裁 Kent Walker 指出,"近年来大数据正在以惊人的指数增长。随着计算机存储成本的下降,存储数据的量激增。截至 2000 年,人类仅存储大约 12EB 的数据,但如今,人们每天都要产生 2EB 的数据。过去两年的时间里产生了世界上 90% 以上的数据。"IDC 报告显示,预计到 2020 年全球数据总量将超过 40ZB(相当于 4 万亿吉字节),这一数据量是 2011 年的 22 倍。2008~2020 年数据的增长规模如图 1-2 所示。

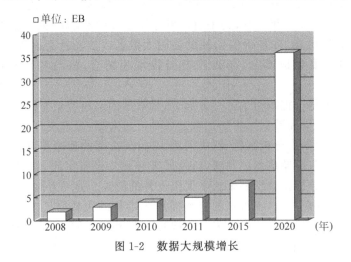

图 1-2 数据大规模增长

2. 数据类型多

从数据组织形式的角度来看,数据类型可以简单地被分为结构化数据、非结构化数据及半结构化数据。结构化数据能够用数据或统一的结构加以表示,如数字、符号等,是传统的关系数据模型、行数据,存储于数据库,可用二维表结构表示。结构化数据包括银行交易数据、商品购买信息数据等格式规范严谨的数据库数据。非结构化数据是指那些无法通过事先定义的数据模型表达或无法存入关系型数据库表中的数据,如办公文档、图片、音频和视频等。半结构化数据介于完全结构化数据和非结构化数据之间,XML、JSON 文档就属于半结构化数据。它一般是自描述的,数据的结构和内容混在一起,没有明显的区分。数据类型如表 1-1 所示。

表 1-1 数据类型

数据类型	数据类型的描述
结构化数据	结构化数据是指可以使用关系型数据库表示和存储,如 MySQL、Oracle、SQL Server 等,表现为二维形式的数据。数据以行为单位,一行数据表示一个实体的信息,每一行数据的属性是相同的
非结构化数据	非结构化数据就是没有固定结构的数据。各种文档、图片、视频/音频等都属于非结构化数据。对于这类数据,一般直接整体进行存储,且存储为二进制的数据格式
半结构化数据	半结构化数据属于同一类实体可以有不同的属性,这些属性可能是数值型的,也可能是文本型的,还可能是字典或者列表。常见的半结构化数据有 XML 和 JSON

大数据环境下的数据类型繁多。在早期,绝大部分的数据信息是以结构化的形式存放在数据库中。这些数据处理起来比较方便,但是,随着计算机技术的快速发展,大数据环境下,半结构化数据和非结构化数据在整个数据量中所占的比例大幅度上升。据统计,在企业数据中,目前已有超过 80% 的数据是以非数据结构化的形式存在的,结构化数据仅占 20%。

多类型的数据对数据的处理能力提出了更高的要求。值得注意的是,由于非结构化数据占据了大数据的统治地位,而其蕴含了无尽的知识和能量,这就对现代数据处理技术提出了更高的要求,从算法到架构,以应对非结构化数据增加带来的挑战。

3. 价值密度低

随着物联网的广泛应用,信息感知无处不在,产生海量数据,但这些数据价值密度较低,如何通过强大的机器学习算法迅速地完成数据的价值"提纯",是大数据时代亟待解决的难题。再者,由于大数据具有大容量和数据类型多的问题,其价值密度低这一特性显得尤为突出。

另外,在价值密度低这个层面,大数据要求人们所处理的数据集是有巨大商业价值或社会价值的。阿里巴巴愿意花巨大代价提高推荐系统的准确性,就是在于其推荐系统的准确率的提高,能大大提高平台的交易量,从而具有非常巨大的商业价值。在全国部署"天眼"系统,提高大数据技术在天眼系统的分量,就是因为天眼系统分析能力的每一小步提升,都能在降低犯罪率、打击犯罪、保障人民群众安全、信用取证等方面有巨大的社会价值。

4. 数据时效性强

数据时效性强意味着对数据的处理速度有更高的要求,以便能够从数据中及时地提取知识和能量。在大数据环境下,随着数据量的剧增和数据类型逐渐多样化,数据中所隐藏的高时效性特征显得越来越突出。在传统的数据分析中,数据处理的工作重点更多地放在对历史数据的挖掘和分析。例如,预测一个季度商场食品的销售量,要从过去几年同一季度同种商品的销售数据来进行分析,并就得出的结果进行预测,最终制订销售计划方案。但是,这种以过长时间间隙的历史数据为基础的数据分析所做出的计划方案,往往会因技术的革新和市场变化的加剧,导致做出的决策分析误差较大。

在这样的背景下,企业必须实时分析所拥有的最新数据,并提取其中有价值的信息,以产生对未来的生产具有指导意义的分析结果。例如,在台风天气中,气象部门应实时汇报台风过境前后的路径走向。这就需要相关技术部门随时收集某一刻最新的台风路径数据进行分析,并及时做好应对措施。

5. 准确性高

准确性是指数据处理结果的准确度。大数据中的内容是与真实世界中的发生息息相关的,研究大数据就是从庞大的网络数据中提取出能够解释和预测现实事件的过程,通过大数据的分析处理,最后能够解释结果和预测未来。在小数据时代,由于小数据集搜集数据比较困难,因而在分析数据时往往更着重于分析方法,这会不可避免地产生主观偏差、准确性不高等现象。大数据时代,通过技术手段分析全部数据,准确性大大提高。

6. 复杂性高

复杂性是指数据本身的复杂性、计算的复杂性和信息系统的复杂性。

数据本身的复杂性表现在图文检索、主题发现、语义分析、情感分析等数据分析工作十分困难,其原因是大数据涉及复杂的类型、复杂的结构和复杂的模式,数据本身具有很高的复杂性。

计算机的复杂性表现在大数据计算不能像处理小样本数据集那样做全局数据的统计分析和迭代计算,在分析大数据时,需要重新审视和研究它的可计算性、计算复杂性和求解算法。大数据样本量巨大,内在关联密切而复杂,价值密度分布极不均衡,这些特征对建立大数据计算范式提出了挑战。对于PB级的数据,即使只有线性复杂性的计算也难以实现,而且,由于数据分布的稀疏性,可能做了许多无效计算。

系统的复杂性表现在大数据对计算机系统的运行效率和能耗提出了苛刻要求,大数据处理系统的效能评价与优化问题具有挑战性,不但要求厘清大数据的计算复杂性与系统效率、能耗间的关系,还要综合度量系统的吞吐率、并行处理能力、作业计算精度、作业单位能耗等多种效能因素。针对大数据的价值稀疏性和访问弱局部性的特点,需要研究大数据的分布式存储和处理架构。

1.2 大数据的分析过程

大数据的分析过程一般包括大数据的采集、大数据的存储方式、大数据分析技术、大数据的展示及应用。

1.2.1 大数据的采集

什么是大数据采集技术？大数据采集技术就是对数据进行 ETL(Extract Transform Load)操作的过程，通过对数据进行提取、转换、加载，最终挖掘数据的潜在价值。数据采集是大数据分析过程中的最基础的环节，它通过移动互联网数据、社交网络数据、传感器数据等方式获得各种类型的结构化、半结构化及非结构化的海量数据。由于采集的数据种类错综复杂，当人们进行数据分析时，必须通过提取技术，从复杂格式的数据中提取出人们需要的数据，对于提取后的数据，由于数据源头的采集可能存在不准确性，因此必须进行数据清洗，对于那些不正确的数据进行过滤、剔除。针对不同的应用场景，还需要对数据进行转换操作，将数据转换成不同的数据格式，最终按照预先定义好的数据仓库模型，将数据加载到数据仓库中。由于数据产生的种类很多、方式不同，对于大数据采集系统，主要分为以下 3 类系统。

(1) 网络数据采集系统。通过网络爬虫和一些网站平台提供的公共 API(如 Twitter 和新浪微博 API)等方式从网站上获取数据。目前常用的网页爬虫系统有 Apache Nutch、Crawler4j、Scrapy 等框架。Apache Nutch 是一个高度可扩展和可伸缩性的分布式爬虫框架，通过提交 MapReduce 任务来抓取网页数据，可以将网页数据存储在 HDFS 中。Nutch 可以分布式多任务进行数据爬取、存储和索引，由多个机器并行做爬取任务，Nutch 大大提高了系统爬取数据的能力。Crawler4j、Scrapy 都是爬虫框架，开发人员可以利用爬虫 API 接口实现数据的爬取，Crawler4j、Scrapy 框架大大提高了开发人员的开发效率，可以很快地完成一个爬虫系统的开发。

(2) 系统日志采集系统。系统日志采集系统就是收集日志数据提供离线和在线的实时分析数据。国内一些大的公司(如淘宝、百度、腾讯)每天都会产生大量的日志数据，通过首先对这些日志信息进行日志采集、收集，然后进行数据分析挖掘，为公司决策和分析提供了可靠的数据保障。目前常用的开源日志收集系统有 Cloudera-flume、Facebook-scribe、Hadoop-chukwa 和 kafka 等。Cloudera-flume 是一个分布式、可靠、可用的服务系统，具有基于流式数据流的简单灵活的架构。scribe 是 Facebook 开源的日志采集系统。scribe 实际上是一个分布式共享队列，它可以从各种数据源上收集日志数据，然后放入它上面的共享队列中。Apache 的 chukwa 是一个较新的开源项目，它提供了很多模块以支持 Hadoop 集群日志分析。kafka 采用 scala 语言编写，使用了多种效率优化机制，整体架构比较新颖(push/pull)，更适合异构集群。

(3) 数据库采集系统。目前还有一些企业会使用传统的关系型数据库 MySQL 和 Oracle 等来存储数据。除此之外，HBase、Redis 和 MongoDB 这样的 NoSQL 数据库也常用于数据的采集。通过数据库采集系统直接与企业业务后台服务器结合，将企业业务后台产生的大量的业务记录写入数据库中，最后由特定的处理分析系统进行分析。

1.2.2 大数据的存储方式

存储系统作为数据中心最核心的数据基础,不再是传统分散的、单一的底层设备。除了要具备高性能、高安全、高可靠等基于大数据应用需求,"应用定义存储"概念被提出。

1. 分布式系统

分布式系统包含多个自主的处理单元,通过计算机网络互联来协作完成分配的任务,其分而治之的策略能够更好地处理大规模数据分析问题。分布式文件系统(Hadoop Distributed File System,HDFS)是一个高度容错性系统,被设计成适用于批量处理,能够提供高吞吐量的数据访问,存储管理需要多种技术的协同工作,其中文件系统为其提供最底层存储能力的支持。该系统源于 Google 在 2003 年 10 月发表的 GFS(Google File System)论文,它其实就是 GFS 的一个克隆版本。Google 公司为了存储海量搜索数据而设计了专用文件系统 GFS,尽管 Google 公布了该系统的一些技术细节,但 Google 并没有将该系统的软件部分作为开源软件发布。

2. NoSQL 数据库

关系型数据库已经无法满足 Web 2.0 的需求,主要表现为:无法满足海量数据的管理需求;无法满足数据高并发的需求;高可扩展性和高可用性的功能太低。NoSQL(Not only SQL)数据库的优势是:可以支持超大规模数据存储;灵活的数据模型可以很好地支持 Web 2.0 应用;具有强大的横向扩展能力,等等。与数据库管理系统(RDBMS)相比,NoSQL 不使用 SQL 作为查询语言,其存储可以不需要固定的表模式,通常也会避免使用 RDBMS 的 JOIN 操作,一般都具备水平可扩展的特性。

NoSQL 的实现具有两个特征:使用硬盘和把随机存储器作存储载体。按照存储格式来分,NoSQL 可以分为 4 类:键值存储数据库、列存储数据库、文档存储数据库和图形数据库。目前比较流行的 NoSQL 数据库有 Casssandra、Luncene、Neo4j、MongoDB 和 HBase。例如,HBase(Hadoop DataBase)是一个高可靠性、高性能、面向列、可伸缩的分布式数据库系统,它使用类似 GFS 的 HDFS 作为底层存储文件,在其上运行 MapRduce 批量处理数据,使用 ZooKeeper 作为协同服务组件。

3. 云数据库

云数据库是基于云计算技术发展的一种共享基础架构的方法,是部署和虚拟化在云计算环境中的数据库。云数据库并非一种全新的数据库技术,而只是以服务的方式提供数据库功能。云数据库所采用的数据模型可以是关系数据库所使用的关系模型(微软的 SQLAzure 云数据库都采用了关系模型),同一个公司也可能提供采用不同数据模型的多种云数据库服务。

4. 大数据存储技术路线

大数据存储技术路线分为以下几种。

第一种是采用大规模并行处理(Massively Parallel Processor,MPP)架构的新型数据库集群,重点面向行业大数据,采用 Shared Nothing 架构,通过列存储、粗粒度索引等多项大数据处理技术,再结合 MPP 架构高效的分布式计算模式,完成对分析类应用的支撑它的。运行环境多为低成本 PC Server,具有高性能和高扩展性的特点,在企业分析类应用领域获得极其广泛的应用,对于企业新一代的数据仓库和结构化数据分析,目前最佳选择是 MPP 数据库。

第二种是基于 Hadoop 的技术扩展和封装,围绕 Hadoop 衍生出相关的大数据技术,应对传统关系型数据库较难处理的数据和场景,如针对非结构化数据的存储和计算等,充分利用 Hadoop 开源的优势,伴随相关技术的不断进步,其应用场景也将逐步扩大。目前最为典型的应用场景就是通过扩展和封装 Hadoop 来实现对互联网大数据存储、分析的支撑。这里面有几十种 NoSQL 技术,也在进一步地细分。对于非结构、半结构化数据处理、复杂的 ETL 流程、复杂的数据挖掘和计算模型,Hadoop 平台更擅长。

第三种是大数据一体机,这是一种专为大数据的分析处理而设计的软、硬件结合的产品,由一组集成的服务器、存储设备、操作系统、数据库管理系统,以及为数据查询、处理、分析用途而特别预先安装及优化的软件组成,高性能大数据一体机具有良好的稳定性和纵向扩展性。

1.2.3 大数据分析技术

数据存储之后,对数据的分布式处理工具有 Hadoop、MapReduce、Storm/JStorm、Samza 和 Spark,以及在此之上的各种不同计算范式,如批处理、流处理和图计算等,包括衍生出编程的计算模型,如 BSP、GAS 等。

MapReduce 一般用于处理大规模数据集,如业务方累计的历史数据,是一种典型的批处理系统。Storm/JStorm 一般用于处理连续不断的数据流,注重数据处理的时效性。Apache Samza 是一个分布式的流处理框架,它使用 Apache Kafka 来传递消息,使用 Apache Hadoop YARN 来提供容错、安全和资源管理等功能。

为了能同时进行批处理和流处理,出现了基于内存的 Spark 计算框架。Spark 是一个高速、通用的集群计算系统,它为 Java、Scala、Python 及 R 语言都提供了应用程序接口,它也是最佳的支持通用执行图的引擎。不仅如此,Spark 也提供了非常丰富的插件工具,其中包括为 SQL 设计的 Spark SQL、结构化的数据处理工具、机器学习库 MLlib、图像处理工具 GraphX 和 Spark Streaming。

Hive 是 Facebook 团队开发的一个可以支持 PB 级别的可伸缩性的数据仓库。这是一个建立在 Hadoop 之上的开源数据仓库解决方案。Pig 是一个基于 Hadoop 的大规模数据分析平台,它提供的 SQL-LIKE 语言称为 Pig Latin,该语言的编译器会把类 SQL 的数据分析请求转换为一系列经过优化处理的 MapReduce 运算。Pig 为复杂的海量数据并行计算提供了一个简单的操作和编程接口。

1.2.4 大数据的展示及应用

可视化技术是利用计算机图形学及图像处理技术,将数据转换为图形或图像形式显示到屏幕上,并进行交互处理的理论、方法和技术。数据可视化是指以图形或图表格式通过人工或以其他方式组织和显示数据。常用的大数据可视化工具有 Echarts、D3.js、Tableau 等。

  Echarts 是百度公司的前端开源工具,是一个使用 JavaScript 实现的开源可视化库,可以流畅地运行在 PC 和移动设备上,底层依赖轻量级的矢量图形库 ZRender,提供直观、交互丰富、可高度个性化定制的数据可视化图表。D3.js 是一个采用 Java 语言编写的开源库,其目标是允许使用标准网页浏览技术(如 HTML 或 CSS)轻松地处理基于数据的文档,D3.js 可以通过使用 HTML、SVG 和 CSS 把数据鲜活形象地展现出来。同时,它提供了强大的可视化组件,可以让使用者以数据驱动的方式去操作 DOM。Tableau 是一种商业智能软件,旨在帮助人们查看和理解数据。Tableau 不仅是一个代码库,也被认为是一组或一系列交互式数据可视化产品。

1.3 大数据的价值、挑战与风险

  纵观整个移动互联网领域,大数据已被认为是继云计算、物联网之后的又一大颠覆性的技术性革命,毋庸置疑,大数据市场是待挖掘的金矿,其价值不言而喻,具有重要的战略意义。

1.3.1 商业价值

  随着大数据的发展,企业也越来越重视数据相关的开发和应用,从而获取更多的市场机会。大数据打破了企业传统数据的边界,改变了过去商业智能仅仅依靠企业内部业务数据的局面,大数据使数据来源更加多样化,不仅包括企业内部数据,也包括企业外部数据,尤其是和消费者相关的数据。

  据 IDC 和麦肯锡的大数据研究结果的总结,大数据主要能在以下 4 个方面挖掘出巨大的商业价值:对顾客群体细分,然后对每个群体量体裁衣般地采取独特的行动;运用大数据模拟实境,发掘新的需求和提高利润;提高大数据成果在各相关部门的分享程度,提高企业决策能力;进行商业模式、产品和服务的创新。

1. 对顾客群体细分,然后对每个群体量体裁衣般地采取独特的行动

  大数据能够帮助企业分析大量数据,从而进一步挖掘市场机会和细分市场,对顾客群体细分,然后对每个群体量体裁衣般地采取独特的行动,获得好的产品概念和创意。大数据分析的关键在于如何去搜集消费者相关的信息,如何获得趋势,挖掘出人们头脑中未来可能会消费的产品概念;用创新的方法解构消费者的生活方式,剖析消费者的生活密码,让吻合消费者未来生活方式的产品研发不再成为问题。大数据分析是发现新客户群体、确定最优供应商、创新产品、理解销售季节性等问题的最好方法。

  在大数据时代之前,要搞清楚海量顾客的购买情况,需要投入惊人的人力、物力、财力,使得这种细分行为毫无商业意义。但是,云存储的海量数据和大数据的分析技术使得对消费者的实时和极端的细分有了成本低、效率极高的可能。企业利用用户在互联网上的访问行为偏好,从而为具有相似特征的用户组提供精确服务,满足用户需求,甚至为每个客户量身定制。这一变革将大大缩减企业产品与最终用户的沟通成本。

2. 运用大数据模拟实境，发掘新的需求和提高利润

现在越来越多的产品中都装有传感器，汽车和智能手机的普及使得可搜集数据呈现爆炸性增长。Blog、Twitter、Facebook 和微博等社交网络也在产生着海量的数据。云计算和大数据分析技术使得商家可以在成本低、效率较高的情况下，实时地对这些数据连同交易行为的数据进行存储和分析。交易过程、产品使用和人类行为都可以数据化。大数据技术可以把这些数据整合起来进行数据挖掘，从而在某些情况下通过模型模拟来判断不同变量的情况下以何种方案投入回报最高。

3. 提高大数据成果在各相关部门的分享程度，提高企业决策能力

大数据能够有效地帮助各个行业用户做出更为准确的商业决策，从而实现更大的商业价值，它从诞生开始就是站在决策的角度出发。虽然不同行业的业务不同，所产生的数据及其所支撑的管理形态也千差万别，但从数据的获取、数据的整合、数据的加工、数据的综合应用、数据的服务和推广、数据处理的生命线流程来分析，所有行业的模式是一致的。

这种基于大数据的决策具有以下特点：

（1）量变到质变。由于数据被广泛挖掘，决策所依据的信息完整性越来越高。

（2）决策技术含量、知识含量大幅度提高。由于云计算的出现，人类没有被海量数据所淹没，能够高效率驾驭海量数据，生产有价值的决策信息。

（3）大数据决策催生了很多过去难以想象的重大解决方案。例如，某些药物的疗效和毒副作用，无法通过技术和简单样本验证，需要几十年海量病历数据分析得出结果；做宏观经济计量模型，需要获得所有企业、居民及政府的决策和行为海量数据，才能得出减税政策最佳方案；反腐倡廉，人类几千年历史都没解决，最近通过微博和"人肉"搜索，贪官在大数据的海洋中无处可藏，人们看到根治的希望。

如果在不同行业的业务和管理层之间增加数据资源体系，通过数据资源体系的数据加工，把今天的数据和历史数据对接，把现在的数据和领导与企业机构关心的指标关联起来，把面向业务的数据转换成面向管理的数据，辅助于领导层的决策，真正实现了从数据到知识的转变，这样的数据资源体系是非常适合管理和决策使用的。

4. 进行商业模式、产品和服务的创新

大数据让企业能够创造新产品和服务，改善现有产品和服务，以及创建全新的业务模式。回顾 IT 历史，似乎每一轮 IT 概念和技术的变革，都伴随着新商业模式的产生。例如个人计算机时代微软公司凭借操作系统获取了巨大财富；互联网时代谷歌公司抓住了互联网广告的机遇；移动互联网时代苹果公司则通过终端产品的销售和应用商店获取了高额利润。

大数据技术可以有效地帮助企业整合、挖掘、分析其所掌握的庞大数据信息，构建系统化的数据体系，从而完善企业自身的结构和管理机制；同时，伴随消费者个性化需求的增长，大数据在各个领域的应用开始逐步显现，已经开始并正在改变着大多数企业的发展途径及商业模式。例如,大数据可以完善基于柔性制造技术的个性化定制生产路径,推动制造业企业的升级改造;依托大数据技术可以建立现代物流体系,其效率远超传统物流企业;利用大数据技术可多维度评价企业信用,提高金融业资金使用率,改变传统金融企业的运营模式等。

1.3.2 社会生活价值

通过大数据技术融合社会应用,使数据参与社会生活的各个方面,挖掘大数据中真正有效的价值,继而促进人们创造更多的社会价值,从而完善社会生活方式,改变人们的未来。

首先,对于个人生活而言,由于大数据的高透明度和实时性,经过数据的精准挖掘,大数据可以为个人提供个性化的医疗服务。过去人们去看病,医生只能对患者的当下身体情况做出判断;而在大数据的帮助下,将来的诊疗可以对一个患者的累计历史数据进行分析,并结合遗传变异、对特定疾病的易感性和对特殊药物的反应等关系,实现个性化的医疗;还可以在患者发生疾病症状前,提供早期的检测和诊断。因此,医疗体系随着大数据的发展而得以完善,极大地方便了人们的生活。

其次,在传统教育模式下,一个教师面对几十个学生进行课堂教育,课后布置同样的作业。然而,学生是各有其特性的,在原来的教育模式下,教师难以真正做到"因材施教"。但是,大数据的出现使得学生在受教育过程中的数据得以完整搜集,包括授课的过程、作业的情况、学生成绩、教师评价等数据。这些数据完全是在学生不自知的情况下被观察、搜集,只需要一定的观测技术与设备的辅助,而不影响学生任何的日常学习与生活,因此它的采集也非常自然、真实。在大数据的支持下,教育将呈现另外的特征:弹性学制、个性化辅导、社区和家庭学习。大数据支撑下的教育,就是要根据每一个人的特点,释放每一个人本来就有的学习能力和天分。

再次,大数据的诞生让社会安全管理更为井然有序。在社会安全管理领域,大数据技术通过对手机数据的挖掘,可以分析实时动态的流动人口来源、出行,实时交通客流信息及拥堵情况。利用短信、微博、微信和搜索引擎,可以收集热点事件,挖掘舆情,还可以追踪造谣信息的源头。美国麻省理工学院曾通过对10万多人手机的通话、短信和空间位置等信息进行处理,提取人们行为的时空规律性,进行犯罪预测。

最后,大数据的发展带动了社会上各行各业的发展。值得一提的是,据盖特纳咨询公司预测,大数据将为全球带来440万个IT新岗位和上千万个非IT岗位。因此,大数据分析人才,特别是深度数据挖掘人才,将会受到人们的追捧。

1.3.3 大数据的挑战与风险

如前所述,大数据正在催生以数据资产为核心的多种商业模式,革新生活模式,产生社会价值,引发积极影响。谷歌、百度、亚马逊等成千上万个互联网公司采集大数据、存储大数据、分析大数据。企业家和科学研究者共同合作,渴望利用大数据技术寻求发展,应用大数据技术投入教育领域,运用大数据技术颠覆医疗服务性行业。目前,大数据还处于发展的初期阶段,大数据在带给人们发展的同时,也让人们面临着更大的挑战与风险。

一方面，与传统数据相比，数据量源源不断地增加，容易导致很多不正确的数据写入数据库中。况且，大数据包括不同的信息来源，多种多样的数据增大了出现混乱的概率。面对各种错综复杂的海量数据，无疑研究者从中分析数据找到确定性结论的难度增大。大数据既包括更密集的信息，也带来更多的错误关系，使人们难以找到研究对象之间真正的关联。因此，大数据技术具有在处理上的风险。另外，由于大数据的容量大（至少PB级以上），存储系统需要有一定的扩展能力。大数据还需要解决实时性问题，尤其是实时监测和网上交易问题特别注重实时性，实时数据的应用很多时候需要设备具有较高的计算性能。

另一方面，一些安全性问题也因大数据的应用而逐渐呈现出来。例如，医疗信息、金融数据以及政府情报等某些特殊行业的应用，都有自己的保密性和安全标准需求，都是IT管理者必须遵从的，这是大数据企业的安全需求。对于个体而言，大数据逐渐深入人们的生活，互联网上到处留下痕迹，这样大大增加了暴露个人隐私的风险。大数据将对安全行业产生深远影响，用户应该如何保护自己的数据安全和个人隐私，逐渐成为安全厂商和用户关心的话题。解决个人数据隐私安全问题，个人数据中心可能不是最好的办法，因为业界对大数据的无限渴求会阻止个人数据中心的实现。但是随着数据量的增长，"数据重要还是隐私重要"将会是一个激烈的辩题。

要充分认识大数据的两面性，既带来机遇和价值，也带来挑战和风险。大数据来源丰富、种类复杂、数据量大，具有不可完全预测的复杂性，未来需要更高的预测优化和前瞻观察。

1.4 大数据的应用

大数据技术已经被视为未来经济生活中的基础，这意味着几乎全部行业都能够在大数据分析技术之上获得经济效益的提升。在销售行业中，通过输入客户的性格、穿戴习惯、所处行业及历史销售数据等信息，销售员将会被大数据分析告知何时给哪一位客户打电话获得订单的概率最高；在品牌形象建立中，能够依据市场情绪的分析，写出与用户能够产生共鸣的文案从而获取消费者的好感；法律行业中能够"阅读"过去数十万判决案例，针对用户输入的案件给出判决概率预测，帮助律师制定辩护策略，而长期来看法律大数据企业很有可能取代大部分初级律师；同样，在零售、广告、医疗等诸多领域，大数据技术都能通过分析数据内在的关系而帮助用户实现购买预测、受众精准投放及病情辅助判断等功能。大数据的应用行业包含电子商务、媒体营销、旅游、物流交通、农业、工业、企业服务、娱乐、汽车、物联网、生命科技、金融科技、房产、教育及政府等诸多产业，如图1-3所示。

1. 电商大数据的应用

电商是最早利用大数据精准营销的行业。大数据技术帮助电子商务行业发现新的商业模式，尤其是购物行为预测分析和购物商品关联分析已经在电商领域得到了很好的应用，并已经帮助电商获得了巨大的利润。收集和分析消费者网上消费行为数据可以帮助商家预测顾客下一步的购物行为。利用顾客留在网站上的行为轨迹数据，分析顾客浏览商品类别，可以帮助商家预测顾客需要哪类商品，并推出相关商品；根据顾客询价情况，商家还可以预测顾客购买力，而对高级顾客推荐名牌商品，对普通顾客推荐物美价廉的商

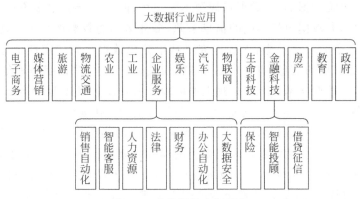

图 1-3 大数据的应用行业

品,以满足不同的顾客对商品的不同心理价位的需求。跟踪顾客经常购物的网店,对此类数据进行分析,可以预测用户的下一次购物行为将可能发生在哪个网店。商品关联分析也是电商大数据应用中的一种。京东商城、当当网、亚马逊、淘宝网等电商网站商品交易数据量非常庞大,其中,淘宝网是目前中国最大的 C2C 电子商务平台,也是国内第一批采用 Hadoop 技术数据平台升级的企业之一,从 2008 年开始,淘宝就开始研究基于 Hadoop 的数据处理平台——云梯(Cloud Ladder),到 2014 年,整个集群达到 1700 个节点,数据总容量 24.3PB。电商数据分析部门通过对这些数据的数据挖掘,构建关联模型,可以更好地组织网站上的商品,减少用户过滤信息的负担,并根据顾客当前的购买行为为顾客提供推荐;借助商品关联分析还可以发现什么样的商品组合顾客多半会一起购买,从而可以向顾客推出商品或者把相关的商品链接放在一起,让顾客快速地浏览需要的商品,节省顾客的时间,满足顾客需求,为顾客提供更好的体验。

2. 医疗大数据的应用

医疗行业是让大数据分析最先发扬光大的传统行业之一。医疗行业拥有大量的病例、病理报告、治愈方案、药物报告等,如果这些数据可以被整理和应用,将会极大地帮助医生和病人。但是,医疗行业中的病理数据,包括病人的病症等,在整个治疗过程中是不断变化的,这使得在诊断疾病时,疾病的确诊和治疗方案的确定非常困难。

借助于大数据平台可以收集不同病例和治疗方案,以及病人的基本特征,可以建立针对疾病特点的数据库。如果未来基因技术发展成熟,可以根据病人的基因序列特点进行分类,建立医疗行业的病人分类数据库。在医生诊断病人时可以参考病人的疾病特征、化验报告和检测报告,再对照疾病数据库来快速帮助病人确诊,明确定位疾病。在制订治疗方案时,医生可以依据病人的基因特点,调取相似基因、年龄、人种、身体情况相同的有效治疗方案,制订出适合病人的治疗方案,帮助更多人及时进行治疗。同时,这些数据也有利于医药行业开发出更加有效的药物和医疗器械。

3. 教育大数据的应用

随着互联网的发展,信息技术已在教育领域有了越来越广泛的应用:考试、课堂、师生互动、校园设备使用、家校关系等。只要互联网技术达到的地方,各个环节都被数据包裹。大数据技术在教育行业的应用能够对学生学习生涯中各个时期的学习行为、考试成绩乃至职业规划进行详细的关联分析和研究。在国外,很多这样的教育信息经过处理后应用到公共教育系统中。例如,美国政府部门在2012年投入一项花费2亿美元的公共教育大数据计划,通过这个项目美国政府希望能够对美国的教育体系进行完善和改革。在大数据背景下,学生在学校的各种表现都可以用数据形式反映出来,其中包括当下学生的行为表现。一方面,可以通过学生之间的行为变化发现内在的联系性。并且,大数据时代可以显示学生的历史行为,各种数据表单都能够记录下来,通过这些数据可以发现学生的学习兴趣、特长爱好等因素。另一方面,由于学生的变化通常情况下都不是很明显,通过云计算完成学生行为大数据的分析,发现学生的问题。而数据从开始到结束是有时间差的,这样就可以通过数据提早发现学生的变化,避免产生不良的结果。也可以利用大数据挖掘学生的内在特征,分析每个学生的学习情况,从而实现个性化教学。

4. 金融大数据的应用

大数据在金融行业的应用较为广泛,上到国际贸易金融,下到个人买卖交易,其中产生的巨量数据都在清楚地告诉人们大数据正在指引着金融行业的发展。大数据在金融行业应用的经典案例有花旗银行利用IBM沃森电脑为财富管理客户推荐产品;美国银行利用客户点击数据集为客户提供特色服务;招商银行对客户刷卡、存取款、电子银行转账、微信评论等行为数据进行分析,每周给客户发送针对性广告信息,里面有顾客可能感兴趣的产品和优惠信息。总体来说,金融行业的发展需要大数据的参与。中国银监会设立的金融消费者保护局最有利地保障了大数据金融的发展。在国外,消费者金融可以帮助客户,提供便利的大数据应用服务,例如对客户日志实施实时检测,进行债权分析等服务。由此可见,利用大数据技术,有助于金融行业提升自身业务水平,从而促进其发展。

5. 农业大数据的应用

农业大数据是指以大数据分析为基础,运用大数据的理念、技术及方法来处理农业生产、销售整个链条中所产生的大量数据,从中得到有用信息以指导农业生产经营、农产品流通和消费的过程。大数据的应用与农业领域的相关科学研究相结合,可以为农业科研、政府决策、涉农企业发展等提供新方法、新思路。农业大数据的应用主要体现在以下几个方面。

(1) 精准生产,预测市场需求。通过汇总农业生产过程中的数据,实现农业生产的供需平衡。农户可以通过大数据平台采集的消费者需求报告,进行市场分析,预测农产品市场需求、辅助农业决策,以此达到规避风险、增产增收、管理透明等预期目标。

(2) 提供栽种管理决策。农业大数据可以利用卫星、无人机采集气候、土壤等数据,为农户提供最优化的栽种管理模式,协助农户有效管理农田,让农户从每一颗种子中获得最高的价值,降低农业成本。

(3) 提供生产过程的管理。农业大数据可通过实时卫星影像数据,分析农作物当前的长势,获得地块信息,预测未来环境趋势走向,得到精确病虫害趋势等,为农户提供精确种植建议及管理指导。

(4) 自动化生产,农业环境监测。通过卫星遥感技术采集农作物生长环境中的各项指标数据,再将其上传至本地或云端数据库,对农业生产的历史数据和实时监控数据进行分析,提高对作物种植面积、进度、产量、环境条件、灾害强度的关联监测能力。农户在作物的生长过程中能够规避气候灾害,采取科学的防治措施及种植方法,将从源头上提高农业生产效率和产量。

(5) 农产品供应链追踪。农业大数据被用来改善各个环节,涵盖农产品生产商、供应商和运输者等,可实现从田间到餐桌每一个过程的追踪。通过GPS定位系统进行实时监控,有助于预防食源性疾病和减少浪费。同时,农业大数据通过深度挖掘和有效整合散落在全国各农业产区的农产品生产和流通数据,为农产品生产和流通提供高效优质的信息服务,从源头上保障食品安全。

6. 旅游大数据的应用

旅游大数据主要由结构化大数据和非结构化大数据两部分组成。结构化大数据主要指各旅游企业的ERP数据、财务系统数据等。非结构化数据则指各类文本、图片、网页、音频和视频等,这些旅游数据成了重要的资源,通过对这些数据的开发和利用,可以预测整个旅游产业的发展趋势。旅游大数据的应用主要体现在以下几个方面。

(1) 旅游需求分析与预测。旅游消费者在互联网上留下的足迹,反映了他们的个人偏好、消费习惯、消费水平等,通过对互联网上产生的海量旅游数据的分析,可以判断旅游市场的消费需求,为游客提供精准化和个性化服务条件。

(2) 精准营销与服务。利用大数据对游客基本属性、行为特征、个人偏好等进行分析,可以对游客消费群体进行分类,并对他们提供个性化需求服务。

(3) 构建多维旅游预警系统。随着经济的发展,个体的旅游需求不断增加,而旅游高峰期一般集中在假期,这种扎堆出行给旅游目的地的安全带来了很大的隐患,必须建立旅游预警系统。传统旅游预警系统的建立基于统计调查数据,数据频度较低,指标体系更新较慢,且预警方法较为单一。通过整合不同来源的大数据,可以建立更加及时准确的旅游预警系统。

(4) 旅游规划和管理。通过对旅游大数据的分析,可以科学地进行市场预测,为旅游规划整体项目设计、旅游路线设计、旅游交通规划、旅游基础服务设施建设等提供依据。旅游管理机构则可整合景区大数据、互联网、运营商等第三方大数据,对旅游地的人流、车流等进行预测,对景区安全进行防范,对景区人力、物力等资源进行合理分配。

7. 气象大数据的应用

气象涉及人类活动的各个方面,包括农牧的发展等。但是,在地球环境日益恶化的今天,气象环境变得恶劣。例如,全球变暖问题,导致冰川融化,海平面上升,威胁到人类未来的居住环境。因此,越来越多的人关注生态环境的保护,以确保人类文明更好地发展。以往,主要是分析环境历史数据从而做出的预测及应对方案。由于实时环境的多变,利用历史数据做出的环境预测是存在误差的。大数据的产生有效地改进了这一弊端。借助于大数据技术,天气预报的准确性和实效性将会大大提高,预报的及时性将会大大提升,同时对于重大自然灾害,如龙卷风,通过大数据计算平台,人们将会更加精确地了解其运动轨迹和危害的等级,有利于帮助人们提高应对自然灾害的能力。天气预报准确度的提升和预测周期的延长将会有利于农业生产的安排。

除了这些经典应用外,大数据在零售、交通、舆情监控方面也被广泛运用。总而言之,实现大数据应用,关键在于实时获取各种不同数据类型的数据流,特别是非结构化数据流,持续采集大数据,并汇总到数据中心,使用有效技术和相关数学工具来分析所采集的数据,实现实时共享,支持业务决策。因而,结合相关数据知识,推动行业发展的应用,是驱动大数据发展的重要力量。

1.5 大数据的处理流程

从大数据的特征和应用行业来看,大数据的来源相当广泛,由此产生的数据类型和应用处理方法千差万别。总体来说,大数据的基本处理流程大都是一致的。大数据的处理流程基本可划分为数据采集、数据处理与集成、数据分析和数据解释4个阶段。即经数据源获取的数据,因为其数据结构不同(包括结构、半结构和非结构数据),用特殊方法进行数据处理和集成,将其转变为统一标准的数据格式方便以后对其进行处理,然后用合适的数据分析方法对这些数据进行处理分析,并将分析的结果利用可视化等技术展现给用户,这就是整个大数据处理的流程,如图1-4所示。

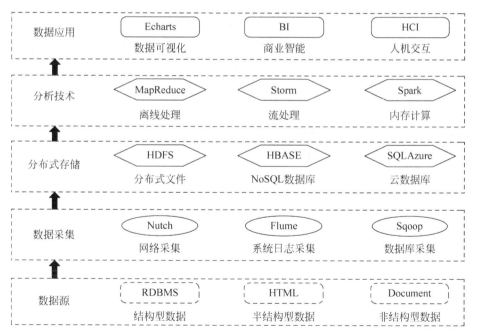

图1-4 大数据的处理流程

1.6 大数据成为人工智能产业的燃料

人工智能技术一直是科学家与技术人员的追求,但其发展并不是一帆风顺的。例如,在最初的自然语言识别技术中,科学家希望通过语法规则使计算机理解语义从而实现智能化,但事实证明这一路径并不可行,其后依据大量数据样本的统计方法才有效提升了自然语言处理的准确度并逐渐达到可用水平。

人工智能实现最大的飞跃是大规模并行处理器的出现,特别是CPU,它具有数千个内核的大规模并行处理单元,这大大加快了现有的人工智能算法的速度。

人工智能应用的数据越多,其获得的结果就越准确。在过去,人工智能由于处理器速度慢、数据量小而不能很好地工作;也没有先进的传感器,并且当时互联网还没有广泛使用,所以很难提供实时数据。如今,人们拥有所需要的一切:快速的处理器、输入设备、网络和大量的数据集。毫无疑问,没有大数据就没有人工智能。人工智能想要发展得更全面、更智能,就必须依托大数据技术,需要大数据的支撑。随着计算机硬件发展及计算能力的提升,庞大的大数据为人工智能提供了多样的、丰富的学习样本。大数据技术的发展为人工智能提供了一定的技术支持,使机器具备了数据处理能力,这样才能解决人工智能的扩展性和成长性问题,人工智能的发展,反过来进一步促进了大数据的发展,二者之间起到了相互促进的作用。

1.7 大数据技术的发展前景

Hadoop是目前发展最成熟的大数据开源平台,吸引了大批互联网企业、传统IT设备商和新创企业参与到项目开发和实际应用过程中。未来很长一段时间内,Hadoop开源生态系统仍将对大数据应用和技术的发展起到积极推动作用。

大数据是新一代信息技术的集中反映,是一个应用驱动性很强的服务领域,是具有无穷潜力的新兴产业领域;目前,其标准和产业格局尚未形成,这是我国实现跨越式发展的宝贵机会。要从战略上重视大数据的开发利用,将它作为转变经济增长方式的有效抓手。因此,大数据技术具有良好的发展前景。

总体来说,大数据技术仍在不断发展中。接下来,将会出现更多的基于大数据研究的应用项目,如机器学习、认知计算及预测分析等。

小结

本章首先介绍了大数据的基本概念、大数据的特征,让读者对大数据有一个基本的了解,接着概述了大数据分析的过程、技术及工具,涉及大数据的采集技术、大数据的存储技术及大数据的分析技术,然后对大数据的商业价值、社会价值,以及面临的挑战与风险也进行了分析,详细描述了大数据在电子商务、医疗、教育、旅游、金融等行业中的应用,最后讨论了大数据与人工智能的关系,以及大数据的应用发展前景。

习题

1. 简述大数据的概念。
2. 简述大数据的基本特征。
3. 简述大数据的分析处理过程。
4. 简述大数据的存储方式。
5. 简述大数据的商业价值和社会生活价值。
6. 以某一行业为例,简述大数据的应用。

第 2 章 大数据集群系统基础

视频讲解

随着云计算、物联网、移动计算、人工智能的发展,导致了大量数据的产生,各种大数据分析系统随之产生,由于数据量之大,对计算的速度和精度要求都比较高,单纯地通过不断增加处理器的数量来增强单个计算机的计算能力已经达不到预想的效果,那么大数据处理的方向逐渐朝着分布式的计算集群来发展,将分布在不同空间的计算机通过网络相互连接组成一个有机的集群,然后将需要处理的大量数据分散到集群中,交由分散系统内的计算机组同时计算,最后将这些计算结果合并得到最终的结果。尽管分散系统内的单个计算机的计算能力不强,但是由于每个计算机只计算一部分数据,而且是多台计算机同时计算,因此就分散系统而言,处理数据的速度会远高于单个计算机。

2.1 大数据集群系统概述

集群技术是指通过高速通信网络将一组相互独立的计算机联系在一起,组成一个计算机系统,该系统中每一台计算机都是一个独立的服务器,运行各自的进程,它们相互之间可以通信,既可以看作是一个个单一的系统,也能够协同起来为用户提供服务。集群中的计算机被称为节点,某一个节点出现问题不能使用,或者功能不够完善难以处理客户的部分请求时,该请求就会被自动转到其他的节点进行处理。节点的运行方式主要受节点设置形式影响,理想的两节点集群中,这两个节点同时运行应用程序,两个服务器都处于活动状态,其中一个节点发生故障,对应的应用程序会立即转移到正常服务器上,此时,一个服务器承担两个节点的工作,服务器的性能会受到影响。

集群就是一组相互独立的计算机,通过高速的网络组成一个计算机系统,每个集群节点都是运行其自己进程的一个独立服务器。对网络用户来讲,后端就像是一个单一的系统,协同向用户提供系统资源、系统服务,通过网络连接成一个组合来共同完一个任务。

集群系统可以在付出较低成本的情况下获得在性能、可靠性、灵活性方面的相对较高的收益,其任务调度则是集群系统中的核心技术,一个客户与集群相互作用时,集群像是一个独立的服务器。

计算机集群简称集群,是一种计算机系统,它通过一组松散集成的计算机软件或硬件连接起来高度紧密地协作完成计算工作,在某种意义上,它们可以被看作是一台计算机。

Hadoop 分布式集群是为了对海量的非结构化数据进行存储和分析而设计的一种特定的集群。其本质上是一种计算集群,也就是对不同的数据进行分配,并对其进行数据的处理。在大数据处理中,Hadoop 之所以能够得到广泛应用,主要在于其进行数据提取、数据变形以及加载等方面的优势非常明显。

2.1.1 集群的分类

集群分为同构与异构两种,它们的区别在于:组成集群系统的计算机之间的体系结构是否相同。计算机集群按功能和结构可以分成以下几类。

(1) 高可用性集群。一般是指当集群中有某个节点失效的情况下,其上的任务会自动转移到其他正常的节点上;还指可以对集群中的某节点进行离线维护后再上线,该过程并不影响整个集群的运行。

(2) 负载均衡集群。负载均衡集群运行时一般通过一个或者多个前端负载均衡器将工作负载分发到后端的一组服务器上,从而达到整个系统的高性能和高可用性。这样的计算机集群有时也被称为服务器群(Server Farm)。一般高可用性集群和负载均衡集群会使用类似的技术,或同时具有高可用性与负载均衡的特点。

(3) 高性能计算集群。高性能计算集群采用将计算任务分配到集群的不同计算节点来提高计算能力,因而主要应用在科学计算领域。比较流行的 HPC 采用 Linux 操作系统和其他一些软件来完成并行运算。这一集群配置通常被称为 Beowulf 集群。这类集群通常运行特定的程序以发挥 HPC Cluster 的并行能力。这类程序一般应用特定的运行库,如专为科学计算设计的 MPI 库。

HPC 集群特别适合于在计算中各计算节点之间发生大量数据通信的计算作业,如一个节点的中间结果或影响到其他节点计算结果的情况。

(4) 网格计算。网格计算或网格集群是一种与集群计算非常相关的技术。网格与传统集群的主要差别是:网格是连接一组相关并不信任的计算机,它的运作更像一个计算公共设施而不是一个独立的计算机;还有,网格通常比集群支持更多不同类型的计算机集合。

网格计算是针对有许多独立作业的工作任务,在计算过程中作业之间无须共享数据。网格主要服务于管理在独立执行工作的计算机之间的作业分配,如资源存储可以被所有节点共享,但作业的中间结果不会影响在其他网格节点上作业的进展。

2.1.2 集群的目的

集群为用户提供一个强大的计算机分布式系统,其目的如下。

(1) 提高性能。一些计算密集型应用,如天气预报、核试验模拟等,需要计算机有很强的运算处理能力,现有的技术,即使普通的大型机器计算也很难胜任,一般都使用计算机集群技术,集中几十台甚至上百台计算机的运算能力来满足要求,提高处理性能一直是集群技术研究的一个重要目标之一。

(2) 降低成本。通常一套较好的集群配置,其软、硬件开销要超过上百万元。但与价值上千万元的专用超级计算机相比已属相当便宜,在达到同样性能的条件下,采用计算机集群比采用同等运算能力的大型计算机具有更高的性价比。

(3) 提高可扩展性。用户若想扩展系统能力,不得不购买更高性能的服务器,才能获得额外所需的 CPU 和存储器。如果采用集群技术,则只需要将新的服务器加入集群中即可,对于客户来看,服务无论从连续性还是性能上都几乎没有变化,好像系统在不知不觉中完成了升级。

(4) 增强可靠性。集群技术使系统在故障发生时仍可以继续工作,将系统停运时间减到最小。集群系统在提高系统的可靠性的同时,也大大减小了故障损失。

2.2 Linux 操作系统

2.2.1 Linux 操作系统简介

Linux 是一个免费的开源操作系统,并且有许多不同的版本,但它们都使用 Linux 内核并且可以安装在各种计算机硬件设备中,如移动平板电脑、路由器、台式计算机、手机等。Linux 出现于 1991 年,是由芬兰赫尔辛基大学学生 Linus Torvalds 和后来加入的众多爱好者共同开发完成的。

2.2.2 Linux 操作系统的特性

Linux 操作系统是一个多用户、多任务、丰富的网络功能,它不仅有可靠的系统安全,而且有良好的可移植性,具有标准的兼容性,良好的用户界面,出色的速度性能,最为重要的是开源。本书使用的 CentOS 主要有以下特点。

(1) 主流。目前的 Linux 操作系统主要应用于生产环境,企业级主流 Linux 系统仍旧是 Redhat Linux 或者 CentOS。

(2) 免费。Redhat Linux 和 CentOS 差别不大,基于 Redhat Linux 提供的可自由使用源代码的企业 CentOS 是一个 Linux 发行版本。

(3) 更新方便。CentOS 独有的 yum 命令支持在线升级,可以即时更新系统,不像 Redhat Linux 那样需要花钱购买支持服务。

2.2.3 Linux 安装与基础操作

Linux 安装环境:CentOS 7.3,官网:http://www.centos.org/。Linux 的根目录下有许多文件(如 bin、sbin、etc、usr、home、root、dev、lib、mnt、boot、tmp、var),其每个目录结构下都会存放相应的文件。

bin(binaries):存放二进制可执行文件。

sbin(super user binaries):存放二进制可执行文件,只有 root 才能访问。

etc(etcetera):存放系统配置文件。

usr(unix shared resources)：存放共享的系统资源。
home：存放用户文件的根目录。
root：超级用户目录。
dev(devices)：存放设备文件。
lib(library)：存放根文件系统中的程序运行所需要的共享库及内核模块。
mnt(mount)：系统管理员安装临时文件系统的安装点。
boot：存放用于系统引导时使用的各种文件。
tmp(temporary)：存放各种临时文件。
var(variable)：存放运行时需要改变数据的文件。

2.2.4 Linux 常用命令

命令格式：

命令 -选项 参数

例如，ls -la /usr ，其中，ls 显示文件和目录列表(list)，而常用参数为-l(long)、-a(all)、-t(time)、-i(inode)，需要注意的是一些隐藏文件及特殊目录。

Linux 命令分为内部命令、外部命令、查看帮助文档三大类。内部命令属于 shell 解析器的一部分，如 cd(切换目录)、pwd(显示当前工作目录)；外部命令是独立于 shell 解析器之外的文件程序，如 ls(显示文件和目录以及目录列表)、mkdir(创建目录)、cp(复制文件或目录)；查看帮助文档如 help cd、man ls。

1. 目录操作类命令

pwd(print working directory)：显示当前工作目录。
cp(copy)：复制文件或目录。
- -r(recursive)：递归处理，将指定目录下的文件与子目录一并复制。

mkdir(make directory)：创建目录。
- -p(parents)：在父目录不存在的情况下先生成父目录。

mv(move)：移动文件或目录、文件或目录改名。
rm(remove)：删除文件。
- -r(recursive)：同时删除该目录下的所有文件。
- -f(force)：强制删除文件或目录。

rmdir(remove directory)：删除空目录。
ln(link)：建立连接文件。
- -s(symbolic)：对源文件建立符号连接，而非硬连接。

2. 文件浏览类命令

cat(catenate)：显示文本文件内容。
more、less：分页显示文本文件内容。

head、tail：查看文本中开头或结尾部分的内容。
haed -n 5 a.log：查看 a.log 文件的前 5 行。
tail -F b.log：循环读取(follow)。
wc(word count)：统计文本的行数、字数、字符数。
- -m：统计文本字符数。
- -w：统计文本字数。
- -l：统计文本行数。

3. 文件查找类命令

find：在文件系统中查找指定的文件。

```
find /etc/ - name " * .c"
```

grep：在指定的文本文件中查找指定的字符串。

4. 进程类命令

top：显示当前系统中耗费资源最多的进程。
ps：显示瞬间的进程状态。
- -e /-A：显示所有进程,环境变量。
- -f：全格式。
- -a：显示所有用户的所有进程(包括其他用户)。
- -u：按用户名和启动时间的顺序来显示进程。
- -x：显示无控制终端的进程。

jobs、bg、fg：任务操作,Ctrl+Z 组合键将当前任务放入后台。
kill：结束一个指定 pid 的进程。
- -l：可查看所有信号名称。

killall：结束指定名称的进程。

5. 文件归档类命令

gzip：压缩(解压)文件或目录,压缩文件扩展名为.gz。
bzip2：压缩(解压)文件或目录,压缩文件扩展名为.bz2。
tar：文件、目录打(解)包。
- -c：建立一个压缩文件的参数指令。
- -x：解开一个压缩文件的参数指令。
- -t：查看 tar 文件包中包含的目录和文件信息。
- -z：是否需要用 gzip 压缩。
- -j：是否需要用 bzip2 压缩。
- -v：压缩的过程中显示文件。
- -f：使用文档名,在 f 之后要立即接文档名。

6. 网络类命令

netstat：显示网络状态信息。
- -a：显示所有连接和监听端口。
- -t(tcp)：仅显示 TCP 相关选项。
- -u(udp)：仅显示 UDP 相关选项。
- -n：拒绝显示别名。
- -p：显示建立相关链接的程序名。

ifconfig：网卡网络配置详细信息。
ip addr：查看 IP 信息。
ping：测试网络的连通性。

7. 系统类命令

dmesg：显示系统设备等信息。
df：显示文件系统磁盘空间的使用情况。
du：显示指定的文件(目录)已使用的磁盘空间的总量。
- -h(human-readable)：文件大小以 KB、MB、GB 为单位显示。
- -s(summarize)：只显示各档案大小的总和。

free：显示当前内存和交换空间的使用情况。
date：时间操作。
cal：日历操作。
clock/hwclock：硬件时钟。

8. 其他类命令

touch：创建空文件。
man：查看文档详细帮助。
shutdown：系统关机。
- -r：关机后立即重新启动。
- -h：关机后不重新启动。

poweroff：关机后关闭电源,等价于 shutdown -h now。
reboot：重新启动,等价于 shutdown -r。
halt：关机后不关电源。
init：系统运行级别。
- 0：停机。
- 1：单用户模式。
- 2：多用户模式,没有 NFS。
- 3：完全多用户模式(标准的运行级)。
- 4：保留。

- 5：X11(xwindow)。
- 6：重新启动。

熟练地掌握这些常用命令,再配合一些快捷键的使用,操作起来会更加方便,特别是巧用 Tab 键,能在操作中节省大量时间,还有 Ctrl+C(停止当前进程)组合键、Ctrl+R(查看命令历史)组合键、Ctrl+L(清屏,与 clear 命令作用相同)组合键、Ctrl+D(终止输入或退出 shell)组合键。

2.3 虚拟化技术

2.3.1 虚拟化技术简介

在计算机中,虚拟化(Virtualization)是一种资源管理技术,是将计算机的各种实体资源,如服务器、网络、内存及存储等,予以抽象、转换后呈现出来,打破实体结构间的不可分割的障碍,使用户以比原本的组态更好的方式来应用这些资源。这些资源的新虚拟部分不受现有资源的架设方式、地域或物理组态所限制。一般所指的虚拟化资源包括计算能力和资料存储。

在实际的生产环境中,虚拟化技术主要用来解决高性能的物理硬件产能过剩和旧的硬件产能过低的重组重用问题,透明化底层物理硬件,从而最大化地利用物理硬件。

早期的计算机大多用于科学计算,计算机不仅价格昂贵,而且硬件资源的利用率低,用户的体验效果也差强人意,从而有了分时系统的提出。为了满足分时系统的需求,克里斯托弗(Christopher Strachey)在 1959 年召开的国际信息处理大会上发表了一篇名为 *Time Sharing in Large Fast Computers*(《大型高速计算机中的时间共享》)的学术报告,在这篇报告中首次提出了虚拟化的概念。

在随后的 10 年中,由于当时工业、科技条件的限制,计算机的硬件资源是相当昂贵的。IBM 在 1956 年推出的首部磁盘储存器件,总容量仅 5MB,但是平均每兆字节需花费 1 万美元。这远远超出了普通大众的承受范围,严重阻碍了人们对计算机的购买力。为了使昂贵的硬件资源得到充分利用,来提高自己的销售额,IBM 最早开发了一种虚拟机技术,能够让用户在一台主机上运行多个操作系统,IBM7044 计算机就是典型的代表。随后,虚拟化技术一直只在大型机上应用,而在 PC、x86 服务器的平台上仍然进展缓慢。

随着科技水平的提高,计算机硬件资源的价格逐渐降低,从 20 世纪 90 年代末开始,x86 计算机由于其成本低廉渐渐代替大型机,为了抢占市场的份额,VMware 需要考虑如何节省客户的开支,来提高自己产品的竞争力。这时,就有了虚拟化技术的再次发展。以 VMware 为代表的虚拟化软件生产商率先实施了以虚拟机监视器为中心的软件解决方案,为虚拟化技术在 x86 计算机环境的发展开辟了道路。

最近的十几年间,诸多厂商(如微软公司、Intel 公司、AMD 公司等)都开始进行虚拟化技术的研究。为了与 VMware 展开直接的竞争,微软公司开发了 Hype-V 技术。微软公司凭借其强大的技术支持,成为 VMware 在小企业市场的主要竞争对手。同时,虚拟化技术的飞速发展也引起了芯片厂商的重视,Intel 公司和 AMD 公司在 2006 年以后都逐步在其 x86 处理器中增加了硬件虚拟化技术。

2008 年以后，云计算技术的发展推动了虚拟化技术成为研究热点。由于虚拟化技术能够屏蔽底层的硬件环境，充分利用计算机的软硬件资源，是云计算技术的重要目标之一，虚拟化对云计算技术的发展产生重大意义是基于 x86 架构的服务器虚拟化技术。

2.3.2 虚拟技术的原理

虚拟机是对真实计算环境的抽象和模拟，虚拟机监视器（Virtual Machine Monitor，VMM）需要为每个虚拟机分配一套数据结构来管理它们的状态，包括虚拟处理器的全套寄存器、物理内存的使用情况、虚拟设备的状态等。

在如图 2-1 所示的典型的分层架构中，提供平台虚拟化的层称为 Hypervisor，它是一种运行在物理服务器和操作系统（Operating System，OS）之间的中间软件层，允许多个 OS 和应用共享一套基础物理硬件，因此也可以看作是虚拟环境中的"元"操作系统。Hypervisor 是所有虚拟化技术的核心，具有规划、部署、管理和优化虚拟基础结构等端到端功能，同时具有非中断地支持多工作负载迁移能力的基本功能。当服务器启动并执行 Hypervisor 时，它会给每一台虚拟机分配适量的内存、CPU、网络和磁盘，并加载所有虚拟机的 GuestOS。有时 Hypervisor 也称为虚拟机监控器（VMM），VMM 既可以与虚拟机在同一台主机上，也可以单独运行在一台主机上，VMM 与虚拟机互不影响。

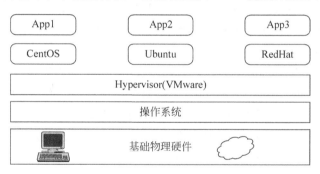

图 2-1 虚拟机在物理机上的结构图

简单来说，虚拟化就是让一台物理计算机能并发运行多个 OS。就拿熟悉的 VMware Workstation 在物理机上的架构来说，物理机的 OS 之上运行了多个进程，VMware 只是其中一个，但是在 VMware 之上又可以运行多个虚拟机，虚拟机的 OS 各自独立，可以是 RedHat、CentOS 和 Ubuntu 等。VMware 可以理解为是一款模拟了一堆硬件的应用软件，在这些硬件上可以运行额外的 OS。1974 年，Popek 和 Goldberg 在一篇论文中定义了"经典虚拟化（classical virtualization）"的基本需求，他们认为，一款真正意义上的 VMM 至少要符合以下三方面的标准。

(1) 等价执行（Equivalent execution）：程序在虚拟化环境中及真实环境中的执行是完全相同的。

(2) 性能（Performance）：指令集中的大部分指令要能够直接运行于 CPU 上。

(3) 安全（Safety）：VMM 要能够完全控制系统资源。

2.3.3 常见的虚拟化软件

1. VirtualBox

VirtualBox 是一款开源免费的虚拟机软件,使用简单、性能优越、功能强大且软件本身并不臃肿。VirtualBox 是由德国软件公司 InnoTek 开发的虚拟化软件,现隶属于 Oracle 旗下,并更名为 Oracle VirtualBox。其宿主机的操作系统支持 Linux、Mac、Windows 三大操作平台,在 Oracle VirtualBox 虚拟机里面,可安装的虚拟系统包括各个版本的 Windows 操作系统、MacOS x(32 位和 64 位都支持)、Linux 内核的操作系统、Open BSD Solaris、IBM OS2 甚至 Android 4.0 等,在这些虚拟的系统里面安装任何软件,都不会对原来的系统造成任何影响。与同类的 VMware Workstation 虚拟化软件相比,VirtualBox 对 Mac 系统的支持要好很多,运行比较流畅,配置比较简单,对于新手来说也不需要太多的专业知识,很容易掌握,并且免费,这一点也比商业化的 VMware Workstation 更吸引人,因此 VirtualBox 更适合预算有限的小公司。

2. VMware Workstation

VMware Workstation 是一款功能强大的商业虚拟化软件,与 VirtualBox 一样,仍然可以在一个宿主机上安装多个操作系统的虚拟机,宿主机的操作系统可以是 Windows 或 Linux,可以在 VMware Workstation 中运行的操作系统有 DOS、Windows 98、Windows XP、Windows 7、Windows 10、Linux、FreeBSD 等。VMware Workstation 虚拟化软件虚拟的各种操作系统仍然是开发、测试、部署新的应用程序的最佳解决方案。VMware Workstation 占的空间比较大,VMware 公司同时还提供一个免费、精简的 Workstation 环境——VMware Player,可在 VMware 官方网站下载使用。对于企业的 IT 开发人员和系统管理员而言,VMware Workstation 在虚拟网络、实时快照、拖曳共享文件夹、支持预启动执行环境(Preboot Execute Environment,PXE)等方面的特点使它成为必不可少的工具。

总体来看,VMware Workstation 的优点在于其计算机虚拟能力,物理机隔离效果非常优秀,它的功能非常全面,倾向于计算机专业人员使用,其操作界面也很人性化。VMware Workstation 的缺点在于其体积庞大,安装耗时较长,并且在运行使用时占用物理机的资源较大。

3. KVM

KVM(Kend-based Virtual Machine)是一种针对 Linux 内核的虚拟化基础架构,它支持具有硬件虚拟化扩展的处理器上的原生虚拟化。最初,它支持 x86 处理器,但现在广泛支持各种处理器和操作系统,包括 Linux、BSD、Solaris、Windows、Haiku、ReactOS 和 AR-OS 等。基于内核的虚拟机(KVM)是针对包含虚拟化扩展(IntelVT 或 AMD-V)的 x86 硬件 Linux 的完全原生虚拟化解决方案。对半虚拟化(paravirtualization)的有限支持也可以通过半虚拟网络驱动程序的形式用于 Linux 和 Windows Guest 系统。

KVM 使用灵活，Guest 操作系统与集成到 Linux 内核中的虚报机管理程序通信，直接寻址硬件，无须修改虚拟化的操作系统，这使得 KVM 成为更快的虚拟机解决方案。KVM 的补丁与 Linux 内核兼容，KVM 在 Linux 内核本身内实现，这进而简化了对虚拟化进程的控制，但是没有成熟的工具可用于 KVM 服务器的管理，KVM 仍然需要改进虚拟网络、虚拟存储的支持，并且增强安全性、高可用性、容错性、电源管理、HPC/实时支持、虚拟 CPU 可伸缩性、跨供应商兼容性、可移植性。

2.3.4 虚拟化技术的优势和劣势

1. 虚拟化技术的优势

虚拟化技术的出现和发展提高了资源的利用率，使得企业能以更低的成本获得更大的收益。从总体上而言，虚拟化的优势体现在以下几个方面。

（1）虚拟化技术可以提高资源利用率。传统的 IT 企业为每一项业务分配一台单独的服务器，服务器的实际处理能力往往远超服务器的平均负载，使得服务器大部分时间都处于空闲状态，造成资源的浪费。而虚拟化技术可以减少必须进行管理的物理资源的数量，隐藏了物理资源的部分复杂性。为了达到资源的最大利用率，虚拟化还把一组硬件资源虚拟化为多组硬件资源，并动态地调整空闲资源，减小服务器的规模。例如，VMware 的用户在使用 VMware 的虚拟基础架构解决方案之后，服务器的利用率通常可由原先的 5%～15% 提升到 60%～80%。

（2）提供相互隔离、高效的应用执行环境。虚拟化技术能够实现较简单的共享机制无法实现的隔离和划分，从而对数据和服务进行可控和安全的访问。例如，用户可以在单台计算机上模拟多个不同、相互之间独立的操作系统，这些虚拟的操作系统可以是 Windows 或 Linux 系统。其中的一个或多个子系统遭受攻击而崩溃时，不会对其他系统造成影响。在使用备份机制后，受到攻击的子系统可以快速恢复。

（3）虚拟化可以简化资源和资源的管理。计算机有硬盘等硬件资源和 Web 服务等软件资源。用户对计算机资源进行访问是通过标准接口来进行的，使用标准接口的好处是用户不用知道虚拟资源的具体实现。底层的基础设施发生变化时，只要标准接口没有发生变化，用户基本上感受不到这种变化。这是因为与用户直接接触的是标准接口，虽然底层的具体实现发生改变，但是用户与虚拟资源进行交互的方式并没有改变。

传统的 IT 服务器资源是硬件相对独立的个体。对每一种资源都要进行相应的维护和升级，会耗费大量的人力和物力。而虚拟化系统降低了用户与虚拟资源之间的耦合度，利用这种松耦合的关系，管理者可以在对用户影响最小的基础上对资源进行管理。此外，虚拟化系统还将资源进行整合，在管理上相对而言比较方便，在升级时也只需添加动作，从而提高工作效率。

（4）虚拟化技术实现软件和硬件的分离。用户在同一个计算机系统上可以运行多个软件系统，不同的软件系统通过虚拟机监视器（Virtual Machine Monitor，VMM）来使用底层的硬件资源，从而实现多个软件系统共享同一个硬件资源，达到软件和硬件的分离。这样，在虚拟化的统一资源池能够运行更多的软件系统，充分利用已有的硬件资源。

2. 虚拟化技术的劣势

任何事物都是有利有弊的，虚拟化技术也不例外。物理计算机上硬件用的时间久了可能会损坏，其上的软件也要定时地更新，防止病毒的感染。虚拟化技术由于是针对实际的计算机来进行的，虚拟化技术方案的部署与使用也有一些劣势。

（1）可能会使物理计算机负载过重。虚拟化技术虽然是在虚拟的环境中运行的，但是其并不是完全虚拟的，依然需要硬件的支持。以服务器虚拟化为例，一台物理计算机上可以虚拟化出多台客户机，每台客户机上又可以安装多个应用程序。若这些应用程序全部运行，就会占用大量的物理计算内存、CPU等硬件系统，从而给物理计算机带来沉重的负担，可能会导致物理计算机负载过重，使各虚拟机上的应用程序运行缓慢，甚至系统崩溃。

（2）升级和维护引起的安全问题。物理计算机的操作系统及操作系统上的各种应用软件都需要不定时地进行升级更新增强其抵抗攻击的能力，每台客户机也都需要进行升级更新。一台物理计算机上安装多台客户机，会导致在客户机上安装补丁速度缓慢。如果客户机上的软件不能及时更新，则会被病毒攻击，带来安全隐患。

（3）物理计算机的影响。传统的物理计算机发生不可逆转的损坏时，若不是作为服务器出现，则只有其自身受到影响。当采用虚拟化技术的物理计算机发生宕机时，其所有的虚拟机都会受到影响，在该计算机上运行的业务也会受到一定程度的影响，甚至损坏。此外，一台物理计算机的虚拟机之间相互通信，通信的过程中可能会导致安全风险。

2.4　CentOS大数据集群系统的组成

社区企业操作系统（Community Enterprise Operating System，CentOS）是Linux发行版之一，它是来自于Red Hat Enterprise Linux依照开放源代码规定释出的源代码所编译而成。由于出自同样的源代码，因此有些要求高度稳定性的服务器以CentOS替代商业版的Red Hat Enterprise Linux使用。二者的不同在于CentOS并不包含封闭源代码软件。

CentOS是RHEL（Red Hat Enterprise Linux）源代码再编译的产物，而且在RHEL的基础上修正了不少已知的Bug，相对于其他Linux发行版，其稳定性值得信赖。

CentOS集群系统拓扑图如图2-2所示。Master为主节点，作为管理服务器。为了避免Master单节点故障，使用双机热备；Standby为备用节点；Slave节点为从节点，作为数据处理服务器。

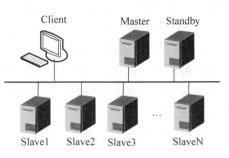

图2-2　CentOS集群系统拓扑图

2.5 大数据集群技术的架构

一般来说,大数据集群的构架主要分为硬件资源层、OS 层、基础设施管理层、文件系统层、资源管理和大数据集群层和大数据应用层,如图 2-3 所示。

图 2-3 大数据集群的架构

在硬件资源层中,这些硬件可能是不同厂商的机器,如 IBM、HP、DELL 或者联想等服务器,也有可能包括不同的构架,如 System X 或者 IBM POWER 等机器。这些机器有可能在机房,也有可能在云端(包括公有云和私有云)。硬件之上,需要安装运行操作系统(OS),一般为 Linux OS,如 RedHat、SUSE、Ubuntu 等。

在基础设施管理层,主要管理资源(更多的是软件资源)以及资源的虚拟化等,如网络资源/设备、计算资源、内存、Slots 等的统一管理和优化分配,此层同时肩负着部署大型 Cluster 的任务,也就是将各个分散的节点通过 Hadoop 软件,统一部署为一个整体。在 Hadoop 集群中,分为管理节点和计算节点。

并行文件系统是大数据集群的重要组成部分,因为大数据有两个主要的特征:一是数据量比较大,如此巨量数据的存储成为了集群需要解决的关键问题之一;二是处理速度快,随着集群技术的发展,并行化的思想尤为明显,并行化的计算产品和工具也层出不穷。所以,并行的文件系统是大数据集群中不可或缺的一部分。例如,在 Hadoop 时代,HDFS 就在 Hadoop 阵营中贡献了中流砥柱的作用。另外一个出色的并行文件系统是 IBM Spectrum Scale,其前身是 IBM GPFS,经过近年来的版本迭代和发展,已经可以支持目前流行的大数据计算模式,如 Spark 等。

在资源管理和大数据集群层,主要部署两方面的组件:一是大数据分析处理组件;二是资源调度和管理组件。在一般情况下,这二者有机地结合在一起,组成一个产品。随着大数据的发展,大数据的分析和处理技术如井喷一般涌现出来,例如,有 Hadoop、Spark、Storm、Dremel/Drill 等大数据解决方案展现出来。需要说明的是,这里所有的方案不是一种技术,而是数种,甚至数十种技术的组合。就拿 Hadoop 来说,后面的关键组件有 MapReduce、HDFS、Hive、HBase、Pig、ZooKeeper 等。资源调度管理主要是维护、分配、管理、监控软硬件资源，包括节点、网络资源、CPU、内存等，根据数据处理的需求来分配资源，并负责回收，如 YARN。

大数据应用层主要通过大数据的可视化来展示复杂的多维数据关系，通过图表展现大数据分析的结果，通过推荐系统提升大数据的应用效果。

2.6 操作实践：大数据集群的部署

2.6.1 集群规划

为了简化操作，使用三台服务器作为集群节点，其中一台为 Master 节点，两台为 Slave 节点。集群节点 IP 规划如下。本书所有的操作都使用 root 用户，切换命令是：su-root。

```
master: 172.30.0.10
slave1: 172.30.0.11
slave2: 172.30.0.12
netmask: 255.255.255.0
gateway:172.30.0.254
```

2.6.2 网络配置

对集群节点进行网络配置，以 Master 节点为例。

（1）设置主机名，操作如下。

```
# hostnamectl set-hostname master
```

或者

```
# vi /etc/hostname
master
```

（2）修改/etc/hosts 文件，操作如下。

```
# vi /etc/hosts
```

添加内容：

```
172.30.0.10    master
172.30.0.11    slave1
172.30.0.12    slave2
```

（3）修改网络配置，操作如下。

```
# vi /etc/sysconfig/network-scripts/ifcfg-ens33
```

添加修改内容：

```
TYPE = Ethernet
BOOTPROTO = none
NAME = ens33
DEVICE = ens33
ONBOOT = yes
IPADDR = 172.30.0.10
PREFIX = 24
GATEWAY = 172.30.0.254
DNS1 = 202.96.128.86
DNS2 = 202.96.134.133
```

(4) 重启网络，并查看网络 IP 地址，操作如下。

```
# systemctl restart network
# ifconfig -a
```

或者

```
# ip addr
```

(5) 关闭并停止 NetworkManager 服务，操作如下。

```
# systemctl disable NetworkManager
# systemctl stop NetworkManager
```

NetworkManager 是用于便携式计算机和其他可移动计算机的理想解决方案，可以自动连接到已知无线网络或并行管理多个网络连接，然后将最快的连接用作默认连接；还可以手动在可用网络之间切换，并使用系统盘中的小程序管理网络连接。NetworkManager 在 Hadoop 环境下不适用，需要关闭。

2.6.3 安全配置

作为一个开放源代码的操作系统，Linux 服务器以其安全、高效和稳定的显著优势得以广泛应用，最重要的安全模块有 SELinux、Iptables 和 firewalld。

视频讲解

SELinux(Security-Enhanced Linux)是美国国家安全局(NSA)对于强制访问控制的实现，是 Linux 历史上最杰出的新安全子系统，SELinux 共有 3 个状态：enforcing(执行中)、permissive(不执行但产生警告)和 disabled(关闭)。

Iptables 是与 Linux 内核集成的 IP 信息包过滤系统。如果 Linux 系统连接到因特网或 LAN、服务器或连接 LAN 和因特网的代理服务器，则该系统有利于在 Linux 系统上更好地控制 IP 信息包过滤和防火墙配置。防火墙在做信息包过滤决定时，有一套遵循和组成的规则，这些规则存储在专用的信息包过滤表中，而这些表集成在 Linux 内核中。

CentOS7 中默认将原来的防火墙 Iptables 升级为 firewalld。防火墙守护 firewalld 服务引入了一个信任级别的概念来管理与之相关联的连接与接口。它支持 IPv4 与 IPv6,并支持网桥,采用 firewall-cmd(command)或 firewall-config(gui)来动态地管理 kernel netfilter 的临时或永久的接口规则,并实时生效而无须重启服务。

为了便于学习,读者可以关闭这些安全配置,设置一些安全选项。

(1) 安全密码控制,操作如下。

```
# vi /etc/login.defs            //适用于新建的用户
```

修改内容:

```
PASS_MAX_DAYS    30             //密码最多使用 30 天,必须更改密码
PASS_MIN_DAYS    0              //密码最少使用 0 天,才能更改密码
PASS_MIN_LEN     5              //可接受的密码长度
PASS_WARN_AGE    7              //密码到期前的警告时间

# chage -M 30 用户名            //适用于已存在的用户,密码 30 天过期
# chage -d 0 用户名             //下次登录时必须更改密码
```

为了降低密码被猜出或暴力破解的风险,应避免长期使用同一个密码。管理员可以在服务器端限制用户密码使用最大有效期天数,对密码已过期的用户,登录时要求重新设置密码,否则拒绝登录。

(2) 设置历史记录、退出自动清空历史记录等,操作如下。

```
# vi /etc/profile
```

添加内容:

```
export HISTSIZE = 5             //历史记录限制为 5 条

# vi ~/.bash_logout
```

添加内容:

```
history -c                      //清空历史记录
```

(3) 设置闲置超时时间,操作如下。

```
# vi /etc/profile
```

添加内容:

```
export TMOUT = 600              //用户将在 10 分钟无操作后自动注销
```

(4) 设置 SElinux,操作如下。

```
# vi /etc/selinux/config
```

修改内容:

```
SELINUX = permissive

# setenforce permissive
# getenforce
```

(5) 设置并停止 firewalld 服务,操作如下。

```
# systemctl disable firewalld
# systemctl stop firewalld
```

2.6.4 时间同步

网络时间协议(Network Time Protocol,NTP)是用来使计算机时间同步化的一种协议,它可以使计算机对其服务器或时钟源(如石英钟、GPS 等)做同步化,并提供高精准度的时间校正,且可以加密确认的方式来防止恶毒的协议攻击。NTP 的目的是在无序的 Internet 环境中提供精确和健壮的时间服务。

NTP 提供准确时间,首先要有准确的时间来源,这一时间应该是国际标准时间(UTC)。NTP 获得 UTC 的时间来源可以是原子钟、天文台、卫星,也可以是 Internet,这样就有了准确而可靠的时间源。时间按 NTP 服务器的等级传播。按照离外部 UTC 源的远近将所有服务器归入不同的 Stratum(层)中。Stratum-1 在顶层,有外部 UTC 接入,而 Stratum-2 则从 Stratum-1 获取时间,Stratum-3 从 Stratum-2 获取时间,以此类推,但 Stratum 层的总数限制在 15 以内。Stratum-1 的时间服务器是整个系统的基础,所有这些服务器在逻辑上形成阶梯式的架构相互连接。计算机主机一般同多个时间服务器连接,利用统计学的算法过滤来自不同服务器的时间,以选择最佳的路径和来源来校正主机时间。即使主机在长时间无法与某一时间服务器相联系的情况下,NTP 服务依然有效运转。为防止对时间服务器的恶意破坏,NTP 使用了识别(Authentication)机制,检查时间服务器与资料的返回路径,以提供对抗干扰的保护机制。NTP 时间同步报文中包含的时间是格林威治时间,是从 1900 年开始计算的。

Chrony 是一个开源的自由软件,它能帮助用户保持系统时钟与时钟服务器同步,因此让用户的时间保持精确。它由 chronyd 和 chronyc 两个程序组成。chronyd 是一个后台运行的守护进程,用于调整内核中运行的系统时钟和时钟服务器同步,它确定计算机增减时间的比率,并对此进行补偿。chronyc 提供了一个用户界面,用于监控性能并进行多样化的配置,它可以在 chronyd 实例控制的计算机上工作,也可以在一台不同的远程计算机上工作。

服务器之间的时间需要同步,但并不是所有机器都可以直接连外网,这时可以用 Chrony 工具解决。解决方法是将其中一台机器设为时间服务器,然后其他服务器和这台时间服务器同步即可。

（1）Master 节点时间同步安装设置，操作如下。

```
[root@master ~]# yum install chrony
[root@master ~]# vi /etc/chrony.conf
```

添加修改内容：

```
local stratum 10
allow 172.30.0.0/24
```

（2）Slave 节点时间同步安装设置，操作如下。

```
[root@slave ~]# yum install chrony
[root@slave ~]# vi /etc/chrony.conf
```

添加修改内容：

```
# server 0.centos.pool.ntp.org iburst
# server 1.centos.pool.ntp.org iburst
# server 2.centos.pool.ntp.org iburst
# server 3.centos.pool.ntp.org iburst
server master iburst
```

（3）设置自动加载并重启 chrony 服务，操作如下。

```
# systemctl enable chronyd.service
# systemctl restart chronyd.service
```

（4）查看 Master 节点时间同步信息，操作如下。

```
[root@master ~]# chronyc sources
210 Number of sources = 4
MS Name/IP address         Stratum Poll Reach LastRx Last sample
===============================================================================
^* 85.199.214.100                1   6   377     3   +4577us[  +10ms] +/-  154ms
^+ static-5-103-139-163.ip.>     1   6   377    10    +43ms[  +49ms] +/-  186ms
^+ ntp.wdc1.us.leaseweb.net      2   6    33    13    -14ms[ -8522us] +/-  352ms
^- gus.buptnet.edu.cn            5   6   167    72   -3955us[ +1588us] +/-  371ms
```

（5）查看 Slave 节点时间同步信息，操作如下。

```
[root@slave ~]# chronyc sources
210 Number of sources = 1
MS Name/IP address         Stratum Poll Reach LastRx Last sample
===============================================================================
^* master                        2   6   377    42   -6960us[  -30ms] +/-  127ms
```

2.6.5 SSH 登录

SSH(Secure Shell)由 IETF 的网络小组(Network Working Group)所制定。SSH 是建立在应用层基础上的安全协议,也是目前较可靠,专为远程登录会话和其他网络服务提供安全性的协议。利用 SSH 协议可以有效防止远程管理过程中的信息泄露问题。SSH 最初是 UNIX 系统上的一个程序,后来迅速扩展到其他操作平台。SSH 在正确使用时可弥补网络中的漏洞。SSH 客户端适用于多种平台,几乎所有 UNIX 平台(包括 HP-UX、Linux、AIX、Solaris、Digital UNIX、Irix)及其他平台,都可运行 SSH。

从客户端来看,SSH 提供两种级别的安全验证。

第一种级别:基于口令的安全验证。

只要用户知道自己的账号和口令,就可以登录到远程主机。所有传输的数据都会被加密,但是不能保证用户正在连接的服务器就是用户想连接的服务器。可能会有别的服务器在冒充真正的服务器,也就是受到"中间人"这种方式的攻击。

第二种级别:基于密钥的安全验证。

需要依靠密钥,也就是用户必须为自己创建一对密钥,并把公用密钥放在需要访问的服务器上。如果用户要连接到 SSH 服务器上,客户端软件就会向服务器发出请求,请求用户用自己的密钥进行安全验证。服务器收到请求之后,先在该服务器上的主目录下寻找用户的公用密钥,然后把它与发送过来的公用密钥进行比较。如果两个密钥一致,服务器就用公用密钥加密"质询"(challenge)并把它发送给客户端软件。客户端软件收到"质询"之后就可以对用户的私人密钥解密,再把它发送给服务器。这种级别不仅加密所有传送的数据,而且防止受到"中间人"这种方式的攻击。

SSH 主要由传输层协议、用户认证协议、连接协议三部分组成。

(1) 传输层协议(SSH-TRANS):提供了服务器认证、数据机密性、信息完整性等方面的保护,也提供压缩功能。SSH-TRANS 通常运行在 TCP/IP 连接上,也可能用于其他可靠数据流上。该协议中的认证基于主机,并且该协议不执行用户认证。更高层的用户认证协议可以设计在此协议之上。

(2) 用户认证协议(SSH-USERAUTH):用于向服务器提供客户端用户鉴别功能。它运行在传输层协议 SSH-TRANS 上面。当 SSH-USERAUTH 开始后,它从低层协议那里接收会话标识符(第一次密钥交换中的哈希值)。会话标识符唯一标识此会话并且适用于标记已证明私钥的所有权。SSH-USERAUTH 也需要知道低层协议是否提供保密性保护。

(3) 连接协议(SSH-CONNECT):将多个加密隧道分成逻辑通道,它运行在用户认证协议上。它提供了交互式登录会话、远程命令执行、转发 TCP/IP 连接和转发 X11 连接。

SSH 是由客户端和服务端的软件组成的,服务端是一个守护进程(daemon),它在后台运行并响应来自客户端的连接请求。服务端一般是 sshd 进程,提供了对远程连接的处理,包括公共密钥认证、密钥交换、对称密钥加密和非安全连接。

SSH 客户端包含 ssh 程序,以及 scp(远程复制)、slogin(远程登录)、sftp(安全 FTP

文件传输)等其他的应用程序。

Hadoop 系统需要使用基于密钥的安全验证的 SSH 登录,Master 节点需要自动登录 Slave 节点,按下面步骤进行安装配置。

(1) 安装 openssh,开启 sshd 服务,操作如下。

```
# yum install openssh
# systemctl enable sshd.service
# systemctl start sshd.service
```

(2) 以 root 用户登录 Master 节点,生成 SSH 密钥对,操作如下。

```
[root@master ~]# ssh-keygen -t dsa
Generating public/private dsa key pair.
Enter file in which to save the key (/root/.ssh/id_dsa): 按 Enter 键
Created directory '/root/.ssh'. 按 Enter 键
Enter passphrase (empty for no passphrase): 按 Enter 键
Enter same passphrase again: 按 Enter 键
Your identification has been saved in /root/.ssh/id_dsa.
Your public key has been saved in /root/.ssh/id_dsa.pub.
The key fingerprint is:
26:f3:cb:a9:8b:e9:7c:ed:a3:93:20:26:14:ef:99:26 root@master
The key's randomart image is:
+--[ DSA 1024]----+
|                 |
| .               |
|  o              |
| ..              |
|. . o o S        |
|.Eo=.   =        |
| oo. . o.        |
|  . o+.oo        |
|   .=.+*=.       |
+-----------------+
```

ssh-keygen 加密方式选为 rsa 和 dsa,默认为 dsa,分别生成含有公用密钥和私用密钥的文件,存放在用户目录的~/.ssh 下。

(3) 把含有公用密钥信息的文件复制到节点机上,操作如下。

```
[root@master ~]# ssh-copy-id -i ~/.ssh/id_dsa.pub master
[root@master ~]# ssh-copy-id -i ~/.ssh/id_dsa.pub slave1
[root@master ~]# ssh-copy-id -i ~/.ssh/id_dsa.pub slave2
```

ssh-copy-id 自动把 id_dsa.pub 中的公用密钥信息添加到 Slave 节点对应用户的~/.ssh/目录下的 authorized_keys 文件中。

（4）使用 SSH 登录节点机，测试如下。

```
[root@master ~]# ssh master
[root@master ~]# exit
登出
Connection to master closed.
[root@master ~]# ssh slave1
[root@slave1 ~]# exit
登出
Connection to slave1 closed.
[root@master ~]# ssh slave2
[root@slave2 ~]# exit
登出
Connection to slave2 closed.
```

小结

集群可以提高系统性能，提高可用性和增强灵活性，能很好地完成复杂的任务。本章首先介绍了集群系统的概念、分类、目的，然后简述了集群相关的技术基础：Linux 操作系统、计算机虚拟化技术及常见的虚拟化软件，接着分析了数据集群技术的架构，最后重点介绍 CentOS 集群系统及部署，以及 SSH 安全登录及配置。

习题

1. 简述大数据集群系统。
2. 简述集群系统的分类。
3. 简述 Linux 操作系统的特性。
4. 简述计算机虚拟化技术以及常见的虚拟化软件。
5. 简述大数据集群技术的架构。
6. 安装 Linux 系统并进行网络配置。
7. 部署 Linux 集群、设置时间同步以及免密钥 SSH 配置。

第 3 章

视频讲解

Hadoop分布式系统

3.1 Hadoop 概述

众所周知，Hadoop 是一个由 Apache 基金会开发的分布式系统基础架构，可安装在一个商用机器集群中，使机器可彼此通信并协同工作，以高度分布式的方式共同存储和处理大量数据。简单地说，Hadoop 是一个海量数据分布式处理的开源软件框架，被部署到一个集群上。

3.1.1 Hadoop 简介

Hadoop 是一个能够让用户轻松架构和使用的分布式计算平台。用户可以轻松地在 Hadoop 上开发和运行处理海量数据的应用程序。

Hadoop 是以一种可靠、高效、可伸缩的方式进行处理的。Hadoop 是可靠的，因为它假设计算元素和存储会失败，所以它维护多个工作数据副本，确保能够针对失败的节点重新分布处理。Hadoop 是高效的，因为它以并行的方式工作，通过并行处理加快处理速度。Hadoop 还是可伸缩的，能够处理 PB 级数据。此外，Hadoop 依赖于社区服务器，因此它的成本比较低，任何人都可以使用。它主要有以下几个优点。

(1) 高可靠性。Hadoop 按位存储和处理数据的能力值得人们信赖。

(2) 高扩展性。Hadoop 是在可用的计算机集簇间分配数据并完成计算任务的，这些集簇可以方便地扩展到数以千计的节点中。

(3) 高效性。Hadoop 能够在节点之间动态地移动数据，并保证各个节点的动态平衡，因此处理速度非常快。

(4) 高容错性。Hadoop 能够自动保存数据的多个副本，并且能够自动将失败的任务重新分配。

3.1.2 Hadoop 的发展历程

2003 年,Google 发表了第一篇关于其云计算核心技术 GFS 的论文。该论文呈现出 Doug Cutting 等在研发 Apache 开源项目 Nutch 搜索引擎时,正面临着如何将其架构扩展到可以处理数 10 亿规模网页的难题。继而,他们于 2004 年编写了一个开放源代码的类似系统——Nutch 分布式文件系统(Nutch Distributed File System,NDFS)。同年,Google 在著名的 OSDI 国际会议上发表了一篇题为 *MapReduce：Simplified Data Processing on Large Clusters* 的论文,简要介绍 MapReduce 的基本设计思想。该论文发表后,MapReduce 编程模型在解决大型分布式并行计算问题上具有极大的可操作性。紧接着,Nutch 团队尝试依据 Google MapReduce 的设计思想,模仿 Google MapReduce 框架的设计思路,用 Java 设计实现出了一套新的 MapReduce 并行处理软件系统,并将其与 Nutch 分布式文件系统 NDFS 结合,用以支持 Nutch 搜索引擎的数据处理。2006 年,他们把 NDFS 和 MapReduce 从 Nutch 项目中分离出来,成为一套独立的大规模数据处理软件系统。有意思的是这个系统的命名是使用 Doug Cutting 小儿子当时牙牙学语称呼自己的玩具小象的名字"Hadoop"。Hadoop 能支持 PB 级海量数据,可扩展性强。可靠、高效、可扩展和开源的特性,使得 Hadoop 技术得到了迅猛发展,并在 2008 年成为 Apache 的顶级项目。

自从 2006 年 Hadoop 正式成为 Apache 开源组织的独立项目后,由于其低成本、高性能,深受广大用户的欢迎,经过短短几年的发展,Hadoop 及其技术在不断地改进完善中,目前已形成一个强大的系统。

下面列举了 Hadoop 在成为独立项目后的发展与演进中的重要事件,以便大家了解 Hadoop 的发展历程。

(1) Hadoop 最初是由 Apache Lucene 项目的创始人 Doug Cutting 开发的文本搜索库。Hadoop 源自于 2002 年的 Apache Nutch 项目——一个开源的网络搜索引擎,并且也是 Lucene 项目的一部分。

(2) 2004 年,Nutch 项目也模仿 GFS 开发了自己的分布式文件系统 NDFS(Nutch Distributed File System),也就是 HDFS 的前身。

(3) 2004 年,谷歌公司又发表了另一篇具有深远影响的论文,阐述了 MapReduce 分布式编程的思想。

(4) 2005 年,Nutch 开源实现了谷歌的 MapReduce。

(5) 2006 年 2 月,Apache Hadoop 项目正式启动以支持 MapReduce 和 HDFS 的独立发展。

(6) 2007 年 4 月,雅虎公司实现了包含 1000 个计算节点的 Hadoop 集群。

(7) 2008 年,淘宝开始投入研究基于 Hadoop 的系统——云梯,并将其用于处理电子商务相关数据。云梯 1 的总容量大概为 9.3PB,包含了 1100 台机器,每天处理约 18 000 道作业,扫描 500TB 数据。

(8) 2008 年 1 月,Hadoop 成为 Apache 顶级项目,获得了业界更为广泛的关注。

(9) 2008 年 2 月,雅虎公司宣布其搜索引擎产品部署在一个拥有 1 万个内核的

Hadoop 集群上。

(10) 2008 年 7 月,Hadoop 打破 1TB 数据排序基准测试记录。雅虎公司的一个 Hadoop 集群用 209 秒完成 1TB 数据的排序,比上一年的纪录保持者保持的 297 秒快了近 90 秒。

(11) 2009 年 5 月,雅虎的团队使用 Hadoop 对 1TB 的数据进行排序只花了 62 秒时间。

(12) 2009 年 7 月,Hadoop Core 项目更名为 Hadoop Common;MapReduce 和 HDFS 成为 Hadoop 项目的独立子项目;Avro 和 Chukwa 成为 Hadoop 新的子项目。

(13) 2010 年 5 月,Avro 数据传输中间件和 HBase 数据库从 Hadoop 项目中脱离出来,成为 Apache 顶级项目。此外,IBM 提供了基于 Hadoop 的大数据分析软件——InfoSphere BigInsights,包括基础版和企业版。

(14) 2010 年 9 月,Hive 数据仓库工具和 Pig 数据分析平台从 Hadoop 项目中脱离出来,成为 Apache 顶级项目。

(15) 2011 年 1 月,ZooKeeper 脱离 Hadoop,成为 Apache 顶级项目。

(16) 2011 年 5 月,Mapr Technologies 公司推出分布式文件系统和 MapReduce 引擎——Mapr Distribution for Apache Hadoop。该项目由 Hortonworks 在 2010 年 3 月提出,HCatalog 主要用于解决数据存储、元数据的问题,主要解决 HDFS 的瓶颈,它提供了一个地方来存储数据的状态信息,这使得数据清理和归档操作可以很容易地进行。

(17) 2011 年 8 月,Cloudera 公布了一项有益于合作伙伴生态系统的计划——创建一个生态系统,以便硬件供应商、软件供应商及系统集成商可以一起探索如何使用 Hadoop 更好地洞察数据。

(18) 2011 年 12 月,Hadoop1.0.0 版本发布,标志着 Hadoop 技术进入成熟期。

(19) 2012 年 5 月,Hadoop 发布 2.0 Alpha 版本,对 MapReduce、HDFS 等部分进行了重大改进,标志着 Hadoop 技术进入一个新的发展阶段。

(20) 2013 年 8 月,Hadoop1.2.1 稳定版发布。

(21) 2014 年 2 月,Spark 逐渐代替 MapReduce 成为 Hadoop 的默认执行引擎,并成为 Apache 基金会顶级项目。

(22) 2015 年 10 月,Cloudera 公布继 HBase 以后的第一个 Hadoop 原生存储替代方案——Kudu。

(23) 2015 年 12 月,Cloudera 发起的 Impala 和 Kudu 项目加入 Apache 孵化器。

(24) 2017 年 12 月,Apache Hadoop 3.0.0 版本发布。

3.1.3 Hadoop 原理及运行机制

Hadoop 的核心由 3 个子项目组成:Hadoop Common、HDFS、和 MapReduce。其中,Hadoop Common 在 Hadoop 0.20 版本以前被称为 Hadoop Core。Hadoop Common 子项目为 Hadoop 整体构架提供基础支撑性功能。Hadoop Common 包括文件系统(File System)、远程过程调用协议(RPC)和数据串行化库(Serialization Libraries)。HDFS 是一个分布式文件系统,具有低成本、高可靠性、高吞吐量的特点。MapReduce 是一个计算模型,用于大数据量的计算。在实际应用环境中,这 3 个核心子项目配合默契,结合其他子项目共同完成用户提交的大数据处理请求。下面主要讲述了 HDFS 和 MapReduce 这两个核心子项目所包含的主要逻辑组件。

1. HDFS 组件

HDFS(Hadoop Distributed FileSystem)是一种专门为 MapReduce 这类框架下的大规模分布式数据处理而设计的文件系统。可以把一个大数据集(100TB)在 HDFS 中存储为单个文件,大多数其他的文件系统无力实现这一点。

HDFS 的组件主要有 Namenode、SecondaryNamenode 及 Datanode。

1) Namenode

Namenode,即元数据节点。元数据节点用来管理文件系统的命名空间。它将所有文件和文件夹的元数据保存在一个文件系统树中。这些信息也会存储在 Namenode 维护的两个本地磁盘文件中,这两个本地磁盘文件是命名空间镜像文件(name space image)和编辑日志文件(editlog)。Namenode 还保存了一个文件包括哪些数据块、分布在哪些数据节点上这些信息。然而这些信息并不存储在硬盘上,而是在系统启动的时候从数据节点收集而来。

2) SecondaryNamenode

SecondaryNamenode,即从元数据节点。在 Hadoop 集群环境上,只有一个 Namenode 节点。那么,一旦 Namenode 节点出现故障,整个系统将会受到影响。为了提高 Namenode 的可靠性,从 Hadoop 0.23 版本开始引入了 Secondary Namenode。但是,SecondaryNamenode 并不是 Namenode 出现问题的时候的备用节点,它和 Namenode 负责不同的事情。SecondaryNamenode 的主要功能是周期性地将元数据节点命名空间的镜像文件和修改日志文件合并,以防日志文件过大。合并过后的命名空间镜像文件也在 SecondaryNamenode 中保存了一份,以便在元数据节点出现故障时,可以恢复数据。

3) Datanode

Datanode,即数据节点。Datanode 是文件系统中真正存储数据的地方,是 HDFS 文件系统中保存数据的节点。HDFS 中的文件通常被分割成多个数据块,以冗余备份的形式存储在多个 Datanode 中。客户端(client)或者元数据信息(Namenode)可以向数据节点请求写入或者读出数据块;而 Datanode 周期性地向 Namenode 汇报其存储的数据块信息。

2. MapReduce 组件

MapReduce 也采用了 Master/Slave(M/S)架构。它主要由 JobClient、JobTracker、TaskTracker 和 Task 组件组成。下面分别对这几个组件进行介绍。

1) JobClient

用户编写的 MapReduce 程序通过 JobClient 提交到 JobTracker 端;同时,用户可通过 Client 提供的一些接口查看作业运行状态。在 Hadoop 内部用"作业"(Job)表示 MapReduce 程序。一个 MapReduce 程序可对应若干个作业,而每个作业会被分解成若干个 Map/Reduce 任务(Task)。

2) JobTracker

JobTracker 主要负责 MapReduce 的资源监控和作业调度。JobTracker 监控所有 TaskTracker 与作业的状态情况，一旦发现失败情况后，其会将相应的任务转移到其他节点；同时，JobTracker 会跟踪任务的执行进度、资源使用量等信息，并将这些信息告诉任务调度器，而调度器会在资源出现空闲时，选择合适的任务使用这些资源。在 Hadoop 中，任务调度器是一个可插拔的模块，用户可以根据自己的需要设计相应的调度器。每一个 Hadoop 集群中只有一个 JobTracker。

3) TaskTracker

TaskTracker 主要负责执行由 JobTracker 分配的任务。TaskTracker 会周期性地通过 Heartbeat 将本节点上资源的使用情况和任务的运行进度汇报给 JobTracker，同时接收 JobTracker 发送过来的命令并执行相应的操作（如启动新任务、结束任务等）。

4) Task

Task 分为 MapTask 和 ReduceTask 两种，均由 TaskTracker 启动，负责具体地执行 Map 任务和 Reduce 任务的程序。

3.2 Hadoop 相关技术及生态系统

Hadoop 最为核心的技术是 HDFS 和 MapReduce。除此之外，为了满足大数据平台更高的存储和运算要求，Hadoop 技术不断拓展，在原来的基础上研发了很多其他技术，构成一个完整的分布式计算系统。Hadoop 生态系统主要包括 HDFS、MapReduce、Spark、Storm、HBase、Hive、Pig、ZooKeeper、Avro、Sqoop、Ambari、HCatalog、Chukwa、Flume、Tez、Phoenix、Mahout、Shark 等，Hadoop 生态系统如图 3-1 所示。

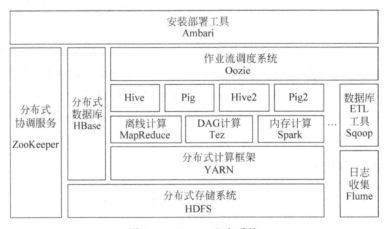

图 3-1 Hadoop 生态系统

（1）HDFS(Hadoop Distributed File System)：Hadoop 分布式文件系统，由早期的 NDFS 演化而来。它是一个高度容错的系统，能检测和应对硬件故障，用于在低成本的通用硬件上运行。并且，其简化了文件的一致性模型，通过流式数据访问，提供高吞吐量应用程序数据访问功能，适合带有大型数据集的应用程序。

(2) MapReduce：一个编程模型和软件框架，用于在大规模计算机集群上编写对大数据进行快速处理的并行化程序。

(3) Spark：一个开源的数据分析集群计算框架，最初由加州大学伯克利分校 AMPLab 开发，建立于 HDFS 之上。Spark 与 Hadoop 一样，用于构建大规模、低延时的数据分析应用。Spark 采用 Scala 语言实现，使用 Scala 作为应用框架，能够对大数据进行分析处理。

(4) Storm：一个分布式的、容错的实时计算系统，由 BackType 开发，后被 Twitter 收购。Storm 属于流处理平台，多用于实时计算并更新数据库。Storm 也可被用于"连续计算"(continuous computation)，对数据流做连续查询，在计算时就将结果以流的形式输出给用户。它还可用于"分布式 RPC"，以并行的方式运行大型的运算。

(5) HBase：一个分布式和面向列的动态模式数据库，不同于一般的关系数据库，它是一个适合于非结构化大数据存储的数据库，支持随机、实时读/写访问。

(6) Hive：Hadoop 中的一个重要子项目，最早由 Facebook 设计，是建立在 Hadoop 基础上的数据仓库架构，它提供了类似于 SQL 的查询语言，通过实现该语言，可以方便地进行数据汇总、特定查询，以及分析存放在 Hadoop 兼容文件系统中的大数据。

(7) Pig：运行在 Hadoop 上，是对大型数据集进行分析和评估的平台。Pig 包括了一个数据分析语言和运行环境，其特点是其结构设计支持真正的并行化处理，因此适合应用于大数据处理环境。与 Hive 一样，Pig 降低了对大型数据集进行分析和评估的门槛。

(8) Oozie：一个管理 Hadoop 作业、可伸缩、可扩展、可靠的工作流调度系统，它内部定义了三种作业：①工作流作业：由一系列动作构成的有向无环图(DAGs)；②协调器作业：按时间频率周期性触发 Oozie 工作流的作业；③Bundle 作业：管理协调器作业。

(9) ZooKeeper：作为一个分布式的服务框架，解决了分布式计算中的一致性问题。在此基础上，ZooKeeper 可用于处理分布式应用中经常遇到的一些数据管理问题，如统一命名服务、状态同步服务、集群管理、分布式应用配置项的管理等。

(10) Avro：由 Doug Cutting 牵头开发，是一个数据序列化系统。类似于其他序列化机制，Avro 可以将数据结构或者对象转换成便于存储和传输的格式，其设计目标是用于支持数据密集型应用，适合大规模数据的存储与交换。

(11) Sqoop：SQL-to-Hadoop 的缩写，是 Hadoop 的周边工具，它的主要作用是在结构化数据存储与 Hadoop 之间进行数据交换，即可以用于传统数据库（如 SQL、Oracle）中的数据导入 HDFS 或者 MapReduce，并将处理后的结果导出到传统数据库中。

(12) Ambari：一个用于安装、管理和监控 Hadoop 集群的 Web 界面工具，它提供一个直观的操作工具和一个健壮的 Hadoop API，可以隐藏复杂的 Hadoop 操作，使集群操作大大简化。

(13) HCatalog：一个用于管理 Hadoop 产生的数据的表存储管理系统。它提供了一个共享的数据模板和数据类型的机制，并对数据表进行抽象，同时支持 Hadooop 不同数据处理工具之间的联系。

(14) Chukwa：开源的数据收集系统，用于监控大规模分布式系统。它构建在 Hadoop 的 HDFS 和 MapReduce 基础之上，继承了 Hadoop 的可伸缩性和鲁棒性。Chukwa

包含一个强大和灵活的工具集，提供了数据的生成、收集、排序、去重、分析和展示等一系列功能，是 Hadoop 使用者、集群运营人员和管理人员的必备工具。

（15）Flume：Flume 是 Cloudera 开发维护的分布式、可靠、高可用的日志收集系统。它将数据从产生、传输、处理并最终写入目标的路径的过程抽象为数据流，在具体的数据流中，数据源支持在 Flume 中定制数据发送方，从而支持收集各种不同协议的数据。

（16）Tez：一个基于 Hadoop YARN 之上的 DAG（有向无环图，Directed Acyclic Graph）计算框架。它把 Map/Reduce 过程拆分成若干个子过程，同时可以把多个 Map/Reduce 任务组合成一个较大的 DAG 任务，减少了 Map/Reduce 之间的文件存储。同时合理组合其子过程，减少任务的运行时间。

（17）Phoenix：一个构建在 Apache HBase 之上的 SQL 中间层，完全使用 Java 编写，提供 HBase scan，并编排执行以生成标准的 JDBC 结果集，直接使用 HBase API、协同处理器与自定义过滤器。对于简单查询来说，其性能量级是毫秒；对于百万级别的行数来说，其性能量级是秒。

（18）Mahout：一种基于 Hadoop 的机器学习和数据挖掘的分布式计算框架算法集，实现了多种 MapReduce 模式的数据挖掘算法。

（19）Shark：即 Hive on Spark，一个专为 Spark 打造的大规模数据仓库系统，兼容 Apache Hive，无须修改现有的数据或者查询，就可以用 100 倍的速度执行 Hive QL。Shark 支持 Hive 查询语言、元存储、序列化格式及自定义函数，与现有 Hive 部署无缝集成，是一个更快、更强大的替代方案。

3.3 操作实践：Hadoop 安装与配置

视频讲解

3.3.1 安装 JDK

由于 Hadoop 是由 Java 语言开发的，所以需要安装 JDK。安装 JDK 的具体步骤如下。

（1）下载 JDK 安装包 jdk-8u141-linux-x64.tar.gz（本书将下载好的安装包放在 home/hadoop/software 目录下）。

（2）卸载 Centos 自带的 OpenJDK（root 权限下）。

查看系统已有的 openjdk，如图 3-2 所示。

```
[root@master ~]# rpm -qa|grep jdk
```

图 3-2　查看 openjdk

卸载上述找到的 openjdk 包，如图 3-3 和图 3-4 所示。

```
[root@master ~]# yum -y remove java-1.8.0-openjdk-headless-1.8.0.102-4.b14.el7.x86_64
```

图 3-3　卸载 openjdk 包（一）

```
[root@master ~]# yum -y remove java-1.7.0-openjdk-headless-1.7.0.111-2.6.7.8.el7.x86_64
```

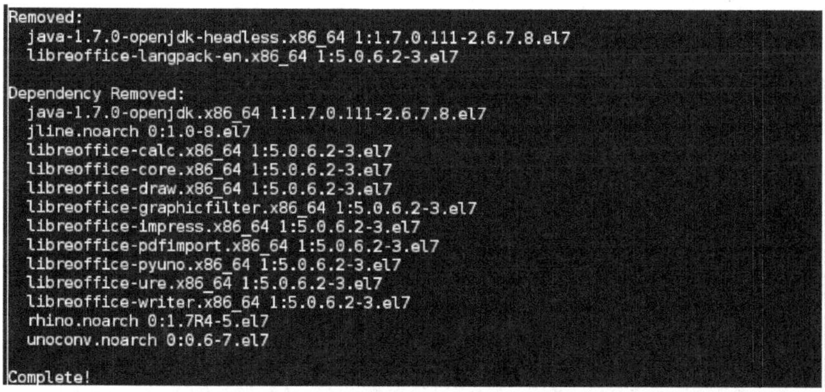

```
Removed:
  java-1.7.0-openjdk-headless.x86_64 1:1.7.0.111-2.6.7.8.el7
  libreoffice-langpack-en.x86_64 1:5.0.6.2-3.el7
Dependency Removed:
  java-1.7.0-openjdk.x86_64 1:1.7.0.111-2.6.7.8.el7
  jline.noarch 0:1.0-8.el7
  libreoffice-calc.x86_64 1:5.0.6.2-3.el7
  libreoffice-core.x86_64 1:5.0.6.2-3.el7
  libreoffice-draw.x86_64 1:5.0.6.2-3.el7
  libreoffice-graphicfilter.x86_64 1:5.0.6.2-3.el7
  libreoffice-impress.x86_64 1:5.0.6.2-3.el7
  libreoffice-pdfimport.x86_64 1:5.0.6.2-3.el7
  libreoffice-pyuno.x86_64 1:5.0.6.2-3.el7
  libreoffice-ure.x86_64 1:5.0.6.2-3.el7
  libreoffice-writer.x86_64 1:5.0.6.2-3.el7
  rhino.noarch 0:1.7R4-5.el7
  unoconv.noarch 0:0.6-7.el7
Complete!
```

图 3-4　卸载 openjdk 包（二）

这时，已有 Openjdk 卸载完了。（注：同样的操作在节点 slave1 和 slave2 上进行。）

（3）传输文件，将文件夹 software 中的 jdk 安装包复制到文件夹/opt/java/里面。

```
[root@master ~]# cd /home/hadoop/software
[root@master software]# ls
hadoop-2.7.3-src.tar.gz jdk-8u141-linux-x64.tar.gz
[root@master software]# cp jdk-8u141-linux-x64.tar.gz/opt/java/
```

将/home/hadoo/下的安装包复制到/opt/目录下，如图 3-5 所示。

```
[root@master opt]# cp /home/hadoop/jdk-8u141-linux-x64.tar.gz /opt/
```

```
[root@master opt]# cp /home/hadoop/jdk-8u141-linux-x64.tar.gz /opt/
[root@master opt]# ls
jdk-8u141-linux-x64.tar.gz  rh
```

图 3-5　复制安装包

解压命令：

```
tar -zxvf 安装包名
[root@master opt]# tar -zxvf jdk-8u141-linux-x64.tar.gz
```

查看解压后的安装包：

```
[root@master opt]# ls
jdk1.8.0_141 jdk-8u141-linux-x64.tar.gz rh
```

同样的安装步骤在其余节点上进行，如图 3-6 和图 3-7 所示。

```
[root@slave1 opt]# ls
jdk1.8.0_141  jdk-8u141-linux-x64.tar.gz  rh
[root@slave1 opt]#
```

图 3-6　查看安装包(一)

```
[root@slave2 opt]# ls
jdk1.8.0_141  jdk-8u141-linux-x64.tar.gz  rh
[root@slave2 opt]#
```

图 3-7　查看安装包(二)

进入目录下查看:

```
[root@master software]# cd /opt/java
[root@master java]# ls
jdk-8u141-linux-x64.tar.gz
[root@master java]# tar -zxvf jdk-8u141-linux-x64.tar.gz
```

安装完毕后,记录下 jdk 的路径。
(4) 将安装的 jdk 路径添加至系统环境变量中。
修改系统环境变量命令:

```
[root@master ~]# gedit /root/.bash_profile
```

在文件末尾加上如下内容,如图 3-8 所示。

```
export JAVA_HOME=/opt/java/jdk1.8.0_141/
export PATH=$JAVA_HOME/bin:$PATH
export PATH=$PATH:$JAVA_HOME/bin:$JRE_HOME/bin
export CLASSPATH=.:$JAVA_HOME/lib/dt.jar:$JAVA_HOME/lib/tools.jar
```

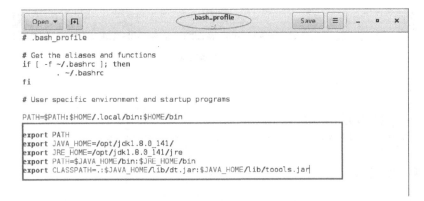

图 3-8　将安装后 jdk 路径添加至系统环境变量中

(5) 关闭 profile 文件,执行下列命令使配置生效:

```
[root@master ~]# source /root/.bash_profile
```

此时，我们就可以通过 java -version 命令检查 jdk 路径是否配置成功。

Master，如图 3-9 所示。

```
root@master ~]# vi /root/.bash_profile
root@master ~]# source /root/.bash profile
root@master ~]# java -version
ava version "1.8.0_141"
Java(TM) SE Runtime Environment (build 1.8.0_141-b15)
Java HotSpot(TM) 64-Bit Server VM (build 25.141-b15, mixed mode)
root@master ~]#
```

图 3-9 Master

Slave1，如图 3-10 所示。

```
root@slave1 ~]# vi /root/.bash_profile
root@slave1 ~]# source /root/.bash profile
root@slave1 ~]# java -version
ava version "1.8.0_141"
Java(TM) SE Runtime Environment (build 1.8.0_141-b15)
Java HotSpot(TM) 64-Bit Server VM (build 25.141-b15, mixed mode)
root@master ~]#
```

图 3-10 Slave1

Slave2，如图 3-11 所示。

```
root@slave2 ~]# vi /root/.bash_profile
root@slave2 ~]# source /root/.bash profile
root@slave2 ~]# java -version
ava version "1.8.0_141"
Java(TM) SE Runtime Environment (build 1.8.0_141-b15)
Java HotSpot(TM) 64-Bit Server VM (build 25.141-b15, mixed mode)
root@master ~]#
```

图 3-11 Slave2

至此，所有节点的 jdk 1.8 安装完毕。

3.3.2 安装 Hadoop

Hadoop2.7.3 的安装步骤如下。

(1) 下载 Hadoop-2.7.3 安装包 hadoop-2.7.3.tar.gz。

(2) 在安装包所在目录下解压文件，如图 3-12 所示。

```
[root@master software]# tar -zxvf hadoop-2.7.3.tar.gz
#复制解压后的 hadoop-2.7.3 文件到/opt/目录下
[root@master ~]# cp -r /home/hadoop/hadoop-2.7.3/opt/
```

```
[root@master ~]# cd /opt/
[root@master ~]# ls -l h*
total 4
drwxr-xr-x. 16  root  root  4096 Jul 27 01:16 hadoop-2.7.3
[root@master ~]#
```

图 3-12 安装 Hadoop

3.3.3 配置 Hadoop

配置 Hadoop 的具体步骤如下。

(1) 在 master 节点上，首先在/home/hadoop/目录下创建 4 个新文件夹，如图 3-13 所示。

```
[root@master ~]# mkdir -p /home/hadoop/hadoopdir/data
[root@master ~]# mkdir -p /home/hadoop/hadoopdir/temp
[root@master ~]# mkdir -p /home/hadoop/hadoopdir/logs
[root@master ~]# mkdir -p /home/hadoop/hadoopdir/pids
[root@master ~]# cd /home/hadoop/hadoopdir/
[root@master hadoopdir]$ ls
data  logs  pids  temp
```

图 3-13　创建新文件夹

(2) 同样在 slave1、slave2 节点上创建 4 个新文件夹，如图 3-14 所示。

```
[root@slave1 ~]# mkdir -p /home/hadoop/hadoopdir/data
[root@slave1 ~]# mkdir -p /home/hadoop/hadoopdir/temp
[root@slave1 ~]# mkdir -p /home/hadoop/hadoopdir/logs
[root@slave1 ~]# mkdir -p /home/hadoop/hadoopdir/pids
[root@slave1 ~]# cd /home/hadoop/hadoopdir/
[root@slave1 hadoopdir]$ ls
data  logs  pids  temp
```

图 3-14　在 Slave1、Slave2 节点上创建新文件夹

进入到解压后的 hadoop-2.7.3/etc/hadoop 目录下，开始配置文件：
① hadoop-env.sh：

```
export JAVA_HOME = /opt/jdk1.8.0_141/
export HADOOP_LOG_DIR = /home/hadoop/hadoopdir/logs
export HADOOP_PID_DIR = /home/hadoop/hadoopdir/pids
```

② Mapred-env.sh：
添加下列语句：

```
export JAVA_HOME = /opt/ jdk1.8.0_141/
export HADOOP_MAPRED_LOG_DIR = /home/hadoop/hadoopdir/logs
export HADOOP_MAPRED_PID_DIR = /home/hadoop/hadoopdir/pids
```

③ yarn-env.sh：

```
export JAVA_HOME = /opt/ jdk1.8.0_141/
YARN_LOG_DIR = /home/hadoop/hadoopdir/logs
```

④ Slaves 文件，如图 3-15 所示。

图 3-15　配置 Slaves 文件

⑤ Core-site.xml：

```xml
<configuration>
    <property>
        <name>fs.defaultFS</name>
        <value>hdfs://master:9000</value>
    </property>
    <property>
        <name>io.file.buffer.size</name>
        <value>131072</value>
    </property>
    <property>
        <name>hadoop.tmp.dir</name>
        <value>file:///home/hadoop/hadoopdir/temp</value>
    </property>
</configuration>
```

⑥ Hdfs-site.xml：

```xml
<configuration>
    <property>
        <name>dfs.namenode.name.dir</name>
        <value>file:///home/hadoop/hadoopdir/name</value>
    </property>
    <property>
        <name>dfs.datanode.data.dir</name>
        <value>file:///home/hadoop/hadoopdir/data</value>
    </property>
    <property>
        <name>dfs.replication</name>
        <value>2</value>
    </property>
    <property>
        <name>dfs.blocksize</name>
        <value>64m</value>
    </property>
    <property>
        <name>dfs.namenode.secondary.http-address</name>
        <value>master:9001</value>
    </property>
    <property>
        <name>dfs.webhdfs.enabled</name>
        <value>true</value>
    </property>
</configuration>
```

⑦ Mapred-site.xml。

首先执行复制命令：

```
[root@master hadoop]# cp /opt/hadoop-2.7.3/etc/hadoop/mapred-site.xml.template /opt/hadoop-2.7.3/etc/hadoop/mapred-site.xml
```

然后打开文件进行添加：

```
<configuration>
    <property>
        <name>mapreduce.framework.name</name>
        <value>yarn</value>
        <final>true</final>
    </property>
    <property>
        <name>mapreduce.jobhistory.address</name>
        <value>master:10020</value>
    </property>
    <property>
        <name>mapreduce.jobtracker.http.address</name>
        <value>master:50030</value>
    </property>
    <property>
        <name>mapred.job.tracker</name>
        <value>http://master:9001</value>
    </property>
    <property>
        <name>mapreduce.jobhistory.webapp.address</name>
        <value>master:19888</value>
    </property>
</configuration>
```

⑧ Yarn-site.xml：

```
<property>
    <name>yarn.nodemanager.aux-services</name>
    <value>mapreduce_shuffle</value>
</property>
<property>
<name>yarn.nodemanager.aux-services.mapreduce_shuffle.class</name>
<value>org.apache.hadoop.mapred.ShuffleHandler</value>
</property>
<property>
<name>yarn.resourcemanager.hostname</name>
<value>master</value>
</property>
<property>
<name>yarn.resourcemanager.scheduler.address</name>
<value>master:8030</value>
</property>
```

```
<property>
<name>yarn.resourcemanager.resource-tracker.address</name>
<value>master:8031</value>
</property>
<property>
<name>yarn.resourcemanager.address</name>
<value>master:8032</value>
</property>
<property>
<name>yarn.resourcemanager.admin.address</name>
<value>master:8033</value>
</property>
<property>
<name>yarn.resourcemanager.webapp.address</name>
<value>master:8088</value>
</property>
```

至此，所有配置文件已完成。

(3) 在 master 机器上，把整个配置好的 hadoop2.7.3 文件夹复制到其他节点。

注意：在这里，为了统一，把 master 根目录下/opt/中的 hadoop-2.7.3 复制到/home/hadoop/目录下。

```
[root@master~]# scp -r /opt/ hadoop-2.7.3 hadoop@master:/home/hadoop/
[root@master~]# scp -r /opt/ hadoop-2.7.3 hadoop@slave1:/home/hadoop/
[root@master~]# scp -r /opt/ hadoop-2.7.3 hadoop@slave2:/home/hadoop/
```

所以，现在所有节点配置好的文件都在/home/hadoop/目录下了，如图 3-16 所示。

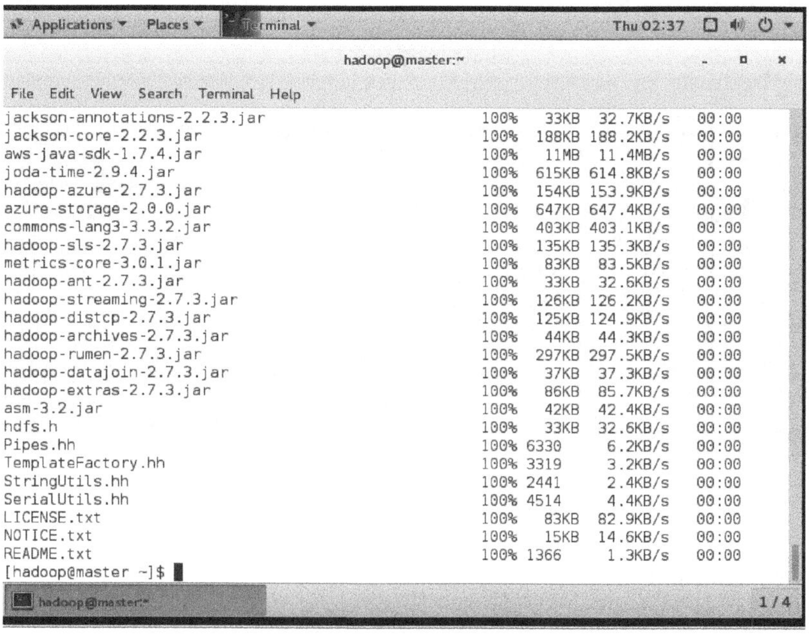

图 3-16　配置文件完成

3.3.4 格式化

进入 hadoop-2.7.3/bin 目录,进行格式化,如图 3-17 和图 3-18 所示。

```
[root@master ~]# cd /home/hadoop/hadoop-2.7.3/bin
[root@master bin]# ./hdfs namenode -format
```

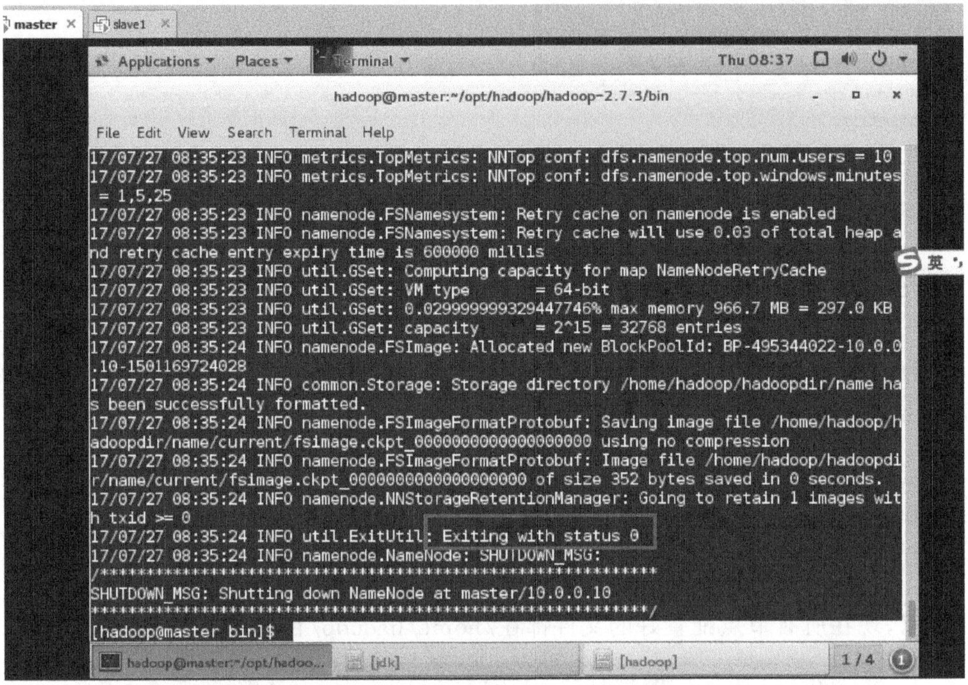

图 3-17 格式化(一)

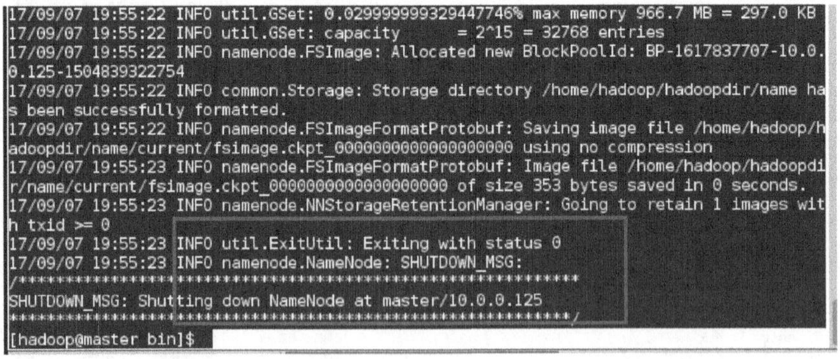

图 3-18 格式化(二)

3.3.5 运行 Hadoop

进入 cd/home/hadoop/hadoop-2.7.3/sbin 目录下。

(1) 启动 hdfs 集群：

```
[root@master sbin]# ./start-dfs.sh
```

(2) 启动 yarn 集群：

```
[root@master sbin]# ./start-yarn.sh
```

这样，hadoop 集群就跑起来了。如果要关闭，在 sbin 目录下进行：

```
[root@master sbin]# ./stop-dfs.sh
[root@master sbin]# ./stop-yarn.sh
```

(3) 使用命令 jps 查看节点：
Master，如图 3-19 所示。

```
[root@master ~]# jps
61024 SecondaryNameNode
61461 Jps
61189 ResourceManager
60814 NameNode
[root@master ~]#
```

图 3-19　Master

Slave1，如图 3-20 所示。

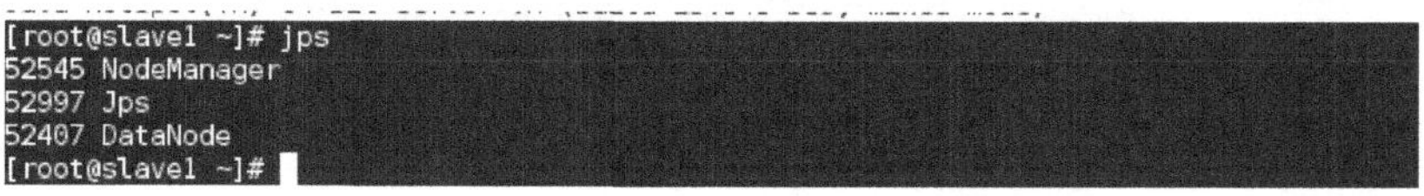

```
[root@slave1 ~]# jps
52545 NodeManager
52997 Jps
52407 DataNode
[root@slave1 ~]#
```

图 3-20　Slave1

Slave2，如图 3-21 所示。

```
[root@slave2 ~]# jps
52768 Jps
52216 DataNode
52347 NodeManager
[root@slave2 ~]#
```

图 3-21　Slave2

如果要关闭，在 sbin 目录下，执行以下命令：

```
[root@master sbin]# ./stop-dfs.sh
[root@master sbin]# ./stop-yarn.sh
```

然后需要用浏览器打开 http://master:50070，即 HDFS 的 Web 页面，可以看到集群信息和 Datenode 相关信息，如图 3-22 所示。

用浏览器打开 http://master:8088，即 YARN 的 Web 页面，可以看到集群相关信息，如图 3-23 所示。

图 3-22 打开 HDFS 的 Web 页面

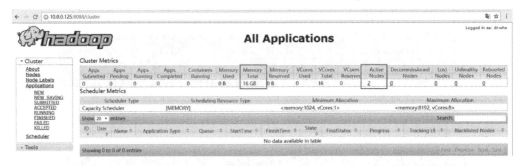

图 3-23

小结

本章首先介绍了 Hadoop 海量数据分布式处理框架、Hadoop 的优点及 Hadoop 的发展历程，然后详细描述了 Hadoop 原理、运行机制，以及 3 个重要的组件（Hadoop Common、HDFS 和 MapReduce），接着简述了 Hadoop 开源技术生态系统的相关组件，最后重点介绍了 Hadoop 安装与配置，包括 JDK 的安装与配置、Hadoop 的安装与配置、Hadoop 的运行。

习题

1. 简述 Hadoop 系统及其优点。
2. 简述 Hadoop 原理及运行机制。
3. 简述 Hadoop 技术生态系统。
4. 学会 JDK 的安装和配置。
5. 掌握 Hadoop 的安装和配置。

第 4 章

HDFS分布式文件系统

分布式文件系统(Distributed File System,DFS)通常指 C/S 架构或网络文件系统，用户数据没有直接连接到本地主机，而是存储在远程存储服务器上。NFS/CIFS 是最为常见的分布式文件系统，这就是我们说的网络存储(Network Attached Storage,NAS)系统。分布式文件系统中，存储服务器的节点数可能是一个，也可以是多个。对于单个节点的分布式文件系统来说，存在单点故障和性能瓶颈问题。除了 NAS 以外，典型的分布式文件系统还有 AFS(Andrew File System)和 HDFS(Hadoop Distributed File System)等。

4.1 HDFS

HDFS 源于 Google 在 2003 年 10 月发表的 GFS(Google File System)论文。它其实就是 GFS 的一个克隆版本。Hadoop 分布式文件系统(Hadoop Distributed File System,HDFS)是 Hadoop 项目的核心子项目，是分布式计算中数据存储管理的基础，是基于流数据模式访问和处理超大文件的需求而开发的，可以运行于廉价的商用服务器上。它所具有的高容错性、高可靠性、高可扩展性、高获得性、高吞吐率等特征为海量数据提供了安全性高的存储，为超大数据集(large data set)的应用处理带来了很多便利。

HDFS 被设计成适合运行在通用硬件(commodity hardware)上的分布式文件系统。它与现有的分布式文件系统有很多共同点，但与其他的分布式文件系统的区别也是很明显的。HDFS 是一个高度容错性的系统，适合部署在廉价的机器上。HDFS 能提供高吞吐量的数据访问，非常适合大规模数据集上的应用。HDFS 放宽了一部分 POSIX 约束，来实现流式读取文件系统数据的目的。HDFS 在最开始是作为 Apache Nutch 搜索引擎项目的基础架构而开发的。

4.1.1 设计前提和设计目标

HDFS 设计的前提和设计目标如下。

(1) 硬件错误。硬件错误是常态而不是异常。HDFS 可能由成百上千的服务器所构成,每个服务器上存储着文件系统的部分数据。面对的现实是构成系统的组件数目是巨大的,而且任一组件都有可能失效,这意味着总是有一部分 HDFS 的组件是不工作的。因此错误检测和快速、自动的恢复是 HDFS 最核心的架构目标。

(2) 流式数据访问。运行在 HDFS 上的应用和普通的应用不同,需要流式访问它们的数据集。HDFS 的设计中更多地考虑到了数据批处理,而不是用户交互处理。与数据访问的低延迟问题相比,更关键的是数据访问的高吞吐量。POSIX 标准设置的很多硬性约束对 HDFS 应用系统不是必需的。为了提高数据的吞吐量,在一些关键方面对 POSIX 的语义做了一些修改。

(3) 大规模数据集。运行在 HDFS 上的应用具有很大的数据集。HDFS 上的一个典型文件大小一般都为吉字节至太字节。因此,HDFS 被调节以支持大文件存储。它能提供较宽的数据传输带宽,能在一个集群里扩展到数百个节点。一个单一的 HDFS 实例应该能支撑数以千万计的文件。

(4) 简单的一致性模型。HDFS 应用需要一个"一次写入多次读取"的文件访问模型。一个文件经过创建、写入和关闭之后就不需要改变。这一假设简化了数据一致性的问题,并且使高吞吐量的数据访问成为可能。Map/Reduce 应用或者网络爬虫应用都非常适合这个模型。目前还有计划在将来扩充这个模型,使之支持文件的附加写操作。

(5) 移动计算比移动数据更划算。一个应用请求的计算,离它操作的数据越近就越高效,在数据达到海量级别的时候更是如此。因为这样就能降低网络阻塞的影响,提高系统数据的吞吐量。将计算移动到数据附近,比之将数据移动到应用显然更好。HDFS 为应用提供了将它们自己移动到数据附近的接口。

(6) 异构软硬件平台间的可移植性。HDFS 在设计的时候就考虑到平台的可移植性。这种特性方便了 HDFS 作为大规模数据应用平台的推广。

但是 HDFS 也有自己的劣势,并不适合所有场合。

(1) 低时延长数据访问。例如,毫秒级的存储数据,毫秒级以内读取数据,HDFS 很难实现,它适合高吞吐率的场景,就是在某一时间内写入大量的数据。

(2) 小文件存储。存储大量小文件(小于 HDFS 系统的 Block 大小的文件),它会占用 Namenode 大量的内存来存储文件、目录和块信息。这样是不可取的,因为 Namenode 的内存总是有限的,小文件存储的寻道时间会超过读取时间,与 HDFS 的设计目标相违背。

(3) 并发写入、文件随机修改。一个文件只能有一个写,不允许多个线程同时写,仅支持数据追加,不支持文件的随机修改。

4.1.2 Namenode 和 Datanode

HDFS 采用 Master/Slave 架构。一个 HDFS 集群是由一个 Namenode 和一定数目的 Datanodes 组成的。Namenode 是一个中心服务器,负责管理文件系统的名字空间(namespace)

以及客户端对文件的访问。集群中的 Datanode 负责管理它所在节点上的存储。HDFS 暴露了文件系统的名字空间,用户能够以文件的形式在上面存储数据,如图 4-1 所示。

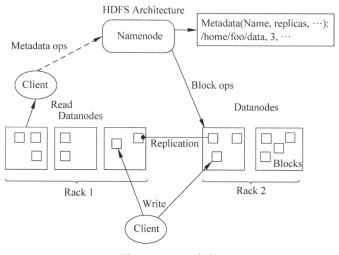

图 4-1 HDFS 架构

从内部看,一个文件其实被分成一个或多个数据块,这些块存储在一组 Datanode 上。Namenode 执行文件系统的名字空间操作,如打开、关闭、重命名文件或目录;它也负责确定数据块到具体 Datanode 节点的映射。Datanode 负责处理文件系统客户端的读/写请求,在 Namenode 的统一调度下进行数据块的创建、删除和复制。

Namenode 和 Datanode 被设计成可以在普通的商用机器上运行。这些机器一般运行着 GNU/Linux 操作系统。HDFS 采用 Java 语言开发,因此任何支持 Java 的机器都可以部署 Namenode 或 Datanode。由于采用了可移植性极强的 Java 语言,使得 HDFS 可以部署到多种类型的机器上。一个典型的部署场景是一台机器上只运行一个 Namenode 实例,而集群中的其他机器分别运行一个 Datanode 实例。这种架构并不排斥在一台机器上运行多个 Datanode,只不过这样的情况比较少见。

集群中单一 Namenode 的结构大大简化了系统的架构。Namenode 是所有 HDFS 元数据的仲裁者和管理者,这样用户数据永远不会流过 Namenode。

4.1.3 文件系统的名字空间

HDFS 支持传统的层次型文件组织结构。用户或者应用程序可以创建目录,然后将文件保存在这些目录中。文件系统名字空间的层次结构和大多数现有的文件系统类似:用户可以创建、删除、移动或重命名文件。当前,HDFS 不支持用户磁盘配额和访问权限控制,也不支持硬链接和软链接,但是 HDFS 架构并不妨碍实现这些特性。

Namenode 负责维护文件系统的名字空间,任何对文件系统名字空间或属性的修改都将被 Namenode 记录下来。应用程序可以设置 HDFS 保存的文件的副本数目。文件副本的数目称为文件的副本系,这个信息也是由 Namenode 保存的。

Namenode 将对文件系统的改动追加保存到本地文件系统上的一个日志文件(edits)

中。当一个 Namenode 启动时,它首先从一个映像文件(fsimage)中读取 HDFS 的状态,接着应用日志文件中的 edits 操作。然后它将新的 HDFS 状态写入(fsimage)中,并使用一个空的 edits 文件开始正常操作。因为 Namenode 只有在启动阶段才合并 fsimage 和 edits,所以久而久之日志文件可能会变得非常庞大,特别是对大型的集群。日志文件主要的缺点是下一次 Namenode 启动会花很长时间。

Secondary Namenode 定期合并 fsimage 和 edits 日志,将 edits 日志文件大小控制在一个限度下。因为内存需求和 Namenode 在一个数量级上,所以通常 Secondary Namenode 和 Namenode 运行在不同的机器上。Secondary Namenode 通过 bin/start-dfs.sh 在 conf/masters 中指定的节点上启动。

Secondary Namenode 的检查点进程启动是由两个配置参数控制的。

(1) fs.checkpoint.period 指定连续两次检查点的最大时间间隔,默认值是 1 小时。

(2) fs.checkpoint.size 定义 edits 日志文件的最大值,一旦超过这个值,会导致强制执行检查点(即使没到检查点的最大时间间隔),默认值是 64MB。

Secondary Namenode 保存最新检查点的目录与 Namenode 的目录结构相同,所以 Namenode 可以在需要的时候读取 Secondary Namenode 上的检查点镜像。

如果 Namenode 上除了最新的检查点以外,所有其他的历史镜像和 edits 文件都丢失了,Namenode 可以引入这个最新的检查点,以下操作可以实现这个功能。

(1) 在配置参数 dfs.name.dir 指定的位置建立一个空文件夹。

(2) 把检查点目录的位置赋值给配置参数 fs.checkpoint.dir。

(3) 启动 Namenode,并加上-importCheckpoint。

Namenode 会从 fs.checkpoint.dir 目录读取检查点,并将它保存在 dfs.name.dir 目录下。如果 dfs.name.dir 目录下有合法的镜像文件,Namenode 会启动失败。Namenode 会检查 fs.checkpoint.dir 目录下镜像文件的一致性,但是不会去改动它。

Secondary Namenode 的工作流程如图 4-2 所示。

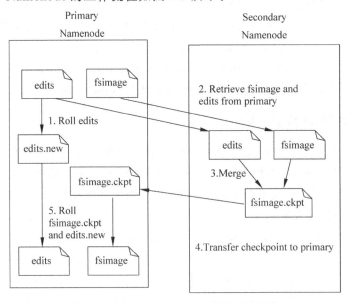

图 4-2 Secondary Namenode 的工作流程

(1) Secondary Namenode 通知 Namenode 切换 edits 文件。
(2) Secondary Namenode 通过 http 请求从 Namenode 获得 fsimage 和 edits 文件。
(3) Secondary Namenode 将 fsimage 载入内存,然后开始合并 edits。
(4) Secondary Namenode 将新的 fsimage 发回给 Namenode。
(5) Namenode 用新的 fsimage 替换旧的 fsimage。

4.1.4 数据复制

HDFS 被设计成能够在一个大集群中不同机器上可靠地存储超大文件。它将每个文件存储成一系列的数据块,除了最后一个,所有的数据块都是同样大小的。为了容错,文件的所有数据块都会有副本。每个文件的数据块大小和副本系数都是可配置的。应用程序可以指定某个文件的副本数目。副本系数可以在文件创建时指定,也可以在之后改变。HDFS 中的文件都是一次性写入的,并且严格要求在任何时候只能有一个写入者,如图 4-3 所示。

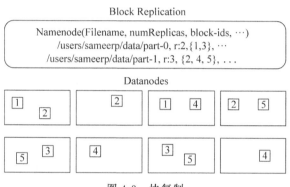

图 4-3 块复制

Namenode 全权管理数据块的复制,它周期性地从集群中的每个 Datanode 接收心跳信号和块状态报告(block report)。接收到心跳信号意味着该 Datanode 节点工作正常。块状态报告包含了一个该 Datanode 上所有数据块的列表。

1. Block 的副本放置策略

副本的存放是 HDFS 可靠性和性能的关键。优化的副本存放策略是 HDFS 区分于其他大部分分布式文件系统的重要特性。这种特性需要做大量的调优,并需要经验的积累。HDFS 采用一种称为机架感知(rack-aware)的策略来改进数据的可靠性、可用性和网络带宽的利用率。目前实现的副本存放策略只是在这个方向上的第一步。实现这个策略的短期目标是验证它在生产环境下的有效性,观察它的行为,为实现更先进的策略打下测试和研究的基础。

大型 HDFS 实例一般运行在跨越多个机架的计算机组成的集群上,不同机架上的两台机器之间的通信需要经过交换机。在大多数情况下,同一个机架内的两台机器间的带宽会比不同机架的两台机器间的带宽大。

通过一个机架感知的过程，Namenode可以确定每个Datanode所属的机架ID。一个简单但没有优化的策略就是将副本存放在不同的机架上，这样可以有效防止当整个机架失效时数据的丢失，并且允许读数据的时候充分利用多个机架的带宽。这种策略设置可以将副本均匀分布在集群中，有利于组件失效情况下的负载均衡。但是，因为这种策略的一个写操作需要传输数据块到多个机架，这增加了写的代价。

在大多数情况下，副本系数是3，HDFS的存放策略是将一个副本存放在本地机架的节点上，一个副本放在同一机架的另一个节点上，最后一个副本放在不同机架的节点上。这种策略减少了机架间的数据传输，这就提高了写操作的效率。机架的错误远远比节点的错误少，所以这个策略不会影响到数据的可靠性和可用性。与此同时，因为数据块只放在两个(不是3个)不同的机架上，所以此策略减少了读取数据时需要的网络传输总带宽。在这种策略下，副本并不是均匀分布在不同的机架上。1/3的副本在一个节点上，2/3的副本在另一个机架上，其他副本均匀分布在剩下的机架中，这一策略在不损害数据可靠性和读取性能的情况下改进了写的性能。

2. 副本选择

为了降低整体的带宽消耗和读取延时，HDFS会尽量让读取程序读取离它最近的副本。如果在读取程序的同一个机架上有一个副本，那么就读取该副本。如果一个HDFS集群跨越多个数据中心，那么客户端也将首先读本地数据中心的副本。

3. 安全模式

Namenode启动后会进入一个称为安全模式的特殊状态。处于安全模式的Namenode是不会进行数据块复制的。Namenode从所有的Datanode接收心跳信号和块状态报告。块状态报告包括了某个Datanode所有的数据块列表。每个数据块都有一个指定的最小副本数。当Namenode检测确认某个数据块的副本数目达到这个最小值，那么该数据块就会被认为是副本安全(safely replicated)的。在一定百分比(这个参数可配置)的数据块被Namenode检测确认安全之后(加上一个额外的30秒等待时间)，Namenode将退出安全模式状态。接下来它会确定还有哪些数据块的副本没有达到指定数目，并将这些数据块复制到其他Datanode上。

4. 文件系统元数据的持久化

Namenode上保存着HDFS的名字空间。任何对文件系统元数据产生修改的操作，Namenode都会使用EditLog的事务日志记录下来。例如，在HDFS中创建一个文件，Namenode就会在Editlog中插入一条记录来表示。同样地，修改文件的副本系数也将在Editlog插入一条记录。Namenode在本地操作系统的文件系统中存储这个Editlog。整个文件系统的名字空间，包括数据块到文件的映射、文件的属性等，都存储在fsimage文件中，这个文件也是放在Namenode所在的本地文件系统上。

Namenode在内存中保存着整个文件系统的名字空间和文件数据块映射(blockmap)的映像。这个关键的元数据结构设计得很紧凑，因而一个有4GB内存的Namenode足够支撑大量的文件和目录。当 Namenode 启动时,它从硬盘中读取 Editlog 和 fsimage,将所有 Editlog 中的事务作用在内存中的 fsimage 上,并将这个新版本的 fsimage 从内存中保存到本地磁盘上,然后删除旧的 Editlog,因为这个旧的 Editlog 的事务都已经作用在 fsimage 上了,这个过程称为一个检查点(checkpoint)。在当前实现中,检查点只发生在 Namenode 启动时,之后将实现支持周期性的检查点。

Datanode 将 HDFS 数据以文件的形式存储在本地的文件系统中,它并不知道有关 HDFS 文件的信息。它把每个 HDFS 数据块存储在本地文件系统的一个单独的文件中。Datanode 并不在同一个目录创建所有的文件,实际上,它用试探的方法来确定每个目录的最佳文件数目,并且在适当的时候创建子目录。在同一个目录中创建所有的本地文件并不是最优的选择,这是因为本地文件系统可能无法高效地在单个目录中支持大量的文件。当一个 Datanode 启动时,它会扫描本地文件系统,产生一个这些本地文件对应的所有 HDFS 数据块的列表,然后作为报告发送到 Namenode,这个报告就是块状态报告。

5. 通信协议

所有的 HDFS 通信协议都是建立在 TCP/IP 协议上。客户端通过一个可配置的 TCP 端口连接到 Namenode,通过 ClientProtocol 协议与 Namenode 交互。而 Datanode 使用 DatanodeProtocol 协议与 Namenode 交互。一个远程过程调用(RPC)模型被抽象出来封装 ClientProtocol 和 Datanodeprotocol 协议。在设计上,Namenode 不会主动发起 RPC,而是响应来自客户端或 Datanode 的 RPC 请求。

6. 健壮性

HDFS 的主要目标是即使在出错的情况下也要保证数据存储的可靠性。常见的 3 种出错情况:Namenode 出错、Datanode 出错和网络割裂(network partitions)。

7. 磁盘数据错误,心跳检测和重新复制

每个 Datanode 节点周期性地向 Namenode 发送心跳信号。网络割裂可能导致一部分 Datanode 与 Namenode 失去联系。Namenode 通过心跳信号的缺失来检测这一情况,并将这些近期不再发送心跳信号的 Datanode 标记为宕机,不会再将新的 I/O 请求发给它们。任何存储在宕机 Datanode 上的数据将不再有效。Datanode 的宕机可能会引起一些数据块的副本系数低于指定值,Namenode 不断地检测这些需要复制的数据块,一旦发现就启动复制操作。在下列情况下,可能需要重新复制:某个 Datanode 节点失效、某个副本遭到损坏、Datanode 上的硬盘错误或者文件的副本系数增大。

8. 集群均衡

HDFS 的架构支持数据均衡策略。如果某个 Datanode 节点上的空闲空间低于特定的临界点,按照均衡策略系统就会自动地将数据从这个 Datanode 移动到其他空闲的 Datanode。当对某个文件的请求突然增加,那么也可能启动一个计划创建该文件新的副本，并且同时重新平衡集群中的其他数据。这些均衡策略目前还没有实现。

9. 数据完整性

从某个 Datanode 获取的数据块有可能是损坏的，损坏可能是由 Datanode 的存储设备错误、网络错误或者软件 bug 造成的。HDFS 客户端软件实现了对 HDFS 文件内容的校验和(checksum)检查。当客户端创建一个新的 HDFS 文件，会计算这个文件每个数据块的校验和，并将校验和作为一个单独的隐藏文件保存在同一个 HDFS 名字空间下。当客户端获取文件内容后，它会检验从 Datanode 获取的数据与相应的校验和文件中的校验和是否匹配，如果不匹配，客户端可以选择从其他 Datanode 获取该数据块的副本。

10. 元数据磁盘错误

fsimage 和 Editlog 是 HDFS 的核心数据结构。如果这些文件损坏了，整个 HDFS 实例都将失效。因而，Namenode 可以配置成支持维护多个 fsimage 和 Editlog 的副本。任何对 fsimage 或 Editlog 的修改，都将同步到它们的副本上。这种多副本的同步操作可能会降低 Namenode 每秒处理的名字空间事务数量。然而这个代价是可以接受的，因为即使 HDFS 的应用是数据密集的，它们也非元数据密集的。当 Namenode 重启时，它会选取最近的完整的 fsimage 和 Editlog 来使用。

Namenode 是 HDFS 集群中的单点故障(single point of failure)所在。如果 Namenode 机器故障，是需要手工进行干预的。目前，自动重启或在另一台机器上做 Namenode 故障转移的功能还没实现。

4.1.5 HDFS 读流程

客户端将要读取的文件路径发送给 namenode，namenode 获取文件的元信息(主要是 block 的存放位置信息)返回给客户端，客户端根据返回的信息找到相应 datanode 逐个获取文件的 block 并在客户端本地进行数据追加合并从而获得整个文件，如图 4-4 所示。

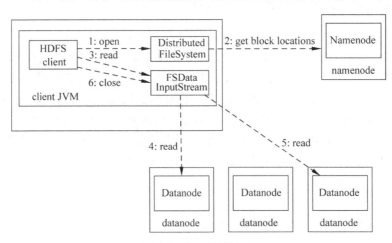

图 4-4　HDFS 读流程

详细流程如下。

（1）HDFS 的客户端调用 Distributed FileSystem.open 方法打开这个文件（HDFS 中 API 的一个对象）。

（2）通过 Distributed FileSystem 发送给 Namenode 请求，同时将用户信息及文件名的信息等发送给 Namenode，并返回给 Distributed FileSystem，该文件包含 block 所在的 Datanode 位置。

（3）HDFS 客户端通过 FSDataInputStream 按顺序读取 Datanode 中的 block 信息（它会选择负载最低的或离客户端最近的一台 Datanode 去读 block）。

（4）FSDataInputStream 按顺序一个一个地读，直到所有的 block 都读取完毕。

（5）当读取完毕后会将 FSDataInputStream 关闭。

4.1.6　HDFS 写流程

客户端要向 HDFS 写数据，首先要与 namenode 通信以确认可以写文件并获得接收文件 block 的 datanode，然后客户端按顺序将文件逐个 block 传递给相应 datanode，并由接收到 block 的 datanode 负责向其他 datanode 复制 block 的副本，如图 4-5 所示。

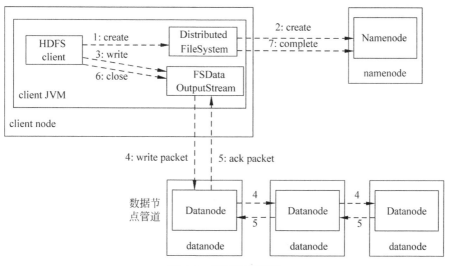

图 4-5　HDFS 写流程

详细流程如下。

（1）HDFS 的客户端调用 Distributed FileSystem.create 方法创建这个文件（HDFS 中 API 的一个对象）。

（2）通过 Distributed FileSystem 发送客户端的请求给 Namenode（Namenode 主要是接收客户端请求）并且会带着文件要保存的位置、文件名、操作的用户名等信息一起发送给 Namenode。

（3）Namenode 会给客户端返回一个 FSDataOutputStream，同时也会返回文件要写入哪些 Datanode 上（负载较低的）。

(4) 通过 FSDataOutputStream 进行写操作，在写之前就做文件的拆分，将文件拆分成多个 block，第一个写操作写在负载比较低的 Datanode 上，并将这个 block 复制到其他的 Datanode 上。

(5) 当所有的 block 副本复制完成后会反馈给 FSDataOutputStream。

(6) 当所有的 block 副本全都复制完成，就可以将 FSDataOutputStream 流关闭。

(7) 通过 Distributed FileSystem 更新 Namenode 中的源数据信息。

4.2 HDFS 操作实践

4.2.1 HDFS Shell

视频讲解

调用文件系统(FS)Shell 命令应使用 bin/hadoop fs < args >或 bin/hdfs dfs < args >的形式。所有的 FS shell 命令使用 URI 路径作为参数。URI 格式是 scheme://authority/path。对 HDFS 文件系统，scheme 是 hdfs；对本地文件系统，scheme 是 file。其中，scheme 和 authority 参数都是可选的，如果未加指定，就会使用配置中指定的默认 scheme。一个 HDFS 文件或目录，如/parent/child，可以表示成 hdfs://namenode:namenodeport/parent/child，或者更简单的/parent/child(假设配置文件中的默认值是 namenode:namenodeport)。大多数 FS Shell 命令的行为和对应的 UNIX Shell 命令类似，不同之处会在下面介绍各命令使用详情时指出。出错信息会输出到 stderr，其他信息输出到 stdout。

(1) 查看帮助：

```
# hdfs dfs - help
Usage: hadoop fs [generic options]
        [- appendToFile < localsrc > ... < dst >]
        [- cat [- ignoreCrc] < src > ...]
        [- checksum < src > ...]
        [- chgrp [- R] GROUP PATH...]
        [- chmod [- R] < MODE[,MODE]... | OCTALMODE > PATH...]
        [- chown [- R] [OWNER][:[GROUP]] PATH...]
        [- copyFromLocal [- f] [- p] [- l] < localsrc > ... < dst >]
        [- copyToLocal [- p] [- ignoreCrc] [- crc] < src > ... < localdst >]
        [- count [- q] [- h] < path > ...]
        [- cp [- f] [- p | - p[topax]] < src > ... < dst >]
        [- createSnapshot < snapshotDir > [< snapshotName >]]
        [- deleteSnapshot < snapshotDir > < snapshotName >]
        [- df [- h] [< path > ...]]
        [- du [- s] [- h] < path > ...]
        [- expunge]
        [- find < path > ... < expression > ...]
```

```
[-get [-p] [-ignoreCrc] [-crc] <src> ... <localdst>]
[-getfacl [-R] <path>]
[-getfattr [-R] {-n name | -d} [-e en] <path>]
[-getmerge [-nl] <src> <localdst>]
[-help [cmd ...]]
[-ls [-d] [-h] [-R] [<path> ...]]
[-mkdir [-p] <path> ...]
[-moveFromLocal <localsrc> ... <dst>]
[-moveToLocal <src> <localdst>]
[-mv <src> ... <dst>]
[-put [-f] [-p] [-l] <localsrc> ... <dst>]
[-renameSnapshot <snapshotDir> <oldName> <newName>]
[-rm [-f] [-r|-R] [-skipTrash] <src> ...]
[-rmdir [--ignore-fail-on-non-empty] <dir> ...]
[-setfacl [-R] [{-b|-k} {-m|-x <acl_spec>} <path>]|[--set <acl_spec> <path>]]
[-setfattr {-n name [-v value] | -x name} <path>]
[-setrep [-R] [-w] <rep> <path> ...]
[-stat [format] <path> ...]
[-tail [-f] <file>]
[-test -[defsz] <path>]
[-text [-ignoreCrc] <src> ...]
[-touchz <path> ...]
[-truncate [-w] <length> <path> ...]
[-usage [cmd ...]]
```

（2）创建目录：

```
# hdfs dfs -mkdir /test
```

（3）列目录：

```
# hdfs dfs -ls /
Found 2 items
drwxr-xr-x   - root supergroup          0 2018-06-23 20:52 /test
drwxrwx---   - root supergroup          0 2018-06-23 14:11 /tmp
```

（4）编辑本地文件：

```
# vi test.txt
Hello Welcome
广东技术师范大学
汕尾职业技术学院
```

（5）上传文件：

```
# hdfs dfs -put test.txt /test
```

(6) 查看文件：

```
# hdfs dfs -ls /test
Found 1 items
-rw-r--r--   3 root supergroup         65 2018-06-23 21:00 /test/test.txt
# hdfs dfs -cat /test/test.txt
Hello Welcome
广东技术师范大学
汕尾职业技术学院
```

4.2.2　HDFS Java API

视频讲解

　　HDFS 设计的主要目的是对海量数据进行存储，也就是说在其上能够存储大量文件(可以存储 TB 级的文件)。HDFS 将这些文件分割之后，存储在不同的 Datanode 上，HDFS 提供了两种访问接口：Shell 接口和 Java API 接口，对 HDFS 里面的文件进行操作，具体每个 Block 放在哪台 Datanode 上面，对于开发者来说是透明的。

　　Java 抽象类 org.apache.hadoop.fs.FileSystem 定义了 Hadoop 的一个文件系统接口。该类是一个抽象类，通过以下两种静态工厂方法可以获取 FileSystem 实例：

```
public static FileSystem.get(Configuration conf) throws IOException
public static FileSystem.get(URI uri, Configuration conf) throws IOException
```

具体方法实现如下。

(1) public boolean mkdirs(Path f) throws IOException：一次性新建所有目录(包括父目录)，f 是完整的目录路径。

(2) public FSOutputStream create(Path f) throws IOException：创建指定 path 对象的一个文件，返回一个用于写入数据的输出流，create()有多个重载版本，允许用户指定是否强制覆盖已有的文件、文件备份数量、写入文件缓冲区大小、文件块大小及文件权限。

(3) public boolean copyFromLocal(Path src, Path dst) throws IOException：将本地文件复制到文件系统。

(4) public boolean exists(Path f) throws IOException：检查文件或目录是否存在。

(5) public boolean delete(Path f, Boolean recursive)：永久性删除指定的文件或目录，如果 f 是一个空目录或文件，那么 recursive 的值就会被忽略。只有 recursive=true 时，一个非空目录及其内容才会被删除。

(6) FileStatus 类封装了文件系统中文件和目录的元数据，包括文件长度、块大小、备份、修改时间、所有者及权限信息。

4.2.3　Eclipse 开发环境

　　Eclipse 是一个开放源代码的、基于 Java 的可扩展开发平台。

Eclipse 是 Java 的集成开发环境(IDE),当然 Eclipse 也可以作为其他开发语言的集成开发环境,如 C、C++、PHP 和 Ruby 等。Eclipse 附带了一个标准的插件集,包括 Java 开发工具(Java Development Kit,JDK)。

配置 Eclipse 开发环境的具体步骤如下。

(1) 下载 Eclipse 插件,参考网站如下:

```
https://github.com/winghc/hadoop2x-eclipse-plugin
https://codeload.github.com/winghc/hadoop2x-eclipse-plugin/zip/master
```

(2) 安装 JDK 和 Eclipse,解压到/opt 目录:

```
# cd /opt
# tar xvzf /root/jdk-8u131-linux-x64.tar.gz
# mv /opt/jdk1.8.0_131 /opt/jdk1.8
# tar xvzf /root/eclipse-java-luna-SR2-linux-gtk-x86_64.tar.gz
```

(3) 复制插件到/opt/eclipse/plugins/目录下:

```
# cp /root/hadoop-eclipse-plugin-2.6.0.jar /opt/eclipse/plugins/
```

(4) 以 root 用户登录图形桌面,运行 eclipse:

```
# /opt/eclipse/eclipse &
```

(5) 新建 MapReduce Project 项目,项目名为 HdfsEx,选择 Configure Hadoop install directory,输入 Hadoop 安装目录"/opt/hadoop",如图 4-6 所示。

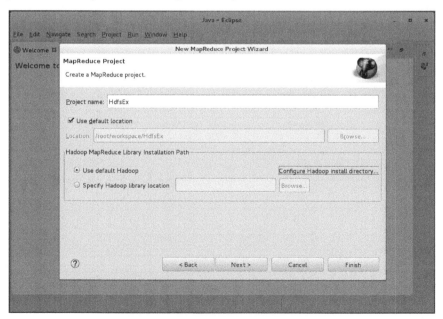

图 4-6　新建 MapReduce Project 项目

(6) 选择 Map/Reduce Locations 选项,右击其下面的窗口,在弹出的菜单中选择 New Hadoop location 选项,如图 4-7 所示。

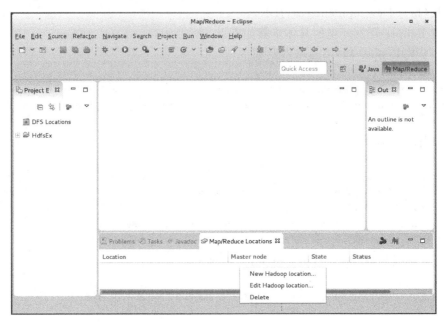

图 4-7　New Hadoop location 选项

(7) 在 New Hadoop location 对话框的 Location name 中输入"master",Map/Reduce(V2) Master 的 Host 的原 localhost 改为 master,DFS Master 的 Port 的原"50040"改为"9000",其他不变,如图 4-8 所示。

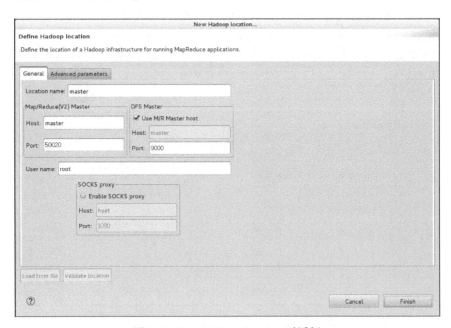

图 4-8　New Hadoop location 对话框

(8) 选择 DFS Locations→master 选项,可以看到 HDFS 上的目录,如图 4-9 所示。

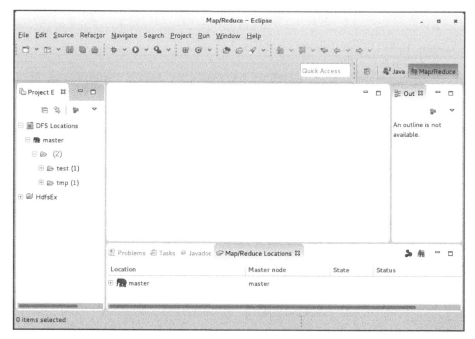

图 4-9 DFS Locations 目录

4.2.4 综合实例

(1) 在项目 HdfsEx 中创建类 OperateHDFS.java,代码如下。

视频讲解

```
import java.io.IOException;
import java.net.URI;
import java.net.URISyntaxException;
import org.apache.commons.lang.StringUtils;
import org.apache.hadoop.conf.Configuration;
import org.apache.hadoop.fs.BlockLocation;
import org.apache.hadoop.fs.FileStatus;
import org.apache.hadoop.fs.FileSystem;
import org.apache.hadoop.fs.FileUtil;
import org.apache.hadoop.fs.Path;
import org.apache.hadoop.fs.PathFilter;
import org.apache.hadoop.hdfs.DistributedFileSystem;
import org.apache.hadoop.hdfs.protocol.DatanodeInfo;
import org.apache.log4j.Logger;
public class OperateHDFS {
    //HDFS 的访问路径
    static final String HDFSUri = "hdfs://master:9000/";
    protected static Logger logger = Logger.getLogger(OperateHDFS.class);
    /**
```

```java
 * 获取文件系统
 * @return FileSystem
 */
public static FileSystem getFileSystem() {
    //读取配置文件
    Configuration conf = new Configuration();
    FileSystem fs = null;
    String hdfsUri = HDFSUri;
    if (StringUtils.isBlank(hdfsUri)) {
        //返回默认文件系统
        try {
            fs = FileSystem.get(conf);
        } catch (IOException e) {
            e.printStackTrace();
            logger.error("", e);
        }
    } else {
        //返回指定的文件系统
        try {
            URI uri = new URI(hdfsUri.trim());
            fs = FileSystem.get(uri, conf);
        } catch (URISyntaxException | IOException e) {
            e.printStackTrace();
            logger.error("", e);
        }
    }
    return fs;
}


/**
 * 创建文件目录
 * @param path
 */
public void mkdir(String path) {
    try {
        FileSystem fs = getFileSystem();
        String hdfsUri = HDFSUri;
        if (StringUtils.isNotBlank(hdfsUri)) {
            path = hdfsUri + path;
        }
        fs.mkdirs(new Path(path));
        fs.close();
    } catch (IllegalArgumentException | IOException e) {
        e.printStackTrace();
        logger.error("", e);
    }
```

```java
    }

    /**
     * 删除文件或者文件目录
     * @param path
     */
    public void rmdir(String path) {
        try {
            FileSystem fs = getFileSystem();
            String hdfsUri = HDFSUri;
            if (StringUtils.isNotBlank(hdfsUri)) {
                path = hdfsUri + path;
            }
            fs.delete(new Path(path), true);
            fs.close();
        } catch (IllegalArgumentException | IOException e) {
            e.printStackTrace();
            logger.error("", e);
        }
    }

    /**
     * 根据Filter获取目录下的文件
     *
     * @param path
     * @param pathFilter
     * @return String[]
     */
    public String[] ListFile(String path, PathFilter pathFilter) {
        String[] files = new String[0];
        try {
            FileSystem fs = getFileSystem();
            String hdfsUri = HDFSUri;
            if (StringUtils.isNotBlank(hdfsUri)) {
                path = hdfsUri + path;
            }
            FileStatus[] status;
            if (pathFilter != null) {
                //根据Filter列出目录内容
                status = fs.listStatus(new Path(path), pathFilter);
            } else {
                //列出目录内容
                status = fs.listStatus(new Path(path));
            }
            Path[] listedPaths = FileUtil.stat2Paths(status);
            //转换String[]
```

```java
            if (listedPaths != null && listedPaths.length > 0) {
                files = new String[listedPaths.length];
                for (int i = 0; i < files.length; i++) {
                    files[i] = listedPaths[i].toString();
                }
            }
            fs.close();
        } catch (IllegalArgumentException | IOException e) {
            e.printStackTrace();
            logger.error("", e);
        }
        return files;
    }

    /**
     * 文件上传至 HDFS
     * @param delSrc
     * @param overwrite
     * @param srcFile
     * @param destPath
     */
    public void copyFileToHDFS(boolean delSrc, boolean overwrite,
            String srcFile, String destPath) {
        Path srcPath = new Path(srcFile);
        String hdfsUri = HDFSUri;
        if (StringUtils.isNotBlank(hdfsUri)) {
            destPath = hdfsUri + destPath;
        }
        Path dstPath = new Path(destPath);
        //实现文件上传
        try {
            FileSystem fs = getFileSystem();
            fs.copyFromLocalFile(srcPath, dstPath);
            fs.copyFromLocalFile(delSrc, overwrite, srcPath, dstPath);
            fs.close();
        } catch (IOException e) {
            e.printStackTrace();
            logger.error("", e);
        }
    }

    /**
     * 从 HDFS 下载文件
     * @param srcFile
     * @param destPath
     */
```

```java
public void getFile(String srcFile, String destPath) {
    String hdfsUri = HDFSUri;
    if (StringUtils.isNotBlank(hdfsUri)) {
        srcFile = hdfsUri + srcFile;
    }
    Path srcPath = new Path(srcFile);
    Path dstPath = new Path(destPath);

    try {
        FileSystem fs = getFileSystem();
        fs.copyToLocalFile(srcPath, dstPath);
        fs.close();
    } catch (IOException e) {
        e.printStackTrace();
        logger.error("", e);
    }
}

/**
 * 获取 HDFS 集群节点信息
 * @return DatanodeInfo[]
 */
public DatanodeInfo[] getHDFSNodes() {
    //获取所有节点
    DatanodeInfo[] dataNodeStats = new DatanodeInfo[0];
    try {
        FileSystem fs = getFileSystem();
        //获取分布式文件系统
        DistributedFileSystem hdfs = (DistributedFileSystem) fs;
        dataNodeStats = hdfs.getDatanodeStats();
    } catch (IOException e) {
        e.printStackTrace();
        logger.error("", e);
    }
    return dataNodeStats;
}

/**
 * 查找某个文件在 HDFS 集群的位置
 * @param filePath
 * @return BlockLocation[]
 */
public BlockLocation[] getFileBlockLocations(String filePath) {
    String hdfsUri = HDFSUri;
    if (StringUtils.isNotBlank(hdfsUri)) {
        filePath = hdfsUri + filePath;
```

```java
        }
        Path path = new Path(filePath);
        //文件块位置列表
        BlockLocation[] blkLocations = new BlockLocation[0];
        try {
            FileSystem fs = getFileSystem();
            //获取文件目录
            FileStatus filestatus = fs.getFileStatus(path);
            //获取文件块位置列表
            blkLocations = fs.getFileBlockLocations(filestatus, 0,
                    filestatus.getLen());
        } catch (IOException e) {
            e.printStackTrace();
            logger.error("", e);
        }
        return blkLocations;
}

/**
 * 文件重命名
 * @param srcPath
 * @param dstPath
 */
public boolean rename(String srcPath, String dstPath) {
    boolean flag = false;
    try {
        FileSystem fs = getFileSystem();
        String hdfsUri = HDFSUri;
        if (StringUtils.isNotBlank(hdfsUri)) {
            srcPath = hdfsUri + srcPath;
            dstPath = hdfsUri + dstPath;
        }
        flag = fs.rename(new Path(srcPath), new Path(dstPath));
    } catch (IOException e) {
        e.printStackTrace();
        logger.error("", e);
    }
    return flag;
}

/**
 * 判断目录是否存在
 * @param srcPath
 * @param dstPath
 */
public boolean existDir(String filePath, boolean create) {
```

```java
            boolean flag = false;
            if (StringUtils.isEmpty(filePath)) {
                return flag;
            }
            try {
                Path path = new Path(filePath);
                //FileSystem 对象
                FileSystem fs = getFileSystem();
                if (create) {
                    if (!fs.exists(path)) {
                        fs.mkdirs(path);
                    }
                }
                if (fs.isDirectory(path)) {
                    flag = true;
                }
            } catch (Exception e) {
                logger.error("", e);
            }
            return flag;
        }

        public static void main(String[] args) {
            String testpath = "testpath";
            OperateHDFS op = new OperateHDFS();
            //创建目录
            op.mkdir(testpath);
            //列文件目录
            String[] filelist = op.ListFile("/", null);
            for (String f : filelist)
                System.out.println(f);
            //删除目录
            op.rmdir(testpath);
            //上传文件
            op.copyFileToHDFS(false, true, "/root/hello.txt", testpath);
            //下载文件
            op.getFile("/test/test.txt", "/root/");
            //查询文件在 HDFS 集群的信息
            BlockLocation[] blkLoc = op.getFileBlockLocations("/test/test.txt");
            for (int i = 0; i < blkLoc.length; i++) {
                System.out.println(blkLoc[i].toString());
                System.out.println("文件在 block 中的偏移量" + blkLoc[i].getOffset()
                    + ", 长度" + blkLoc[i].getLength());
            }
        }
    }
```

（2）运行程序结果如图 4-10 所示。

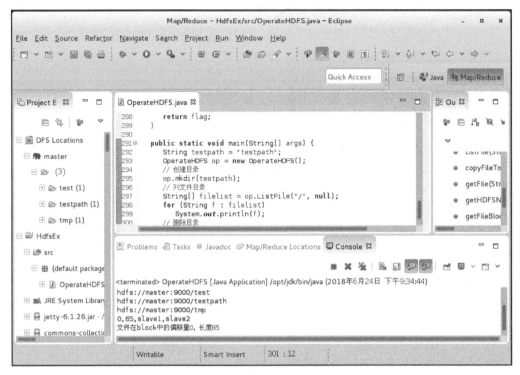

图 4-10　程序运行结果图

小结

Hadoop 实现了一个高可扩展的分布式文件系统——HDFS，HDFS 作为 Hadoop 底层基础设施，为云计算提供高可靠、高性能的存储服务。HDFS 在很大程度上借鉴了 Google GFS 文件系统的设计思想，具有高度容错、支持大数据集等诸多特性。本章介绍了 HDFS 设计的前提和设计目标，重点介绍了 HDFS 架构模块功能及运行机制，以及 HDFS 读/写文件流程、HDFS Shell 和 HDFS JAVA API，最后给出一个综合实例，使用 Java API 对 HDFS 进行目录创建、目录删除、文件上传和文件下载等操作。

习题

1. 简述 HDFS 的体系架构。
2. 简述 HDFS 读数据的流程。
3. 简述 HDFS 写数据的流程。
4. 简述 Block 副本的存放策略。
5. 编写程序实现对 HDFS 文件的读/写等。

第 5 章

MapReduce分布式计算

5.1 MapReduce 简介

Hadoop MapReduce 是一个快速、高效、简单用于编写并行处理大数据程序及应用在大集群上的编程框架。其前身是 Google 公司的 MapReduce,它是 Google 公司的核心计算模型,将复杂的、运行于大规划集群上的并行计算过程高度地抽象到了两个函数:Map 和 Reduce。它适合用 MapReduce 来处理的数据集(或任务),需要满足一个基本要求:待处理的数据集可以分解成许多小的数据集,而且每一个小数据集都可以完全并行地进行处理。概念 Map(映射)和 Reduce(归纳)及它们的主要思想,都是从函数式编程语言中借来的,同时包含了从矢量编程语言中借来的特性。Hadoop MapReduce 极大地方便了编程人员在不会分布式并行编程的情况下,将自己的程序运行在分布式系统上。

5.1.1 MapReduce 架构

和 HDFS 一样,MapReduce 也是采用 Master/Slave 的架构,其架构如图 5-1 所示。它主要由 Client、JobTracker、TaskTracker 及 Task 4 个部分组成。

(1) Client 会在用户端通过 Client 类将应用配置参数打包成 jar 文件存储到 hdfs,并把路径提交到 Jobtracker,然后由 JobTracker 创建每一个 Task(即 MapTask 和 ReduceTask),并将它们分发到各个 TaskTracker 服务中去执行。

(2) JobTracker 负责资源监控和作业调度。JobTracker 监控所有 TaskTracker 与 job 的健康状况,一旦发现失败,就将相应的任务转移到其他节点。同时,JobTracker 会跟踪任务的执行进度、资源使用量等信息,并将这些信息告诉任务调度器,而调度器会在资源出现空闲时,选择合适的任务使用这些资源。在 Hadoop 中,任务调度器是一个可插拔的模块,用户可以根据自己的需要设计相应的调度器。

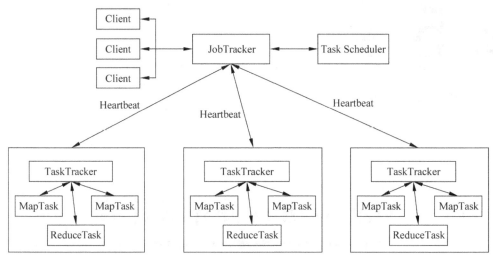

图 5-1 MapReduce 架构图

(3) TaskTracker 会周期性地通过 Heartbeat 将本节点上资源的使用情况和任务的运行进度汇报给 JobTracker,同时接收 JobTracker 发送过来的命令并执行相应的操作(如启动新任务、结束任务等)。TaskTracker 使用 slot 等量划分本节点上的资源量。slot 代表计算资源(CPU、内存等)。一个 Task 获取一个 slot 后才有机会运行,而 Hadoop 调度器的作用就是将各个 TaskTracker 上的空闲 slot 分配给 Task 使用。slot 分为 Map slot 和 Reduce slot 两种,分别供 MapTask 和 ReduceTask 使用。TaskTracker 通过 slot 数目(可配置参数)限定 Task 的并发度。

(4) TaskTracker 分为 MapTask 和 ReduceTask 两种,均由 TaskTracker 启动。HDFS 以固定大小的 block 为基本单位存储数据,而对于 MapReduce 而言,其处理单位是 split。split 是一个逻辑概念,只包含一些元数据信息,如数据起始位置、数据长度、数据所在节点等。它的划分方法完全由用户自行决定。但需要注意的是,split 的多少决定了 MapTask 的数目,因为每个 split 只会交给一个 MapTask 处理。split 和 block 的关系如图 5-2 所示。

MapTask 的执行过程如图 5-3 所示。由图 5-3 可知,MapTask 先将对应的 split 迭代解析成一个个 key/value 对,依次调用用户自定义的 map() 函数进行处理,最终将临时结果存放到本地磁盘上,其中临时数据被分成若干个 partition,每个 partition 将被一个 ReduceTask 处理。

ReduceTask 的执行过程如图 5-4 所示。该过程分为以下 3 个阶段。

(1) 从远程节点上读取 MapTask 中间结果(称为"shuffle 阶段")。

(2) 按照 key/value 对进行排序(称为"sort 阶段")。

(3) 依次读取<key, value list>,调用用户自定义的 reduce() 函数处理,并将最终结果保存到 HDFS 上(称为"reduce 阶段")。

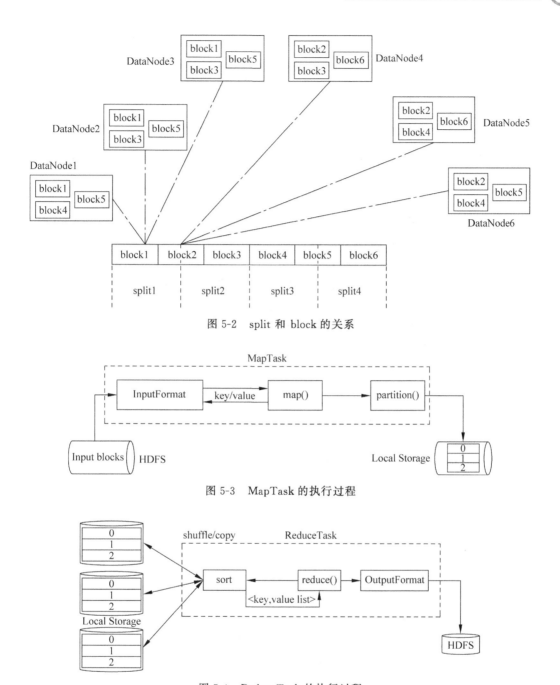

图 5-2　split 和 block 的关系

图 5-3　MapTask 的执行过程

图 5-4　ReduceTask 的执行过程

5.1.2　MapReduce 的原理

MapReduce 采用的是"分而治之"的策略，当我们处理大规模的数据时，将这些数据拆解成多个部分，并利用集群的多个节点同时进行数据处理，然后对各个节点得到的中间结果进行汇总，经过进一步的计算，得到最终结果。

一个MapReduce作业(job)通常会把输入的数据集切分为若干个独立的数据块,由map任务(task)以完全并行的方式处理它们。框架会对map的输出先进行排序,然后把结果输入给reduce任务。通常,作业的输入和输出都会被存储在文件系统中,整个框架负责任务的调度和监控,以及重新执行已经失败的任务。

通常,MapReduce框架的计算节点和存储节点是运行在一组相同的节点上,也就是说,运行MapReduce框架和运行HDFS文件系统的节点通常是在一起的。这种配置允许框架在那些已经存好的数据的节点上高效地调度任务,这可以使整个集群的网络带宽被非常高效地利用。

MapReduce框架由一个主节点(ResourceManager)、多个子节点(运行NodeManager)和MRAppMaster(每个任务一个)共同组成。应用程序至少应该指明输入/输出的位置(路径),并通过实现合适的接口或抽象类提供map和reduce函数,再加上其他作业的参数,就构成了作业配置(job configuration)。Hadoop的job client提交作业(jar包/可执行程序等)和配置信息给ResourceManager,后者负责分发这些软件和配置信息给slave、调度任务且监控它们的执行,同时提供状态和诊断信息给job-client。

虽然Hadoop框架是用Java实现的,但MapReduce应用程序不一定要用Java来写,也可以使用Ruby、Python、C++等来编写。

MapReduce框架的流程如图5-5所示。

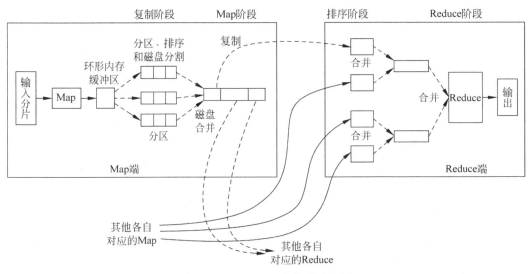

图5-5 MapReduce框架的流程图

针对上面的流程可以分为两个阶段来描述。

1. Map阶段

(1) InputFormat根据输入文件产生键值对,并传送到Mapper类的map函数中。

(2) Map输出键值对到一个没有排序的缓冲内存中。

(3) 当缓冲内存达到给定值或map任务完成,在缓冲内存中的键值对就会被排序,然后输出到磁盘中的溢出文件中。

（4）如果有多个溢出文件，那么就会整合这些文件到一个文件中，且是排序的。

（5）这些排序过的、在溢出文件中的键值对会等待 Reducer 类的获取。

2．Reduce 阶段

（1）Reducer 类获取 Mapper 类的记录，然后产生另外的键值对，最后输出到 HDFS 中。

（2）Reducer 类中的 reduce 方法针对每个 key 调用一次。

（3）Shuffle：相同的 key 被传送到同一个的 Reducer 类中。

（4）当有一个 Mapper 类完成后，Reducer 类就开始获取相关数据，所有的溢出文件会被排到一个内存缓冲区中。

（5）当 Reducer 所有相关的数据都传输完成后，所有溢出文件就会被整合和排序。

（6）当内存缓冲区满了后，就会在本地磁盘产生溢出文件。

（7）Reducer 类输出到 HDFS。

5.1.3 MapReduce 的工作机制

1．MapReduce 运行图

MapReduce 运行图如图 5-6 所示。

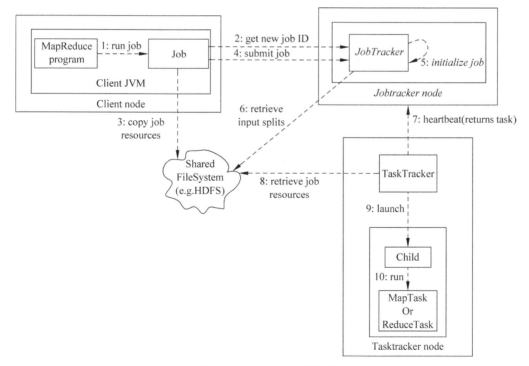

图 5-6　MapReduce 运行图

2. 运行解析

1) 作业的提交

(1) 此方法调用 submit()。在 submit()方法里面连接 JobTracker,即生成一个内部 JobSummitter(实际上是 new JobClient()),在 new JobClient()里面生成一个 JobSubmissionProtocol 接口(JobTracker 实现了此接口)对象 jobSubmitClient(是它连接或对应着 JobTracker),在 submit()方法里面也调用 JobClient.submitJobInternal(conf)方法返回一个 RunningJob(步骤 1)。

(2) 参数 true 说明要调用方法 jobClient.monitorAndPrintJob()即检查作业的运行情况(每秒一次),如果有变化就报告给控制台。

jobClient.submitJobInternal()所实现的提交作业过程如下。

① 向 JobTracker 请求一个新的 job ID(步骤 2)。

② 检查作业的输出路径,如果未指定或已存在,则不提交作业并抛错误给程序。

③ 计算并生成作业的输入分片,如果路径不存在,则不提交作业并抛错误给程序。

④ 将运行作业所需要的资源(包括作业 jar 文件、配置文件和计算所得的输入分片)复制到 JobTracker 的文件系统中以 job ID 命名的目录下(即 HDFS 中)。作业 jar 副本较多(mapred.submit.replication = 10)(步骤 3)。

⑤ 告知 JobTracker 作业准备执行(真正地提交作业 jobSubmitClient.submitJob())(步骤 4)。

2) 作业的初始化

(1) JobTracker 接收到对其 submitJob()方法的调用后,将其放入内部队列,交由 job scheduler 进行调度,并对其进行初始化,包括创建一个正在运行作业的对象——封装任务和记录信息(步骤 5)。

(2) 为了创建任务运行列表,Job Scheduler 首先从共享文件系统中获取已计算好的输入分片信息(步骤 6),然后为每个分片创建一个 map 任务。

(3) 创建的 reduce 任务数量由 Job 的 mapred.reduce.task 属性决定(setNumReduceTasks()设置),schedule 创建相应数量的 reduce 任务。任务在此时被指定 ID。

(4) 除了 map 和 reduce 任务,还有 setupJob 和 cleanupJob 需要建立:由 TaskTrackers 在所有 map 开始前和所有 reduce 结束后分别执行,这两个方法在 OutputCommitter 中(默认是 FileOutputCommitter)。setupJob()创建输出目录和任务的临时工作目录; cleanupJob()删除临时工作目录。

3) 作业的分配

(1) 每个 TaskTracker 定期发送心跳给 JobTracker,告知自己还活着,并附带消息说明自己是否已准备好接受新任务。JobTracker 以此来分配任务,并使用心跳的返回值与 TaskTracker 通信(步骤 7)。JobTracker 利用调度算法先选择一个 job,然后再选此 job 的一个 task 分配给 TaskTracker。

(2) 每个 TaskTracker 会有固定数量的 map 和 reduce 任务槽,数量由 TaskTracker

核的数量和内存大小来决定。JobTracker 会先将 TaskTracker 的所有的 map 槽填满，然后才填此 TaskTracker 的 reduce 任务槽。

（3）JobTracker 分配 map 任务时会选取与输入分片最近的 TaskTracker，分配 reduce 任务用不着考虑数据本地化。

4）任务的执行

（1）TaskTracker 分配到一个任务后，首先从 HDFS 中把作业的 jar 文件及运行所需要的全部文件（DistributedCache 设置的）复制到 TaskTracker 本地（步骤 8）。

（2）TaskTracker 为任务新建一个本地工作目录，并把 jar 文件的内容解压到这个文件夹下。

（3）TaskTracker 新建一个 TaskRunner 实例来运行该任务（步骤 9）。

（4）TaskRunner 启动一个新的 JVM 来运行每个任务（步骤 10），以便客户的 map/reduce 不会影响 TaskTracker。

5）进度和状态的更新

一个作业和它的每个任务都有一个状态，包括作业或任务的运行状态（running、successful、failed）、map 和 reduce 的进度、计数器值、状态消息或描述。

map 进度标准是处理输入所占比例；reduce 是 copy\merge\reduce 整个进度的比例。

Child JVM 有独立的线程，每隔 3 秒检查任务更新标志，如果有更新，就会报告给此 TaskTracker。

TaskTracker 每隔 5 秒给 JobTracker 发心跳。

JobTracker 合并这些更新，产生一个表明所有运行作业及其任务状态的全局试图。

JobClient.monitorAndPrintJob() 每秒查询这些信息。

6）作业的完成

当 JobTracker 收到最后一个任务（this will be the special job cleanup task）的完成报告后，便把 job 状态设置为 successful。

job 得到完成信息便从 waitForCompletion() 返回。

最后，JobTracker 清空作业的工作状态，并指示 TaskTracker 也清空作业的工作状态（如删除中间输出）。

3．失败解析

1）任务失败

（1）子任务失败。当 map 或 reduce 子任务中的代码抛出异常，JVM 进程会在退出之前向父进程 TaskTracker 发送错误报告，TaskTracker 会将此（任务尝试）task attempt 标记为 failed 状态，释放一个槽以便运行另外一个任务。

（2）JVM 失败。JVM 突然退出，即 JVM 错误，这时 TaskTracker 会注意到进程已经退出，标记为 failed。

① 任务失败有重试机制，重试次数 map 任务设置是由 mapred.map.max.attempts 属性控制的，reduce 是由 mapred.reduce.max.attempts 属性控制的。

② 一些 job 可以完成总体任务的一部分就能够接受,这个百分比由 mapred. map. failures. precent 和 mapred. reduce. failures. precent 参数控制。

③ 任务尝试(task attempt)是可以中止(killed)的。

2) TaskTracker 失败

作业运行期间,TaskTracker 会通过心跳机制不断与系统 JobTracker 通信,如果某个 TaskTracker 运行缓慢或失败及出现故障,TaskTracker 就会停止或很少向 JobTracker 发送心跳,JobTracker 会注意到此 TaskTracker 发送心跳的情况,从而将此 TaskTracker 从等待任务调度的 TaskTracker 池中移除。

(1) 如果是 map 并且成功完成,JobTracker 会安排此 TaskTracker 上成功运行的 map 任务返回。

(2) 如果是 reduce 并且成功完成,则数据直接使用,因为 reduce 只要执行完就会把输出写到 HDFS 上。

(3) 如果它们属于未完成的作业,那么 reduce 阶段无法获取该 TaskTracker 上的本地 map 输出文件,任何任务都需要重新调度。

另外,即使 TaskTracker 没有失败,如果它上面的失败任务远远高于集群的平均失败任务数,也会被列入黑名单。可以通过重启从 JobTracker 的黑名单中移除。

3) JobTracker 失败

JobTracker 失败应该说是最严重的一种失败方式了,而且在 Hadoop 中存在单点故障的情况下是相当严重的,因为在这种情况下作业最终将失败,尽管这种故障的概率极小。未来版本可以通过启动多个 JobTracker,在这种情况只运行一个主的 JobTracker,通过一种机制来确定哪个是主的 JobTracker。

5.2 MapReduce 操作实践

5.2.1 MapReduce WordCount 编程实例

视频讲解

词频统计是最简单也是最能体现 MapReduce 思想的程序之一,可以称为 MapReduce 版"Hello World",该程序的完整代码可以在 Hadoop 安装包的 src/examples 目录下找到。词频统计主要完成的功能是:统计一系列文本文件中每个单词出现的次数。本节通过分析源代码帮助读者厘清 MapReduce 程序的基本结构。

1. WordCount 代码分析

MapReduce 框架自带的示例程序 WordCount 只包含 Mapper 类和 Reduce 类,其他全部使用默认类。下面为 WordCount 源代码分析。

1) Mapper 类

Map 过程需要继承 org. apache. hadoop. mapreduce 包中的 Mapper 类,并重写 map 方法。

```java
public static class TokenizerMapper extends
    Mapper<Object, Text, Text, IntWritable> {
private final static IntWritable one = new IntWritable(1);
private Text word = new Text();
//map方法,划分一行文本,读一单词写出的一<单词,1>
public void map(Object key, Text value, Context context)
    throws IOException, InterruptedException {
  StringTokenizer itr = new StringTokenizer(value.toString());
  while (itr.hasMoreTokens()) {
    word.set(itr.nextToken());
    context.write(word, one);                //写出<单词,1>
  }
}
}
```

2) Reduce 类

```java
public static class IntSumReducer extends
    Reducer<Text, IntWritable, Text, IntWritable> {
private IntWritable result = new IntWritable();
public void reduce(Text key, Iterable<IntWritable> values, Context context)
    throws IOException, InterruptedException {
  int sum = 0;
  for (IntWritable val : values) {
    sum += val.get();          //相当于<Hello,1><Hello,1>将两个1相加
  }
  result.set(sum);
  context.write(key, result);
}
}
```

3) 主函数

```java
public static void main(String[] args) throws Exception {     //主方法,函数入口
  Configuration conf = new Configuration();                   //实例化配置文件类
  String[] otherArgs =
      new GenericOptionsParser(conf, args).getRemainingArgs();
  if (otherArgs.length != 2) {
    System.err.println("Usage: wordcount <in> <out>");
    System.exit(2);
  }
  Job job = new Job(conf, "word count");                      //实例化Job类
  job.setJarByClass(WordCount.class);                         //设置主类名
  job.setMapperClass(TokenizerMapper.class);                  //指定使用上述自定义Map类
  job.setCombinerClass(IntSumReducer.class);
  job.setReducerClass(IntSumReducer.class);
  job.setOutputKeyClass(Text.class);
  job.setOutputValueClass(IntWritable.class);
```

```java
    FileInputFormat.addInputPath(job, new Path(otherArgs[0]));
    FileOutputFormat.setOutputPath(job, new Path(otherArgs[1]));   //设置输出结果文件位置
    System.exit(job.waitForCompletion(true) ? 0 : 1);                      //提交任务并监控任务状态
}
```

4) 提交 WordCount

```java
public class WordCount {
  public static class TokenizerMapper extends
      Mapper< Object, Text, Text, IntWritable > {
    private final static IntWritable one = new IntWritable(1);
    private Text word = new Text();
    //map方法,划分一行文本,读一单词写出的一<单词,1>
    public void map(Object key, Text value, Context context)
        throws IOException, InterruptedException {
      StringTokenizer itr = new StringTokenizer(value.toString());
      while (itr.hasMoreTokens()) {
        word.set(itr.nextToken());
        context.write(word, one);         //写出<单词,1>
      }
    }
  }

  public static class IntSumReducer extends
      Reducer< Text, IntWritable, Text, IntWritable > {
    private IntWritable result = new IntWritable();
    public void reduce(Text key, Iterable< IntWritable > values, Context context)
        throws IOException, InterruptedException {
      int sum = 0;
      for (IntWritable val : values) {
        sum += val.get();          //相当于< Hello,1 >< Hello,1 >将两个1相加
      }
      result.set(sum);
      context.write(key, result);
    }
  }

  public static void main(String[] args) throws Exception {   //主方法,函数入口
    Configuration conf = new Configuration();                  //实例化配置文件类
    String[] otherArgs =
        new GenericOptionsParser(conf, args).getRemainingArgs();
    if (otherArgs.length != 2) {
      System.err.println("Usage: wordcount < in >< out >");
      System.exit(2);
    }
    Job job = new Job(conf, "word count");                  //实例化Job类
    job.setJarByClass(WordCount.class);                     //设置主类名
    job.setMapperClass(TokenizerMapper.class);              //指定使用上述自定义Map类
```

```
        job.setCombinerClass(IntSumReducer.class);
        job.setReducerClass(IntSumReducer.class);
        job.setOutputKeyClass(Text.class);
        job.setOutputValueClass(IntWritable.class);
        FileInputFormat.addInputPath(job, new Path(otherArgs[0]));
        FileOutputFormat.setOutputPath(job, new Path(otherArgs[1])); //设置输出结果文件位置
        System.exit(job.waitForCompletion(true) ? 0 : 1);             //提交任务并监控任务状态
    }
}
```

2. 打包运行

(1) 在 Eclipse 中选择 File→Export 选项,导出 jar 包。

(2) 新建两个文本文件并上传到 HDFS:

```
[root@master ~]# hdfs dfs -mkdir /input
[root@master ~]# hdfs dfs -rm /input/*
[root@master ~]# hdfs dfs -put sw*.txt /input
```

(3) 运行 jar 包:

```
[root@master ~]# hadoop jar WordCountTest.jar WordCountTest /input /output
```

(4) 查看运行结果:

```
[root@master ~]# hdfs dfs -ls /output
[root@master ~]# hdfs dfs -text /output/part-r-00000
```

5.2.2 MapReduce 倒排索引编程实例

1. 简介

倒排索引是文档检索系统中最常用的数据结构,被广泛地应用于全文搜索引擎。它主要用来存储某个单词(或词组)在一个文档或一组文档中的存储位置的映射,即提供了一种根据内容来查找文档的方式。由于不是根据文档来确定文档所包含的内容,而是进行了相反的操作,因而称为倒排索引(inverted index)。通常情况下,倒排索引由一个单词(或词组)及相关的文档列表组成,文档列表中的文档或者标识文档的 ID 号,或者指定文档所在位置的 URI。

2. 分析与设计

本节实现的倒排索引主要关注的信息为单词、文档 URI 及词频。下面根据 MapReduce 的处理过程给出倒排索引的设计思路。

1) Map 过程

首先使用默认的 TextInputFormat 类对输入文件进行处理,得到文本中每行的偏移

量及其内容。显然,Map过程首先必须分析输入的<key,value>对,得到倒排索引中需要的3个信息：单词、文档URI和词频。这里存在两个问题：第一,<key,value>对只能有两个值,作为key或value值；第二,通过一个Reduce过程无法同时完成词频统计和生成文档列表,所以必须增加一个Combine过程完成词频统计。

这里将单词和URI组成key值,将词频作为value,这样做的好处是可以利用MapReduce框架自带的Map端排序,将同一个文档的相同单词的词频组成列表,传递给Combine过程,实现类似于WordCount的功能。

2) Combine过程

经过map方法处理后,Combine过程将key值相同的value值累加,得到一个单词在文档中的词频。如果单词在文档中词频的输出作为Reduce过程的输入,在Shuffle过程时将面临一个问题：所有具有相同单词的记录(由单词、URI和词频组成)应该交由同一个Reduce处理,但当前的key值无法保证这一点,所以必须修改key值和value值。这次将单词作为key值,URI和词频组成value值。这样做的好处是可以利用MapReduce框架默认的HashPartitioner类完成Shuffle过程,将所有相同的单词发送给同一个Reduce处理。

3) Reduce过程

经过上述两个过程后,Reduce过程只需要将key值相同的value值组合成倒排索引文件所需的格式即可,剩下的事情就可以直接交给MapReduce框架进行处理了。

4) 需要解决的问题

本节设计的倒排索引在文件数目上没有限制,但是单个文件不宜过大(具体值与默认HDFS块大小及相关配置有关),要保证每个文件对应一个split。否则,由于Reduce过程没有进一步统计词频,最终结果可能会出现词频未统计完全的单词。因此可以通过重写Inputformat类将每一个文件作为一个split,避免上述情况；或者执行两次MapReduce,第一次用于统计词频,第二次用于生成倒排索引。除此之外,还可以利用复合键值对等实现包含更多信息的倒排索引。

3. 倒排索引完整源代码

```
import java.io.IOException;
import java.util.StringTokenizer;
import org.apache.hadoop.conf.Configuration;
import org.apache.hadoop.fs.Path;
import org.apache.hadoop.io.IntWritable;
import org.apache.hadoop.io.Text;
import org.apache.hadoop.mapreduce.Job;
import org.apache.hadoop.mapreduce.Mapper;
import org.apache.hadoop.mapreduce.Reducer;
import org.apache.hadoop.mapreduce.lib.input.FileInputFormat;
import org.apache.hadoop.mapreduce.lib.input.FileSplit;
import org.apache.hadoop.mapreduce.lib.output.FileOutputFormat;
```

```java
import org.apache.hadoop.util.GenericOptionsParser;

public class InvertedIndex {
    public static class InvertedIndexMapper extends Mapper<Object, Text, Text, Text> {
        private Text keyInfo = new Text();
        private Text valueInfo = new Text();
        private FileSplit split;

        public void map(Object key, Text value, Context context) throws IOException, InterruptedException {
            split = (FileSplit)context.getInputSplit();
            StringTokenizer itr = new StringTokenizer(value.toString());

            while(itr.hasMoreTokens()) {
                keyInfo.set(itr.nextToken() + ":" + split.getPath().toString());
                valueInfo.set("1");
                context.write(keyInfo, valueInfo);
            }
        }

    }
    public static class InvertedIndexCombiner extends Reducer<Text, Text, Text, Text> {
        private Text info = new Text();
        public void reduce(Text key, Iterable<Text> values, Context context) throws IOException, InterruptedException {
            int sum = 0;
            for(Text value : values) {
                sum += Integer.parseInt(value.toString());
            }
            int splitIndex = key.toString().indexOf(":");
            info.set(key.toString().substring(splitIndex + 1) + ":" + sum);
            key.set(key.toString().substring(0, splitIndex));
            context.write(key, info);
        }
    }
    public static class InvertedIndexReducer extends Reducer<Text, Text, Text, Text> {
        private Text result = new Text();
        public void reducer(Text key, Iterable<Text> values, Context context) throws IOException, InterruptedException {
            String fileList = new String();
            for(Text value : values) {
                fileList += value.toString() + ";";
            }
            result.set(fileList);
            context.write(key, result);
        }
    }
    public static void main(String[] args) throws Exception{
        //TODO Auto-generated method stub
```

```
        Configuration conf = new Configuration();
        String[] otherArgs = new GenericOptionsParser(conf, args).getRemainingArgs();
        if(otherArgs.length != 2) {
            System.err.println("Usage: wordcount <in> <out>");
            System.exit(2);
        }
        Job job = new Job(conf, "InvertedIndex");
        job.setJarByClass(InvertedIndex.class);
        job.setMapperClass(InvertedIndexMapper.class);
        job.setMapOutputKeyClass(Text.class);
        job.setMapOutputValueClass(Text.class);
        job.setCombinerClass(InvertedIndexCombiner.class);
        job.setReducerClass(InvertedIndexReducer.class);

        job.setOutputKeyClass(Text.class);
        job.setOutputValueClass(Text.class);

        FileInputFormat.addInputPath(job, new Path(otherArgs[0]));
        FileOutputFormat.setOutputPath(job, new Path(otherArgs[1]));
        System.exit(job.waitForCompletion(true) ? 0 : 1);
    }
}
```

程序运行和 WordCount 同样原理。

小结

本章首先阐述了 MapReduce 架构,然后介绍了 MapReduce 的工作原理和 MapReduce 的工作机制,最后重点介绍了基于 MapReduce 架构的 WordCount 编程实例和倒排索引编程实例。

习题

1. 简述 MapReduce 架构。
2. 简述 MapReduce 的工作原理。
3. 简述 MapReduce 的工作机制。
4. 编写 MapReduce WordCount。
5. 实现 MapReduce 倒排索引编程。

第 6 章

HBase分布式数据库应用

数据库是按照数据结构来组织、存储和管理数据的仓库。关系型数据库是指采用了关系模型来组织数据的数据库。关系模型是在 1970 年由 IBM 的研究员 E. F. Codd 博士首先提出的,在之后的几十年中,关系模型的概念得到了充分的发展,时至今日,关系型数据库仍旧非常重要。关系型数据库非常适合结构化数据,但对于数据急剧增长和非结构化数据,关系型数据库很难适应。非关系型数据库(NoSQL)的产生就是为了解决大规模数据集合多重数据种类带来的挑战,尤其是大数据应用难题,MongoDB、Cassandra 和 HBase 是比较有名的非关系型数据库。

6.1 HBase 简介

HBase(Hadoop Database)是一个高可靠性、高性能、面向列、可伸缩的分布式存储系统,利用 HBase 技术可在廉价 PC Server 上搭建起大规模结构化存储集群。HBase 是 Google Bigtable 的开源实现,类似 Google Bigtable 利用 GFS 作为其文件存储系统,HBase 利用 Hadoop HDFS 作为其文件存储系统; Google 运行 MapReduce 来处理 Bigtable 中的海量数据,HBase 同样利用 Hadoop MapReduce 来处理 HBase 中的海量数据; Google Bigtable 利用 Chubby 作为协同服务,HBase 利用 ZooKeeper 作为对应。HBase 与 Hadoop 紧密集成,如图 6-1 所示。

6.1.1 HBase 架构

HBase 采用 Master/Slave 架构搭建集群,它隶属于 Hadoop 生态系统,由 Client、HMaster、HRegionServer、ZooKeeper 等组成,它将数据存储于 HDFS 中。HMaster 主要负责利用 ZooKeeper 为 HRegionServer 分配 HRegion。ZooKeeper 是一个高可靠、高

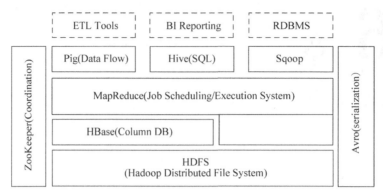

图 6-1 Hadoop 生态系统

可用、持久化的分布式协调系统。Client 使用 HBase 的远程过程调用协议（Remote Procedure Call Protocol，RPC）机制与 HMaster 和 HRegionServer 进行通信，对于管理类操作，Client 与 HMaster 进行 RPC；对于数据读/写类操作，Client 与 HRegionServer 进行 RPC。HBase 架构如图 6-2 所示。

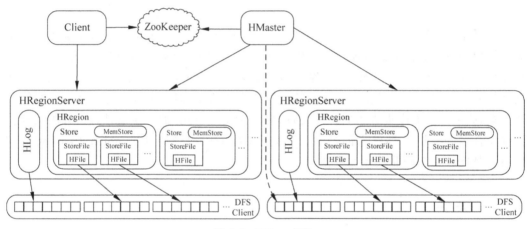

图 6-2 HBase 架构

1. ZooKeeper

ZooKeeper 是一个开放源代码的分布式应用程序协调服务，是 Google 的 Chubby 一个开源的实现，是 Hadoop 和 HBase 的重要组件。分布式的 HBase 依赖于 ZooKeeper 集群，所有节点和客户端必须能够正常访问 ZooKeeper。HBase 默认管理一个单点的 ZooKeeper 集群，HBase 可以把 ZooKeeper 当作自己的一部分来启动和关闭进程。ZooKeeper 也可以直接使用本地配置文件 zoo.cfg，不依赖 HBase 的启动与关闭，独立运行 ZooKeeper。HBase RegionServer 向 ZooKeeper 注册，提供 HBase RegionServer 状态信息；HMaster 启动时会将 HBase 系统表加载到 ZooKeeper Cluster，通过 ZooKeeper Cluster 可以获取当前系统表 Meta 的存储所对应的 HRegionServer 信息。

2. HMaster

HBase 中可以启动多个 HMaster，通过 ZooKeeper 的 Master Election 机制保证总有一个 HMaster 运行。HMaster 管理 HRegionServer，实现其负载均衡；管理和分配 HRegion，如在 HRegion split 时分配新的 HRegion；在 HRegionServer 退出时迁移其内的 HRegion 到其他 HRegionServer 上；实现数据定义和数据操作；管理 NameSpace 和 Table 的元数据；进行权限控制（ACL）管理等。

3. HRegionServer

HRegionServer 主要负责响应用户 I/O 请求，在 HDFS 文件系统中读/写数据，是 HBase 中最核心的模块。HRegionServer 内部管理了一系列 HRegion 对象，每个 HRegion 对应了 Table 中的一个 Region，HRegion 由多个 HStore 组成。每个 HStore 对应了 Table 中的一个 Column Family 的存储，每个 Column Family 其实就是一个集中的存储单元，因此最好将具备共同 I/O 特性的 column 放在一个 Column Family 中，这样做比较高效。

HStore 存储是 HBase 存储的核心，由两部分组成：一部分是 MemStore；另一部分是 StoreFiles。MemStore 是 Sorted Memory Buffer，用户写入的数据首先会放入 MemStore，当 MemStore 满了以后会 Flush 成一个 StoreFile（底层实现是 HFile）；当 StoreFile 文件数量增长到一定阈值时，会触发 Compact 合并操作，将多个 StoreFiles 合并成一个 StoreFile，合并过程中会进行版本合并和数据删除。因此，可以看出 HBase 其实只有增加数据，所有的更新和删除操作都是在后续的 Compact 过程中进行的，这使得用户的写操作只要进入内存中就可以立即返回，保证了 HBase I/O 的高性能。当 StoreFiles Compact 后，会逐步形成越来越大的 StoreFile，当单个 StoreFile 大小超过一定阈值后，会触发 split 操作，同时把当前 Region split 成两个 Region，父 Region 会下线，新 split 出的两个子 Region 会被 HMaster 分配到相应的 HRegionServer 上，使得原先一个 Region 的压力得以分流到两个 Region 上。

4. HFile

实际的存储文件功能是由 HFile 类实现的，它被专门创建用于有效存储 HBase 数据，基于 Hadoop 的 TFile 类，并模仿 Google 的 BigTable 架构使用 SSTable 格式。曾在 HBase 中使用过的 Hadoop 的 MapFile 类被证明性能不够好。

HFile 文件格式的详细信息如图 6-3 所示，文件长度为变长，仅 File Info/Trailer 定长，Trailer 中有指向其他数据块的指针，它是持久化数据到文件结束时写入，写入后即确定其成为不可变的数据存储文件。Index 块记录 Data 块和 Meta 块的偏移量。Data 块和 Meta 块实际上都是可选的，但对于大多数 HFile，用户都可以找到 Data 块。块大小是由 HColumnDescriptor 配置的，而该配置可以在创建时由用户指定或使用默认值。

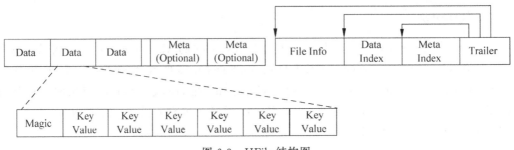

图 6-3　HFile 结构图

5. Write-Ahead-Log

Write-Ahead-Log(WAL)是 HBase 的 HRegionServer 在处理数据插入和删除的过程中用来记录操作内容的一种日志。客户端初始化一个更改数据的动作,如 put、delete 和 incrementColumnvalue(incr),每一种改动都被包装进 KeyValue 对象利用远程调用发送到 ReginoServer 对应这次改动的 Region 中。数据在 HRegionServer 中首先被写入 WAL,然后被写入 MemStore,最后当 MemStore 达到一定的大小或到达指定的时刻之后,数据被异步地持久化到文件系统上。在这之前数据是存储在内存中的,在这段时间里如果 HRegionServer 崩溃了,内存的数据就没有了,这时若有 WAL,就可以恢复数据。

每个 HRegionServer 中都有一个 HLog 对象,HLog 是一个实现 WAL 的类,在每次用户操作写入 MemStore 的同时,也会写一份数据到 HLog 文件中,HLog 文件定期会滚动出新的文件,并删除旧的文件(已持久化到 StoreFile 中的数据)。当 HRegionServer 意外终止后,HMaster 会通过 ZooKeeper 感知到 HMaster 首先会处理遗留的 HLog 文件,将其中不同 Region 的 Log 数据进行拆分,分别放到相应的 Region 目录下;然后将失效的 Region 重新分配,领取到这些 Region 的 HRegionServer 在 Load Region 的过程中,会发现有历史 HLog 需要处理,因此会重置 HLog 中的数据到 MemStore 中;最后刷新到 StoreFiles,完成数据恢复,如图 6-4 所示。

6. System Tables

HBase 内部保留名为 namespace 和 meta 的 System Tables。namespace 指对一组表的逻辑分组,类似关系型数据库中的 DataBase,方便对表在业务上划分。HBase 从 0.98.0 和 0.95.2 两个版本开始支持 namespace 级别的授权操作,HBase 全局管理员可以创建、修改和回收 namespace 的授权。

namespace 下面存储了 HBase 的 namespace、meta 和 acl 3 个表,这里的 meta 表跟 0.94 版本的.META.是一样的,自 0.96 版本之后就已经将-ROOT-表去掉了,直接从 ZooKeeper 中找到 meta 表的位置,然后通过 meta 表定位到 Region。namespace 中存储了 HBase 中的所有 namespace 信息,包括预置的 HBase 和 default。acl 则是表的用户权限控制。

如果自定义了一些 namespace,就会在/hbase/data 目录下新建一个 namespace 文件夹,该 namespace 下的表都将刷新写入该目录下。

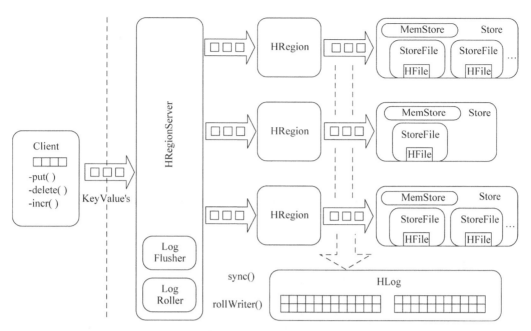

图 6-4　Write-Ahead-Log

通过 describe 命令查看 namespace 和 meta 表结构信息，命令如下。

```
hbase(main):001:0> describe 'hbase:namespace'
Table hbase:namespace is ENABLED
hbase:namespace
COLUMN FAMILIES DESCRIPTION
{NAME => 'info', BLOOMFILTER => 'ROW', VERSIONS => '10', IN_MEMORY => 'true',
KEEP_DELETED_CELLS => 'FALSE', DATA_BLOCK_ENCODING => 'NONE',
TTL => 'FOREVER', COMPRESSION => 'NONE', CACHE_DATA_IN_L1 => 'true',
MIN_VERSIONS => '0', BLOCKCACHE => 'true', BLOCKSIZE => '8192',
REPLICATION_SCOPE => '0'}
1 row(s) in 0.0380 seconds

hbase(main):002:0> describe 'hbase:meta'
Table hbase:meta is ENABLED
hbase:meta, {TABLE_ATTRIBUTES => {IS_META => 'true', REGION_REPLICATION => '1',
coprocessor$1 => '|org.apache.hadoop.hbase.coprocessor
.MultiRowMutationEndpoint|536870911|'}
COLUMN FAMILIES DESCRIPTION
{NAME => 'info', BLOOMFILTER => 'NONE', VERSIONS => '10', IN_MEMORY => 'true',
KEEP_DELETED_CELLS => 'FALSE', DATA_BLOCK_ENCODING => 'NONE',
TTL => 'FOREVER', COMPRESSION => 'NONE', CACHE_DATA_IN_L1 => 'true',
MIN_VERSIONS => '0', BLOCKCACHE => 'true', BLOCKSIZE => '8192',
REPLICATION_SCOPE => '0'}
1 row(s) in 0.0310 seconds
```

namespace 和 meta 可以被看作是两张普通的表,它们有自己的表结构,并且表结构相似。通过 scan 命令查看 namespace 和 meta 的数据,命令如下。

```
hbase(main):003:0> scan 'hbase:namespace'
ROW              COLUMN + CELL
default          column = info:d, timestamp = 1508569563660, value = \x0A\x07default
hbase            column = info:d, timestamp = 1508569564673, value = \x0A\x05hbase
2 row(s) in 0.0360 seconds

hbase(main):004:0> scan 'hbase:meta'
ROW                      COLUMN + CELL
hbase:namespace,,1508569562076.f4a9da80dd4368ddac9fdef5234f91ed.
column = info:regioninfo, timestamp = 1508571228172,
value = {ENCODED => f4a9da80dd4368ddac9fdef5234f91ed,
NAME => 'hbase:namespace,,1508569562076.f4a9da80dd4368ddac9fdef5234f91ed.',
STARTKEY => '', ENDKEY => ''}
hbase:namespace,,1508569562076.f4a9da80dd4368ddac9fdef5234f91ed.
column = info:seqnumDuringOpen, timestamp = 1508571228172,
value = \x00\x00\x00\x00\x00\x00\x00\x0C
hbase:namespace,,1508569562076.f4a9da80dd4368ddac9fdef5234f91ed.
column = info:server, timestamp = 1508571228172, value = slave1:16020
hbase:namespace,,1508569562076.f4a9da80dd4368ddac9fdef5234f91ed.
column = info:serverstartcode, timestamp = 1508571228172, value = 1508570920576
1 row(s) in 0.0370 seconds
```

RowKey 就是 Region Name,它的命名形式是 TableName、StartKey 和 TimeStamp.Encoded。其中,Encoded 是 TableName、StartKey 和 TimeStamp 的 md5 值,示例如下。

```
# echo -n "hbase:namespace,,1508569562076" | md5sum
f4a9da80dd4368ddac9fdef5234f91ed  -
```

Column Family:info 里面包含 Column:regioninfo、seqnumDuringOpen、server、serverstartcode。其中,regioninfo 就是 Region 的详细信息,包括 StartKey、EndKey 和 TimeStamp.Encoded 等;server 存储的是管理这个 Region 的 HRegionServer 的地址。所以当 Region 被拆分、合并或重新分配的时候,都需要来修改这张表的内容。

6.1.2 HBase 的存储

HBase 是基于列式存储的。列式存储是相对于传统关系型数据库的行式存储来说的,列式存储如图 6-5 所示。列式存储的优点是投影效率高、查询时只有涉及的列被读取、任何列都能作为索引,但其缺点是选择完成时被选择的列要重新组装。

HBase 以表的形式存储数据。表由行、列(列族)组成,列族(Column Family)由多个列组成,如表 6-1 所示。一个列族的所有列存储在同一底层的存储文件中,这个存储文件称为 HFile。列族需要在表创建时就定义好,并且不能修改得太频繁,数量也不能太多,列族名必须由可打印字符组成,这与其他名称或值的命名规范有显著不同。与列族的数

```
                基于列式表(Column-based)
        ┌──────┬──────────┬──────────┬──────────┬──────────┐
        │Row ID│Date/Time │ Material │ Customer │ Quantity │
        ├──────┼──────────┼──────────┼──────────┼──────────┤
        │  1   │   953    │    2     │    3     │    1     │
        │  2   │   862    │    6     │    2     │    2     │
        │  3   │   931    │    3     │    6     │    1     │
        │  4   │   915    │    5     │    5     │    3     │
        │  5   │   886    │    2     │    1     │    4     │
        │  6   │   896    │    3     │    4     │    1     │
        └──────┴──────────┴──────────┴──────────┴──────────┘

        基于列式存储(Column-based store)
        ┌─┬─┬─┬─┬───┬───┬───┬───┬───┬───┬───┬───┬───┬───┬───┬───┬───┬───┐
        │1│2│3│4│...│953│862│931│915│...│ 2 │ 6 │ 5 │ 1 │...│ 1 │ 2 │ 3 │...
        └─┴─┴─┴─┴───┴───┴───┴───┴───┴───┴───┴───┴───┴───┴───┴───┴───┴───┘
```

图 6-5　基于列式存储模式

量有限制相反,列的数量没有限制。

表 6-1　Scores Table

Row Key	Timestamp	Student	Courses	
		name	python	math
610215	ts9			86
	ts8			80
	ts6	Kate	87	
610213	ts4		91	
	ts3	Alice		88
590108	ts2			90
	ts1	Kane	83	

Row Key 是用来检索记录的主键,访问 HBase Table 中的行有 3 种方式:通过单个 Row Key 访问、通过 Row Key 的 Range 访问和全表扫描访问。Row Key 可以是任意字符串(最大长度是 64KB),Row Key 在 HBase 内部保存为字节数组,数据存储时按照 Row Key 的字典排序(Byte Order)存储。设计 Row Key 时,将经常一起读取的行存储到一起。每个单元格(Cell)由 Row Key、Family、Column、Timestamp 确定,Cell 中的数据没有类型和长度的限定,以字节码存储。时间戳的类型是 64 位整型。时间戳可以由 HBase 自动赋值,是精确到毫秒的当前系统时间;时间戳也可以由客户显式赋值。如果应用程序要避免数据版本冲突,就必须自己生成具有唯一性的时间戳。每个 Cell 中,不同版本的数据按照时间倒序排序,最新的数据排在最前面。

6.2　HBase 集群部署

HBase 集群一般由一台 HMaster 节点和多台 HRegionServer 节点组成,ZooKeeper 使用 HBase 自带系统。HBase 部署需要 Hadoop 环

视频讲解

境,部署 Hadoop 参照前面章节。

安装 HBase 需要 hbase-1.3.1-bin.tar.gz 软件包,下载网址为 mirrors.aliyun.com。先下载软件包,然后解压到/opt 目录下,解压后子目录名带有版本号,对其改名,使其简洁。

6.2.1　HBase 参数配置

HBase 需要添加环境变量和修改 HBase 参数,修改的配置文件有 hbase-env.sh、hbase-site.xml 和 regionservers,3 台节点机安装配置一样。

(1) 修改 Master 节点和 Slave 节点的/etc/hosts 文件,添加内容如下。

```
172.30.0.10    master
172.30.0.11    slave1
172.30.0.12    slave2
```

(2) 修改 Master 节点和 Slave 节点的/root/.bash_profile 文件,添加内容如下。

```
export HBASE_HOME = /opt/hbase
export PATH = $PATH:$HBASE_HOME/bin
```

(3) 修改配置文件 hbase-env.sh,修改如下。

```
# export JAVA_HOME = /usr/java/jdk1.6.0/
# export HBASE_MANAGES_ZK = true
# 上面两行改为:
export JAVA_HOME = /opt/jdk1.8/
export HBASE_MANAGES_ZK = true
```

注意:HBASE_MANAGES_ZK=true 表示使用自带的 ZooKeeper 系统,HBase 把 ZooKeeper 当作自己的一部分来启动和关闭进程。

(4) 修改配置文件 hbase-site.xml,添加内容如下。

```
<?xml version = "1.0"?>
<?xml - stylesheet type = "text/xsl" href = "configuration.xsl"?>
<configuration>
  <property>
    <name>hbase.rootdir</name>
    <value>hdfs://master:9000/hbase</value>
  </property>
  <property>
    <name>hbase.cluster.distributed</name>
    <value>true</value>
  </property>
  <property>
    <name>hbase.zookeeper.quorum</name>
    <value>master,slave1,slave2</value>
```

```
      </property>
      <property>
        <name>hbase.zookeeper.property.clientPort</name>
        <value>2181</value>
      </property>
      <property>
        <name>hbase.zookeeper.property.datadir</name>
        <value>/opt/hbase/zk_data/</value>
      </property>
</configuration>
```

注意：hbase.zookeeper.quorum 设置为 3 台节点，对于 ZooKeeper，使用节点越多，集群容灾能力越强，一般使用奇数台。使用奇数的目的是只有两台服务器失败的情况下才不可用，而使用偶数在一台服务器失败情况下就会不可用。

（5）修改 regionservers 文件，添加内容如下。

```
master
slave1
slave2
```

6.2.2　HBase 运行与测试

HBase 运行需要先启动 Hadoop 服务，再启动 HBase 服务。

（1）启动 Hadoop，操作如下。

```
[root@master ~]# start-dfs.sh
[root@master ~]# start-yarn.sh
[root@master ~]# mr-jobhistory-daemon.sh start historyserver
```

（2）启动 HBase 服务，操作如下。

```
[root@master ~]# start-hbase.sh
```

（3）在 slave1 节点启动 Backup Master 服务，操作如下。

```
[root@slave1 ~]# hbase-daemon.sh start master
```

如果 Master 服务停止，则 Backup Master 服务自动切换为 Master 服务。

（4）打开浏览器，输入"http://master:16010"，查看 Master 信息，如图 6-6 所示。

（5）打开浏览器，输入"http://slave1:16010"，查看 Backup Master 信息，如图 6-7 所示。

（6）打开浏览器，输入"http://slave1:16030"，查看 Region Server 信息，如图 6-8 所示。

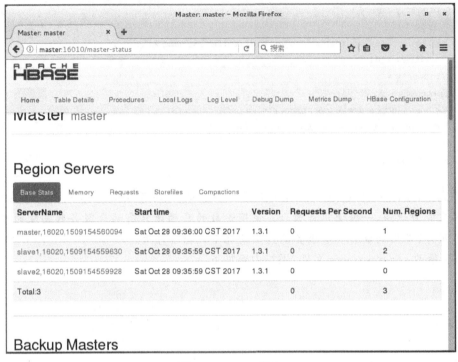

图 6-6　Master 信息

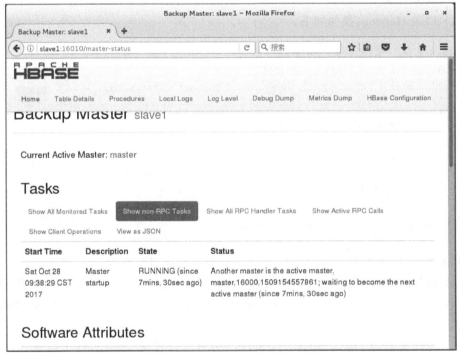

图 6-7　Backup Master 信息

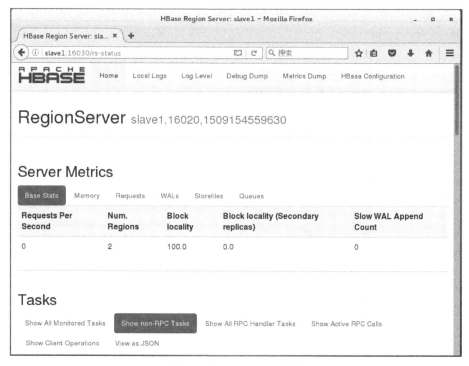

图 6-8 Region Server 信息

(7) 启动 hbase shell,操作如下。

```
# hbase shell
hbase(main):001:0> list
TABLE
0 row(s) in 0.0220 seconds

=> []
hbase(main):002:0> quit
```

6.3 HBase Shell 操作命令

HBase 提供了丰富的访问接口,HBase Shell 是最常用的交互方式,其提供的基本命令如表 6-2 所示。

视频讲解

表 6-2 HBase Shell 命令

命　　令	描　　述
create	创建表
alter	修改列族(Column Family)模式
describe	显示表详细信息
list	列出 HBase 中存在的所有表

续表

命令	描述
count	统计表中行的数量
put	向指向的表单元添加值
get	获取行或单元(Cell)的值
delete	删除指定对象的值(可以为表、行、列对应的值,另外也可以指定时间戳的值)
deleteall	删除指定行的所有元素值
incr	增加指定表的行或列的值
scan	通过对表的扫描来获取单元值
truncate	重新创建指定表
disable	使表无效
enable	使表有效
exists	测试表是否存在
drop	删除表
tools	列出 HBase 所支持的工具
status	返回 HBase 集群的状态信息
shutdown	关闭 HBase 集群(与 exit 不同)
version	返回 HBase 版本信息
exit	退出 HBase Shell

启动 HBase Shell,输入 help 获取帮助,命令如下。

```
# hbase shell
hbase(main):001:0> help
```

帮助信息如下。

```
输入"help "命令""获取命令帮助(例如,help "get",双引号不能省)
输入"help "命令组""获取命令组(例如,help "general")
命令组:
  组名: general
  命令: status, table_help, version, whoami
  组名: ddl
  命令: alter, alter_async, alter_status, create, describe, disable, disable_all,
        drop, drop_all, enable, enable_all, exists, get_table, is_disabled,
        is_enabled, list, locate_region, show_filters
  组名: namespace
  命令: alter_namespace, create_namespace, describe_namespace, drop_namespace,
        list_namespace, list_namespace_tables
  组名: dml
  命令: append, count, delete, deleteall, get, get_counter, get_splits, incr, put,
        scan, truncate, truncate_preserve
  组名: tools
  命令: assign, balance_switch, balancer, balancer_enabled,
        catalogjanitor_enabled, catalogjanitor_run, catalogjanitor_switch,
        close_region, compact, compact_rs, flush, major_compact, merge_region,
```

```
            move, normalize, normalizer_enabled, normalizer_switch, split,
            splitormerge_enabled, splitormerge_switch, trace, unassign,
            wal_roll, zk_dump
   组名: replication
   命令: add_peer, append_peer_tableCFs, disable_peer, disable_table_replication,
            enable_peer, enable_table_replication, get_peer_config,
            list_peer_configs, list_peers, list_replicated_tables, remove_peer,
            remove_peer_tableCFs, set_peer_tableCFs, show_peer_tableCFs
   组名: snapshots
   命令: clone_snapshot, delete_all_snapshot, delete_snapshot,
            delete_table_snapshots, list_snapshots, list_table_snapshots,
            restore_snapshot, snapshot
   组名: configuration
   命令: update_all_config, update_config
   组名: quotas
   命令: list_quotas, set_quota
   组名: security
   命令: grant, list_security_capabilities, revoke, user_permission
   组名: procedures
   命令: abort_procedure, list_procedures
   组名: visibility labels
   命令: add_labels, clear_auths, get_auths, list_labels, set_auths, set_visibility
```

6.3.1 general 操作

general 操作有以下两种。

（1）查询服务器状态，命令如下。

```
hbase(main):001:0> status
1 active master, 0 backup masters, 3 servers, 0 dead, 0.6667 average load
```

（2）查询 HBase 版本，命令如下。

```
hbase(main):002:0> version
1.3.1, r930b9a55528fe45d8edce7af42fef2d35e77677a, Thu Apr  6 19:36:54 PDT 2017
```

6.3.2 namespace 操作

HBase 系统定义了两个默认的 namespace。HBase 为系统命名空间，包括 namespace 和 meta 表；default 为用户默认命名空间，用户表在这里创建。

（1）列出命名空间，命令如下。

```
hbase(main):001:0> list_namespace
NAMESPACE
default
hbase
2 row(s) in 0.0460 seconds
```

(2) 查看命名空间下的表，命令如下。

```
hbase(main):002:0> list_namespace_tables 'hbase'
TABLE
meta
namespace
2 row(s) in 0.0730 seconds
```

(3) 创建命名空间，命令如下。

```
hbase(main):003:0> create_namespace 'my_ns'
```

(4) 查看命名空间，命令如下。

```
hbase(main):004:0> describe_namespace 'my_ns'
DESCRIPTION
{NAME => 'my_ns'}
1 row(s) in 0.0210 seconds
```

(5) 删除命名空间，命令如下。

```
hbase(main):005:0> drop_namespace 'my_ns'
```

6.3.3 DDL 操作

1. 创建表

例如，创建 scores 表，命令如下。

```
hbase(main):001:0> create 'scores','student','courses'
0 row(s) in 1.3940 seconds
 => HBase::Table - scores
hbase(main):002:0> list
TABLE
scores
1 row(s) in 0.0190 seconds
 => ["scores"]
```

2. 获取表信息

例如，获取 scores 表信息，命令如下。

```
hbase(main):001:0> describe 'scores'
Table scores is ENABLED
scores
COLUMN FAMILIES DESCRIPTION
{NAME => 'courses', BLOOMFILTER => 'ROW', VERSIONS => '1', IN_MEMORY => 'false',
```

```
KEEP_DELETED_CELLS => 'FALSE', DATA_BLOCK_ENCODING => 'NONE', TTL => 'FOREVER',
COMPRESSION => 'NONE', MIN_VERSIONS => '0', BLOCKCACHE => 'true',
BLOCKSIZE => '65536', REPLICATION_SCOPE => '0'}
{NAME => 'student', BLOOMFILTER => 'ROW', VERSIONS => '1', IN_MEMORY => 'false',
KEEP_DELETED_CELLS => 'FALSE', DATA_BLOCK_ENCODING => 'NONE', TTL => 'FOREVER',
COMPRESSION => 'NONE', MIN_VERSIONS => '0', BLOCKCACHE => 'true',
BLOCKSIZE => '65536', REPLICATION_SCOPE => '0'}
2 row(s) in 0.0570 seconds
```

3. 增加表的列族

例如，修改 scores 表，增加 department 列族，命令如下。

```
hbase(main):001:0> disable 'scores'
hbase(main):002:0> alter 'scores', { NAME => 'department', VERSIONS => 3 }
Updating all regions with the new schema...
1/1 regions updated.
Done.
0 row(s) in 2.2240 seconds
hbase(main):003:0> enable 'scores'
```

4. 删除表的列族

例如，修改 scores 表，删除 department 列族，命令如下。

```
hbase(main):001:0> disable 'scores'
hbase(main):002:0> alter 'scores', { NAME => 'department', METHOD => 'delete' }
Updating all regions with the new schema...
1/1 regions updated.
Done.
0 row(s) in 2.1880 seconds
hbase(main):003:0> enable 'scores'
```

5. 查询表是否存在

例如，查询 scores 表是否存在，命令如下。

```
hbase(main):001:0> exists 'scores'
Table scores does exist
0 row(s) in 0.0170 seconds
```

6. 判断表是否 enable/disable

例如，判断 scores 表是否 enable/disable，命令如下。

```
hbase(main):022:0> is_enabled 'scores'
true
0 row(s) in 0.0220 seconds

hbase(main):023:0> is_disabled 'scores'
false
0 row(s) in 0.0340 seconds
```

7. 删除表

例如,删除 scores 表,命令如下。

```
hbase(main):001:0> disable 'scores'
hbase(main):002:0> drop 'scores'
hbase(main):003:0> exists 'scores'
Table scores does not exist
0 row(s) in 0.0240 seconds
```

6.3.4 DML 操作

1. 插入记录

例如,插入表 6-1 的数据到 scores 表中,命令如下。

```
hbase(main):001:0> put 'scores','590108','student:name','Kane'
hbase(main):002:0> put 'scores','590108','courses:python','83'
hbase(main):003:0> put 'scores','590108','courses:math','90'
hbase(main):004:0> put 'scores','610213','student:name','Alice'
hbase(main):005:0> put 'scores','610213','courses:math','88'
hbase(main):006:0> put 'scores','610213','courses:python','91'
hbase(main):007:0> put 'scores','610215','student:name','Kate'
hbase(main):008:0> put 'scores','610215','courses:python','87'
hbase(main):009:0> put 'scores','610215','courses:math','80'
hbase(main):010:0> put 'scores','610215','courses:math','86'
```

2. 读取表数据

例如,读取 scores 表中数据,命令如下。

```
hbase(main):001:0> get 'scores','610213'
COLUMN               CELL
courses:math         timestamp=1509023238421, value=88
courses:python       timestamp=1509023238441, value=91
student:name         timestamp=1509023238397, value=Alice
1 row(s) in 0.1590 seconds

hbase(main):002:0> get 'scores','610213','courses'
```

```
COLUMN                  CELL
courses:math            timestamp = 1509023238421, value = 88
courses:python          timestamp = 1509023238441, value = 91
1 row(s) in 0.0380 seconds

hbase(main):003:0 > get 'scores','610213','courses:math'
COLUMN                  CELL
courses:math            timestamp = 1509023238421, value = 88
1 row(s) in 0.0160 seconds
```

3. 扫描表数据

例如,扫描 scores 表中数据,命令如下。

```
hbase(main):001:0 > scan 'scores'
ROW                     COLUMN + CELL
590108                  column = courses:math, timestamp = 1509023238364, value = 90
590108                  column = courses:python, timestamp = 1509023238338, value = 83
590108                  column = student:name, timestamp = 1509023238277, value = Kane
610213                  column = courses:math, timestamp = 1509023238421, value = 88
610213                  column = courses:python, timestamp = 1509023238441, value = 91
610213                  column = student:name, timestamp = 1509023238397, value = Alice
610215                  column = courses:math, timestamp = 1509023240013, value = 86
610215                  column = courses:python, timestamp = 1509023238623, value = 87
610215                  column = student:name, timestamp = 1509023238506, value = Kate
3 row(s) in 0.0410 seconds

hbase(main):002:0 > scan 'scores',{LIMIT = > 2}
ROW                     COLUMN + CELL
590108                  column = courses:math, timestamp = 1509023238364, value = 90
590108                  column = courses:python, timestamp = 1509023238338, value = 83
590108                  column = student:name, timestamp = 1509023238277, value = Kane
610213                  column = courses:math, timestamp = 1509023238421, value = 88
610213                  column = courses:python, timestamp = 1509023238441, value = 91
610213                  column = student:name, timestamp = 1509023238397, value = Alice
2 row(s) in 0.0310 seconds

hbase(main):003:0 > scan 'scores',{STARTROW = > '610213',ENDROW = > '610215'}
ROW                     COLUMN + CELL
610213                  column = courses:math, timestamp = 1509023238421, value = 88
610213                  column = courses:python, timestamp = 1509023238441, value = 91
610213                  column = student:name, timestamp = 1509023238397, value = Alice
1 row(s) in 0.0210 seconds

hbase(main):004:0 > scan 'scores',{COLUMN = > 'courses:python'}
ROW                     COLUMN + CELL
590108                  column = courses:python, timestamp = 1509023238338, value = 83
610213                  column = courses:python, timestamp = 1509023238441, value = 91
610215                  column = courses:python, timestamp = 1509023238623, value = 87
3 row(s) in 0.0380 seconds
```

4. 统计记录数

例如,统计 scores 表的记录数,命令如下。

```
hbase(main):001:0> count 'scores'
3 row(s) in 0.0380 seconds
=> 3
```

5. 删除列

例如,删除 scores 表中的列 english,命令如下。

```
hbase(main):001:0> delete 'scores','610213','courses:python'
0 row(s) in 0.0390 seconds
hbase(main):002:0> get 'scores','610213','courses'
COLUMN                CELL
courses:math          timestamp=1509023238421, value=88
1 row(s) in 0.0200 seconds
```

6. 删除所有行

例如,删除 scores 表中的 Row Key 为 610215 的所有行,命令如下。

```
hbase(main):001:0> deleteall 'scores','610215'
0 row(s) in 0.0100 seconds
hbase(main):002:0> scan 'scores'
ROW                   COLUMN+CELL
590108                column=courses:math, timestamp=1509023238364, value=90
590108                column=courses:python, timestamp=1509023238338, value=83
590108                column=student:name, timestamp=1509023238277, value=Kane
610213                column=courses:math, timestamp=1509023238421, value=88
610213                column=student:name, timestamp=1509023238397, value=Alice
2 row(s) in 0.0260 seconds
```

7. 删除表中所有数据

例如,删除 scores 表中所有数据,命令如下。

```
hbase(main):001:0> truncate 'scores'
Truncating 'scores' table (it may take a while):
 - Disabling table...
 - Truncating table...
0 row(s) in 3.9880 seconds
hbase(main):002:0> scan 'scores'
ROW                   COLUMN+CELL
0 row(s) in 0.2320 seconds
```

6.3.5 授权

HBase 的权限控制是通过 AccessController Coprocessor 协处理器框架实现的，可实现对用户的 RWXCA 的权限控制。HBase 权限与命令对照如表 6-3 所示。

表 6-3 HBase 权限与命令对照表

ACLs	权 限	备 注
READ('R')	Get、Scan、Exists calls	R 表示读权限
WRITE('W')	Put、Delete、LockRow、UnlockRow、IncrementColumnValue、CheckAndDelete、CheckAndPut、Flush、Compact	W 表示写权限
EXEC('X')		表示协处理器端点所需的执行权限
CREATE('C')	Create、Alter、Drop	表示创建权限
ADMIN('A')	Enable、Disable、Snapshot、Restore、Clone、Split、MajorCompact、Grant、Revoke、Shutdown	表示 admin（管理员）权限

在 HBase 中启动授权机制需要修改配置文件 hbase-site.xml，添加内容如下。

```
<property>
  <name>hbase.security.authorization</name>
  <value>true</value>
</property>
<property>
  <name>hbase.coprocessor.master.classes</name>
  <value>org.apache.hadoop.hbase.security.access.AccessController</value>
</property>
<property>
  <name>hbase.coprocessor.region.classes</name>
  <value>org.apache.hadoop.hbase.security.token.TokenProvider,
      org.apache.hadoop.hbase.security.access.AccessController</value>
</property>
<property>
  <name>hbase.superuser</name>
  <value>hbase,root,administrator</value>
</property>
```

1. namespace 授权/撤权

例如，授予用户 kate 对命名空间 my_ns 写权限、撤销其权限，命令如下。

```
hbase(main):001:0> grant 'kate', 'W', '@my_ns'
hbase(main):002:0> revoke 'kate', '@my_ns'
```

2. Table 授权/撤权

例如,授予用户 Alice 对命名空间 my_ns 的表 my_tb 读/写权限、撤销其权限,命令如下。

```
hbase(main):001:0> grant 'Alice', 'RW', 'my_ns:my_tab'
hbase(main):002:0> revoke 'Alice', 'my_ns:my_tab'
```

6.4 HBase 过滤器

视频讲解

HBase 为筛选数据提供了一组过滤器,通过这个过滤器可以在 HBase 中数据的多个维度(行、列、数据版本)上进行对数据的筛选操作,也就是说,过滤器最终能够筛选的数据能够细化到具体的一个存储单元格上(由行键、列族、时间戳定位)。通常来说,通过行键、值来筛选数据的应用场景较多。

Get 和 Scan 两类都支持过滤器,这些类提供的基本 API 不能对行键、列名和列值进行过滤,但过滤器可以实现。过滤器最基本的接口是 Filter,同时用户可以通过继承 Filter 类来实现自己的需求。所有的过滤器都在服务器端生效,称为谓词下推(predicate push down),这样可以保证被过滤掉的数据不会被传送到客户端。用户可以在客户端代码实现过滤的功能(但会影响系统性能)。图 6-9 描述了过滤器怎样在客户端进行配置,怎样在网络传输中被序列化,怎样在服务端执行。

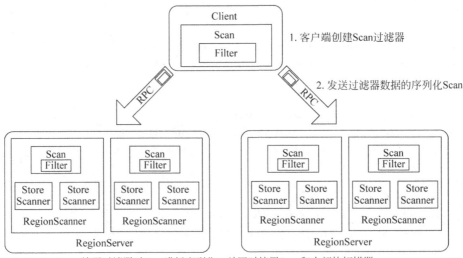

图 6-9 过滤器的工作过程

在过滤器层次结构的最底层是 Filter 接口和 FilterBase 抽象类,它们实现了过滤器的空壳和骨架。大部分实体过滤类一般都直接继承 FilterBase,也有一些间接继承该类,用户定义一个所需要的过滤实例,同时把定义好的过滤器实例传递给 Get 和 Scan 实例,即 setFilter(filter)。在实例化过滤器的时候,用户需要提供一些参数来设定过滤器的用途。其中一组特殊的过滤器,它们继承 CompareFilter,需要用户提供至少两个特定的参数,这两个参数会被基类用于执行它的任务。

1. 行过滤器(RowFilter)

例如,扫描 scores 表中 RowKey 为 610213 的记录,命令如下。

```
hbase(main):001:0> scan 'scores',FILTER=>"RowFilter( = ,'binary:610213')"
ROW                   COLUMN + CELL
610213                column = courses:math, timestamp = 1509024275914, value = 88
610213                column = courses:python, timestamp = 1509024275932, value = 91
610213                column = student:name, timestamp = 1509024275897, value = Alice
1 row(s) in 0.0420 seconds
```

2. 列族过滤器(FamilyFilter)

例如,只扫描 scores 表中列族为 student 的记录,命令如下。

```
hbase(main):002:0> scan 'scores',FILTER=>"FamilyFilter( = ,'binary:student')"
ROW                   COLUMN + CELL
590108                column = student:name, timestamp = 1509024275766, value = Kane
610213                column = student:name, timestamp = 1509024275897, value = Alice
610215                column = student:name, timestamp = 1509024275957, value = Kate
3 row(s) in 0.0300 seconds
```

3. 列名过滤器(QualifierFilter)

例如,扫描 scores 表中列名为 math 的记录,命令如下。

```
hbase(main):003:0> scan 'scores',FILTER=>"QualifierFilter( = ,'binary:math')"
ROW                   COLUMN + CELL
590108                column = courses:math, timestamp = 1509024275872, value = 90
610213                column = courses:math, timestamp = 1509024275914, value = 88
610215                column = courses:math, timestamp = 1509024277353, value = 86
3 row(s) in 0.0330 seconds
```

4. 值过滤器(ValueFilter)

例如,扫描 scores 表中包含 Ka 的记录,命令如下。

```
hbase(main):004:0> scan 'scores',FILTER=>"ValueFilter( = ,'substring:Ka')"
ROW                   COLUMN + CELL
590108                column = student:name, timestamp = 1509024275766, value = Kane
610215                column = student:name, timestamp = 1509024275957, value = Kate
2 row(s) in 0.0400 seconds
```

扫描 scores 表中 courses 大于等于 90 的记录，命令如下。

```
hbase(main):005:0>
scan 'scores',{COLUMNS =>['courses'],FILTER =>"(ValueFilter(>= ,'binary:90'))"}
ROW                   COLUMN + CELL
590108                column = courses:math, timestamp = 1509024275872, value = 90
610213                column = courses:python, timestamp = 1509024275932, value = 91
2 row(s) in 0.0300 seconds
```

5. 前缀过滤器（PrefixFilter）

例如，扫描 scores 表中包含 610 开头的记录，命令如下。

```
hbase(main):006:0> scan 'scores',FILTER =>"PrefixFilter('610')"
ROW                   COLUMN + CELL
610213                column = courses:math, timestamp = 1509024275914, value = 88
610213                column = courses:python, timestamp = 1509024275932, value = 91
610213                column = student:name, timestamp = 1509024275897, value = Alice
610215                column = courses:math, timestamp = 1509024277353, value = 86
610215                column = courses:python, timestamp = 1509024275972, value = 87
610215                column = student:name, timestamp = 1509024275957, value = Kate
2 row(s) in 0.0290 seconds
```

6. 列前缀过滤器（ColumnPrefixFilter）

例如，扫描 scores 表中列名为 py 开头的记录，命令如下。

```
hbase(main):007:0> scan 'scores',FILTER =>"ColumnPrefixFilter('py')"
ROW                   COLUMN + CELL
590108                column = courses:python, timestamp = 1509024275793, value = 83
610213                column = courses:python, timestamp = 1509024275932, value = 91
610215                column = courses:python, timestamp = 1509024275972, value = 87
3 row(s) in 0.0240 seconds
```

7. 行键过滤器（KeyOnlyFilter）

例如，扫描 scores 表中的所有行，value 为空，命令如下。

```
hbase(main):008:0> scan 'scores',FILTER =>"KeyOnlyFilter( )"
ROW                   COLUMN + CELL
590108                column = courses:math, timestamp = 1509024275872, value =
590108                column = courses:python, timestamp = 1509024275793, value =
590108                column = student:name, timestamp = 1509024275766, value =
610213                column = courses:math, timestamp = 1509024275914, value =
610213                column = courses:python, timestamp = 1509024275932, value =
610213                column = student:name, timestamp = 1509024275897, value =
610215                column = courses:math, timestamp = 1509024277353, value =
610215                column = courses:python, timestamp = 1509024275972, value =
```

```
610215                  column = student:name, timestamp = 1509024275957, value =
3 row(s) in 0.0310 seconds
```

8. 首次行键过滤器(FirstKeyOnlyFilter)

例如,扫描 scores 表中的首次行键记录,命令如下。

```
hbase(main):009:0 > scan 'scores',FILTER = >"FirstKeyOnlyFilter( )"
ROW                     COLUMN + CELL
590108                  column = courses:math, timestamp = 1509024275872, value = 90
610213                  column = courses:math, timestamp = 1509024275914, value = 88
610215                  column = courses:math, timestamp = 1509024277353, value = 86
3 row(s) in 0.0290 seconds
```

9. 单列值过滤器(SingleColumnValueFilter)

例如,扫描 scores 表中的 student:name 为 Alice 的记录,命令如下。

```
hbase(main):010:0 > scan 'scores', {COLUMNS = >['student'],
  FILTER = >"SingleColumnValueFilter('student','name', = ,'binary:Alice')"}
ROW                     COLUMN + CELL
610213                  column = student:name, timestamp = 1509024275897, value = Alice
1 row(s) in 0.0310 seconds
```

扫描 scores 表中的 student:lastname 为 Maly 的记录,命令如下。

```
hbase(main):011:0 > scan 'scores', {COLUMNS = >['student'],
  FILTER = >"SingleColumnValueFilter('student','lastname', = ,'binary:Maly')"}
ROW                     COLUMN + CELL
590108                  column = student:name, timestamp = 1509024275766, value = Kane
610213                  column = student:name, timestamp = 1509024275897, value = Alice
610215                  column = student:name, timestamp = 1509024275957, value = Kate
3 row(s) in 0.0380 seconds
```

由于 student:lastname 本身就不存在,如果默认设置或 setFilterIfMissing(false)时,则其他行都显示;如果 setFilterIfMissing(true),则都不显示。另外,setLatestVersionOnly(true)时,只显示最后一个版本。

10. 单列排除过滤器(SingleColumnValueExcludeFilter)

例如,扫描 scores 表中的 student:name 为 Alice 的记录,但不包含 student:name 列,命令如下。

```
hbase(main):012:0 > scan 'scores', FILTER = >
"SingleColumnValueExcludeFilter('student','name', = ,'binary:Alice')"
ROW                     COLUMN + CELL
```

```
610213              column = courses:math, timestamp = 1509024275914, value = 88
610213              column = courses:python, timestamp = 1509024275932, value = 91
1 row(s) in 0.0230 seconds
```

11. 包含结束过滤器(InclusiveStopFilter)

例如,扫描 scores 表中的记录,直到行键为 610213 停止,命令如下。

```
hbase(main):013:0> scan 'scores', FILTER = >"InclusiveStopFilter('610213')"
ROW                 COLUMN + CELL
590108              column = courses:math, timestamp = 1509024275872, value = 90
590108              column = courses:python, timestamp = 1509024275793, value = 83
590108              column = student:name, timestamp = 1509024275766, value = Kane
610213              column = courses:math, timestamp = 1509024275914, value = 88
610213              column = courses:python, timestamp = 1509024275932, value = 91
610213              column = student:name, timestamp = 1509024275897, value = Alice
2 row(s) in 0.0210 seconds
```

12. 列计数过滤器(ColumnCountGetFilter)

例如,扫描 scores 表中的记录,列数超过 1 条停止,命令如下。

```
hbase(main):040:0> scan 'scores', FILTER = >"ColumnCountGetFilter(1)"
ROW                 COLUMN + CELL
590108              column = courses:math, timestamp = 1509024275872, value = 90
1 row(s) in 0.0230 seconds
```

6.5 HBase 编程

HBase 数据操作访问可以通过 HTableInterface 或 HTableInterface 的 HTable 类来完成,二者都支持 HBase 的主要操作。HBase 提供几个 Java API 接口,方便编程调用。

(1) HBaseConfiguration:通过此类可以对 HBase 进行配置。

(2) HBaseAdmin:提供一个接口来管理 HBase 数据库中的表信息。它提供创建表、删除表等方法。

(3) HTableDescriptor:包含了表的名称及其对应列族。HTableDescriptor 提供的方法如下。

```
void addFamily(HColumnDescriptor)                    //添加一个列族
HColumnDescriptor removeFamily(byte[] column)        //移除一个列族
byte[] getName()                                     //获取表的名称
byte[] getValue(byte[] key)                          //获取属性的值
void setValue(String key, String value)              //设置属性的值
```

(4) HColumnDescriptor：维护关于列的信息。HColumnDescriptor 提供的方法如下。

```
byte[] getName()                          //获取列族的名称
byte[] getValue()                         //获取对应的属性的值
void setValue(String key,String value)    //设置对应的属性的值
```

(5) HTable：用户与 HBase 表进行通信，此方法对于更新操作来说是非线程安全，如果启动多个线程尝试与单个 HTable 实例进行通信，那么写缓冲器可能会崩溃。

(6) Put：用于对单个行执行添加操作。

(7) Get：用于获取单个行的相关信息。

(8) Result：存储 Get 或 Scan 操作后获取的单行值。

(9) ResultScanner：客户端获取值的接口。

6.5.1　HBase 表操作编程

编写 Java 应用程序，使用 HBase 提供的 Java API，实现 HBase 表的创建、数据查询和表的删除。

(1) 编写代码。运行 Eclipse 开发工具，打开集成环境界面，如图 6-10 所示，添加 $HBase/lib 下的所有 jar 包。创建 HBaseEx 项目，并在 HBaseEx 项目中新建 HBaseEx1.java 类，代码如下。

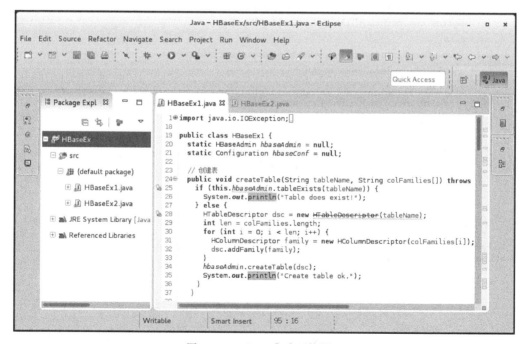

图 6-10　Eclipse 集成环境界面

```java
import java.io.IOException;
import java.util.ArrayList;
import java.util.Iterator;
import java.util.List;
import org.apache.hadoop.conf.Configuration;
import org.apache.hadoop.hbase.HBaseConfiguration;
import org.apache.hadoop.hbase.HColumnDescriptor;
import org.apache.hadoop.hbase.HTableDescriptor;
import org.apache.hadoop.hbase.KeyValue;
import org.apache.hadoop.hbase.client.Get;
import org.apache.hadoop.hbase.client.HBaseAdmin;
import org.apache.hadoop.hbase.client.HTable;
import org.apache.hadoop.hbase.client.Put;
import org.apache.hadoop.hbase.client.Result;
import org.apache.hadoop.hbase.client.ResultScanner;
import org.apache.hadoop.hbase.client.Scan;

public class HBaseEx1 {
    static HBaseAdmin hbaseAdmin = null;
    static Configuration hbaseConf = null;

    //创建表
    public void createTable(String tableName, String colFamilies[])
    throws IOException {
        if (this.hbaseAdmin.tableExists(tableName)) {
            System.out.println("Table does exist!");
        } else {
            HTableDescriptor dsc = new HTableDescriptor(tableName);
            int len = colFamilies.length;
            for (int i = 0; i < len; i++) {
                HColumnDescriptor family = new HColumnDescriptor(colFamilies[i]);
                dsc.addFamily(family);
            }
            hbaseAdmin.createTable(dsc);
            System.out.println("Create table ok.");
        }
    }

    //添加记录
    public void addRecord(String tableName, String rowkey, String family,
    String qualifier, String value ) throws IOException {
        HTable table = new HTable(this.hbaseConf, tableName);
        Put put = new Put(rowkey.getBytes());
        put.add(family.getBytes(), qualifier.getBytes(), value.getBytes());
        table.put(put);
        System.out.println("Add data ok.");
    }
```

```java
//获取一行记录
public Result getOneRecord(String tableName, String rowkey)
throws IOException {
  HTable table = new HTable(this.hbaseConf, tableName);
  Get get = new Get(rowkey.getBytes());
  Result rs = table.get(get);
  return rs;
}

//获取所有记录
public List<Result> getAllRecord(String tableName) throws IOException {
  HTable table = new HTable(this.hbaseConf, tableName);
  Scan scan = new Scan();
  ResultScanner scanner = table.getScanner(scan);
  List<Result> list = new ArrayList<Result>();
  for (Result r : scanner) {
    list.add(r);
  }
  scanner.close();
  return list;
}

//删除表
public void dropTable(String tableName) throws IOException {
  this.hbaseAdmin.disableTable(tableName);
  this.hbaseAdmin.deleteTable(tableName);
}

public static void main(String[] args) throws IOException {
  //初始化
  Configuration conf = new Configuration();
  conf.set("hbase.zookeeper.quorum", "master,slave1,slave2");
  hbaseConf = HBaseConfiguration.create(conf);
  hbaseAdmin = new HBaseAdmin(hbaseConf);
  HBaseEx1 hbaseEx1 = new HBaseEx1();

  //创建表
  String tableName = "demo";
  String colFamilies[] = { "article", "author" };
  hbaseEx1.createTable(tableName, colFamilies);

    //添加记录
  hbaseEx1.addRecord(tableName, "rowkey1", "article","title", "Animal");
  hbaseEx1.addRecord(tableName, "rowkey1", "author","name", "Cat");
  hbaseEx1.addRecord(tableName, "rowkey1", "author","nickname", "Tom");
  hbaseEx1.addRecord(tableName, "rowkey2", "article","title", "Cartoon");
  hbaseEx1.addRecord(tableName, "rowkey2", "author","name", "Mouse");
  hbaseEx1.addRecord(tableName, "rowkey2", "author","nickname", "Jerry");
```

```java
    //查询一条记录
    Result rs1 = hbaseEx1.getOneRecord(tableName, "rowkey1");
    for (KeyValue kv : rs1.raw()) {
      System.out.println("rowkey: " + new String(kv.getRow()));
      System.out.println("family: " + new String(kv.getFamily()));
      System.out.println("qualifier:" + new String(kv.getQualifier()));
      System.out.println("timestamp:" + kv.getTimestamp());
      System.out.println("value: " + new String(kv.getValue()));
    }

    //查询所有记录
    System.out.println("Get all records:");
    List<Result> list = null;
    list = hbaseEx1.getAllRecord(tableName);
    Iterator<Result> it = list.iterator();
    while (it.hasNext()) {
      Result rs2 = it.next();
      for (KeyValue kv : rs2.raw()) {
        System.out.print("rowkey: " + new String(kv.getRow()));
        System.out.print(" family: " + new String(kv.getFamily()));
        System.out.print(" qualifier:" + new String(kv.getQualifier()));
        //System.out.print(" timestamp:" + kv.getTimestamp());
        System.out.println(" value: " + new String(kv.getValue()));
      }
    }

    //删除表
    hbaseEx1.dropTable(tableName);
  }
}
```

(2)导出 HBaseEx1.jar 运行包,其运行及结果如下。

```
# java - jar HBaseEx1.jar
Create table ok.
Add data ok.
Add data ok.
Add data ok.
Add data ok.
Add data ok.
Add data ok.
row key: rowkey1
family: article
qualifier:title
timestamp:1509071287719
value: Animal
row key: rowkey1
```

```
family: author
qualifier:name
timestamp:1509071287739
value: Cat
rowkey: rowkey1
family: author
qualifier:nickname
timestamp:1509071287750
value: Tom
Get all records:
rowkey: rowkey1 family: article qualifier:title value: Animal
rowkey: rowkey1 family: author qualifier:name value: Cat
rowkey: rowkey1 family: author qualifier:nickname value: Tom
rowkey: rowkey2 family: article qualifier:title value: Cartoon
rowkey: rowkey2 family: author qualifier:name value: Mouse
rowkey: rowkey2 family: author qualifier:nickname value: Jerry
```

6.5.2　HBase 过滤查询编程

编写 Java 应用程序，使用 HBase 提供的过滤类进行查询。例如，查询 scores 表中 rowkey 为 610 开头，成绩大于或等于 86 的记录。

视频讲解

（1）编写代码。打开 Eclipse 开发工具，添加 $HBase/lib 下的所有 jar 包。在 HBaseEx 项目中新建 HBaseEx2.java 类，代码如下。

```java
import java.io.IOException;
import java.util.ArrayList;
import java.util.List;
import org.apache.hadoop.conf.Configuration;
import org.apache.hadoop.hbase.HBaseConfiguration;
import org.apache.hadoop.hbase.KeyValue;
import org.apache.hadoop.hbase.client.HBaseAdmin;
import org.apache.hadoop.hbase.client.HTable;
import org.apache.hadoop.hbase.client.Result;
import org.apache.hadoop.hbase.client.ResultScanner;
import org.apache.hadoop.hbase.client.Scan;
import org.apache.hadoop.hbase.filter.BinaryComparator;
import org.apache.hadoop.hbase.filter.CompareFilter.CompareOp;
import org.apache.hadoop.hbase.filter.Filter;
import org.apache.hadoop.hbase.filter.FilterList;
import org.apache.hadoop.hbase.filter.QualifierFilter;
import org.apache.hadoop.hbase.filter.RegexStringComparator;
import org.apache.hadoop.hbase.filter.RowFilter;
import org.apache.hadoop.hbase.util.Bytes;

public class HBaseEx2 {
    static HBaseAdmin hbaseAdmin = null;
    static Configuration hbaseConf = null;
```

```
    public static void main(String[] args) throws IOException {
      Configuration conf = new Configuration();
      conf.set("hbase.zookeeper.quorum", "master,slave1,slave2");
      hbaseConf = HBaseConfiguration.create(conf);
      hbaseAdmin = new HBaseAdmin(hbaseConf);

      HTable table = new HTable(hbaseConf, "scores");
      List<Filter> filters = new ArrayList<Filter>();
      Filter filter1 = new RowFilter(CompareOp.EQUAL,
                    new RegexStringComparator("610"));
      filters.add(filter1);
      Filter filter2 = new QualifierFilter(CompareOp.GREATER_OR_EQUAL,
                    new BinaryComparator(Bytes.toBytes("86")));
      filters.add(filter2);
      FilterList filterList = new FilterList(filters);
      Scan scan = new Scan();
      scan.setFilter(filterList);
      ResultScanner scanner = table.getScanner(scan);
      for(Result res : scanner){
        for (KeyValue kv : res.raw()) {
          System.out.print("rowkey: " + new String(kv.getRow()));
          System.out.print(" family: " + new String(kv.getFamily()));
          System.out.print(" qualifier:" + new String(kv.getQualifier()));
          //System.out.print(" timestamp:" + kv.getTimestamp());
          System.out.println(" value: " + new String(kv.getValue()));
        }
      }
      scanner.close();
    }
}
```

(2) 导出 HBaseEx2.jar 运行包,其运行及结果如下。

```
# java - jar HBaseEx2.jar
rowkey: 610213 family: courses qualifier:math value: 88
rowkey: 610213 family: courses qualifier:python value: 91
rowkey: 610213 family: student qualifier:name value: Alice
rowkey: 610215 family: courses qualifier:math value: 86
rowkey: 610215 family: courses qualifier:python value: 87
rowkey: 610215 family: student qualifier:name value: Kate
```

小结

HBase 是建立在 Hadoop 文件系统之上的分布式面向列的数据库。它是一个横向扩展的开源项目。它提供对数据的随机实时读/写访问,并作为 Hadoop 文件系统的一部分。HBase 是一个数据模型,类似于 Google 的大表设计,可以提供快速随机访问海量的结构化数据。它利用了 Hadoop 的文件系统(HDFS)提供的容错能力。本章详细介绍了 HBase 架构及各组件的功能，介绍了 HBase 集群部署与配置，重点讲解了 HBase Shell 命令的使用，包括 DDL 和 DDM 的操作，详细讲解了 HBase 过滤器的使用，最后讲解了 Java API 编程和 HBase 过滤器编程，提供了两个典型实例。

习题

1. 简述 Write-Ahead-Log 的作用。
2. 简述 HBase 数据是如何存储的。
3. ZooKeeper 在 HBase 中的作用是什么？
4. 简述 HBase 过滤器的工作过程。
5. 编写程序实现对 HBase 数据的查询、添加和删除。

核 心 篇

第 7 章

视频讲解

YARN资源分配

近年来,随着互联网的快速发展和大数据的来临,数据量呈现爆炸式增长。根据国际数据公司 IDC 的监测统计,2011 年全球数据总量达到 1.8ZB,并且以每两年翻番的速度在增长。预计到 2020 年,全球数据总量将达到 40ZB。大数据环境下,如此增长迅速、庞大复杂的数据资源,给传统的数据分析、处理技术带来了巨大的挑战。传统单台高性能服务器的数据处理能力已经不能满足大量的网络服务和越来越多的数据密集型应用的需求,取而代之的是商业服务器集群,它已经成为主要的数据分析平台。

数据密集型应用的集群计算框架的出现有效地缓解了大规模数据处理问题。但是,这些计算框架都只面向某一特定领域的应用。基于这个特点,互联网公司往往需要部署和运行多个计算框架,从而为每个应用选择最优化的计算框架。因此,资源统一管理和调度系统作为集群平台被提出来。集群资源统一管理和调度系统需要同时支持多种不同计算框架,如何管理集群计算资源和不同计算框架间的资源公平分配,成为关键技术难点。

因此,了解典型的计算框架有利于大数据处理技术的发展。其中,YARN 作为 Hadoop 开源项目开发者提出并实现的下一代计算框架,对目前的大数据处理技术有较大贡献。本章主要介绍 YARN 及其实例。

7.1 统一资源管理和调度平台引例

数据量的剧增让人们越来越重视数据资源的存储和管理的问题。云计算、大数据经常意味着需要调动数据中心大量的资源,如何能够快速地匹配合适资源,需要一个聪明的"大脑"。无疑,统一资源管理和调度平台就是这个"大脑",在大数据处理平台中起着重要作用。

7.1.1 背景

随着互联网的高速发展,基于数据密集型应用的计算框架不断出现。从支持离线处理的 MapReduce 到支持在线处理的 Storm,从迭代式计算框架 Spark 到流式处理框架 S4 等计算框架,无一不是为了某一领域的计算应用。但是,由于各种框架诞生于不同的公司或实验室,因此它们各有所长,各自解决了某一类应用问题。而在大部分互联网公司中,这几种框架可能都会采用。考虑到资源利用率、运维成本、数据共享等因素,公司一般希望将所有这些框架部署到一个公共的集群中,让它们共享集群的资源,并对资源进行统一使用。这样,便诞生了资源统一管理与调度平台。其中,典型代表是 Mesos 和 YARN。

7.1.2 特点

统一资源管理和调度平台具有如下特点。

1. 支持多种计算框架

资源统一管理和调度平台应该提供一个全局的资源管理器。所有接入的框架要先向该全局资源管理器申请资源,申请成功之后,再由框架自身的调度器决定资源交由哪个任务使用。也就是说,整个大的系统是个双层调度器,第一层是统一管理和调度平台提供的;第二层是框架自身的调度器。

资源统一管理和调度平台应该提供资源隔离。不同框架中的不同任务往往需要的资源(内存、CPU、网络 I/O 等)不同,它们运行在同一个集群中,会相互干扰,为此,应该提供一种资源隔离机制避免任务之间由于资源争用导致效率下降。

2. 扩展性

现有的分布式计算框架都会将系统扩展性作为一个非常重要的设计目标,例如 Hadoop,好的扩展性意味着系统能够随着业务的扩展线性扩展。资源统一管理和调度平台融入多种计算框架后,不应该破坏这种特性,也就是说,统一管理和调度平台不应该制约框架进行水平扩展。

3. 容错性

同扩展性类似,容错性也是当前分布式计算框架的一个重要设计目标,统一管理和调度平台在保持原有框架的容错特性基础上,自己本身也应具有良好的容错性。

4. 高资源利用率

如果采用静态资源分配,也就是每个计算框架分配一个集群,往往由于作业自身的特点或作业提交频率等原因,集群利用率很低。而将各种框架部署到同一个大的集群中,进行统一管理和调度后,由于各种作业交错及作业提交频率大幅度升高,因此为资源利用率的提升增加了机会。

7.1.3 典型的统一资源调度平台

当前,较为出名的统一资源调度平台有两个:Mesos 和 YARN。下面对这两个典型代表进行简单介绍。

1. Mesos

Mesos 是 Apache 下的开源分布式资源管理框架,它被称为分布式系统的内核。Mesos 诞生于 UC Berkeley 的一个研究项目,现已成为 Apache Incubator 中的项目,当前有一些公司使用 Mesos 管理集群资源,如 Twitter。

Mesos 的架构如图 7-1 所示。总体上看,Mesos 是一个 master/slave 结构,其中,master 是非常轻量级的,仅保存了 framework(各种计算框架称为 framework)和 mesos slave 的一些状态,而这些状态很容易通过 framework 和 slave 重新注册而重构,因而很容易使用 zookeeper 解决 Mesos master 的单点故障问题。Mesos master 实际上是一个全局资源调度器,采用某种策略将某个 slave 上的空闲资源分配给某一个 framework,各种 framework 通过自己的调度器向 Mesos master 注册,以接入 Mesos 中,而 Mesos slave 的主要功能是汇报任务的状态和启动各个 framework 的 executor。

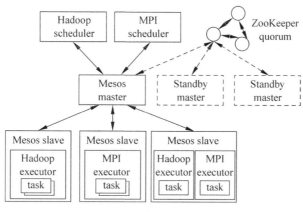

图 7-1 Mesos 的架构图

2. YARN

YARN 是一种新的 Hadoop 资源管理器。YARN 的引入为集群在利用率、资源统一管理和数据共享等方面带来了巨大好处。关于 YARN 在这里只做简单的介绍,在 7.2 节中会有详细的描述。

3. Mesos 和 YARN 比较

Mesos 与 YARN 主要在框架担任的角色、调度机制等方面有明显不同。

(1) 框架担任的角色。

在 Mesos 中,各种计算框架是完全融入 Mesos 中的,也就是说,如果想在 Mesos 中添加一个新的计算框架,首先需要在 Mesos 中部署一套该框架。而在 YARN 中,各种框架作为 Client 端的 Library 使用,仅是用户编写的程序的一个库,不需要事先部署一套该框架。从这点上说,YARN 运行和使用起来更加方便。

(2) 调度机制。

两种系统都采用了双层调度机制,即第一层是源管理系统(Mesos/YARN)将资源分配给应用程序(或框架);第二层应用程序将收到的资源进一步分配给内部的任务。但是资源分配器智能化程度不同,Mesos 是基于 resource offer 的调度机制,包含非常少的调度语义。它只是简单地将资源推给各个应用程序,由应用程序选择是否接收资源,而 Mesos 本身并不知道各个应用程序资源需求。YARN 则不同,应用程序的 ApplicationMaster 会把各个任务的资源要求汇报给 YARN,YARN 根据需要为应用程序分配资源。

7.2 YARN 简介

Apache Hadoop YARN (Yet Another Resource Negotiator,另一种资源协调者)是一种新的 Hadoop 资源管理器,它是一个通用资源管理系统,可为上层应用提供统一的资源管理和调度,它的引入为集群在利用率、资源统一管理和数据共享等方面带来了巨大好处。

Hadoop 的 MapReduce 计算模型较为简单。从业界使用分布式系统的变化趋势和 Hadoop 框架的长远发展来看,MapReduce 的 JobTracker/TaskTracker 机制需要大规模地调整来修复它在可扩展性、内存消耗、线程模型以及可靠性和性能上的缺陷。在过去的几年中,Hadoop 开发团队做了一些 bug 的修复,但是这些修复的成本越来越高,这表明对原框架做出改变的难度越来越大。为从根本上解决旧 MapReduce 框架的性能瓶颈,促进 Hadoop 框架的更长远发展。从 0.23.0 版本开始,Hadoop 的 MapReduce 框架完全重构,发生了根本上的变化。为了 Hadoop 技术的进一步提升,Hadoop 开源项目的开发者提出并实现了全新的计算框架 YARN。Apache Hadoop 2.0 包含 YARN,它将资源管理和处理组件分开,而且基于 YARN 的架构不受 MapReduce 约束。

7.2.1 YARN 架构

YARN 是基于 MapReduce 的基础上演进而来的。其中,最核心的改进是拆分了原 MapReduce 框架中 JobTracker 的两个主要功能,也就是将 JobTracker 的资源管理和作业调度两个功能拆分到独立的进程中。

YARN 的架构如图 7-2 所示。

相对于第一代 Hadoop,YARN 把 Hadoop 中的资源控制、任务调度和具体任务计算的 JobTracker/TaskTracker 架构变为下述的 4 个功能组件,让资源调度和任务调度更加细粒化。

(1) 集群唯一的 ResourceManager。

(2) 每个任务对应的 ApplicationMaster。

(3) 每个机器节点上的 NodeManager。

(4) 运行在每个 NodeManager 上针对某个任务的 Container。

通过上述 4 个功能组件的合作,解决了第一代 Hadoop 中 JobTracker 负责所有资源的调度和任务的调度的重任。

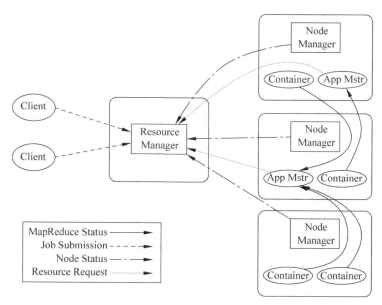

图 7-2 YARN 的架构图

YARN 总体上仍然是 Master/Slave 结构,在整个资源管理框架中,ResourceManager 为 Master,NodeManager 为 Slave。ResourceManager 负责对各个 NodeManager 上的资源进行统一管理和调度。当用户提交一个应用程序时,需要提供一个用以跟踪和管理这个程序的 ApplicationMaster,它负责向 ResourceManager 申请资源,并要求 NodeManager 启动可以占用一定资源的任务。由于不同的 ApplicationMaster 被分布到不同的节点上,因此它们之间不会相互影响。本节将对 YARN 的基本组成结构进行介绍。

1. ResourceManager

ResourceManager(RM)是一个全局的资源管理器,负责整个系统的资源管理和分配。它主要由两个组件构成:调度器(Scheduler)和应用程序管理器(Applications Manager,ASM)。

(1) 调度器。调度器根据容量、队列等限制条件(如每个队列分配一定的资源,最多执行一定数量的作业等),将系统中的资源分配给各个正在运行的应用程序。

需要注意的是,该调度器是一个"纯调度器",它不再从事任何与具体应用程序相关的工作,如不负责监控或跟踪应用的执行状态等,也不负责重新启动因应用执行失败或硬件故障而产生的失败任务,这些均交由应用程序相关的 ApplicationMaster 完成。调度器仅根据各个应用程序的资源需求进行资源分配,而资源分配单位用一个抽象概念"资源容器"(Resource Container,简称 Container)表示,Container 是一个动态资源分配单位,它将内存、CPU、磁盘、网络等资源封装在一起,从而限定每个任务使用的资源量。此外,该调度器是一个可插拔的组件,用户可根据自己的需要设计新的调度器。YARN 提供了多种直接可用的调度器,如 Fair Scheduler 和 Capacity Scheduler 等。

(2) 应用程序管理器。应用程序管理器负责管理整个系统中的所有应用程序,包括应用程序的提交，与调度器协商资源以启动 ApplicationMaster，监控 ApplicationMaster 运行状态并在失败时重新启动它等。

2. ApplicationMaster

用户提交的每个应用程序均包含一个 ApplicationMaster(AM)，主要功能包括以下几个。
（1）与 RM 调度器协商以获取资源（用 Container 表示）。
（2）将得到的任务进一步分配给内部的任务。
（3）与 NM 通信以启动/停止任务。
（4）监控所有任务运行状态，并在任务运行失败时重新为任务申请资源以重启任务。

当前 YARN 自带了两个 AM 实现：一个是用于演示 AM 编写方法的实例程序 distributedshell，它可以申请一定数目的 Container 以并行运行一个 Shell 命令或 Shell 脚本；另一个是运行 MapReduce 应用程序的 AM——MRAppMaster，将在第 8 章对其进行介绍。此外，一些其他的计算框架对应的 AM 正在开发中，如 Open MPI、Spark 等。

3. NodeManager

NodeManager(NM)是每个节点上的资源和任务管理器，一方面，它会定时地向 RM 汇报本节点上的资源使用情况和各个 Container 的运行状态；另一方面，它接收并处理来自 AM 的 Container 启动/停止等各种请求。

4. Container

Container 是 YARN 中的资源抽象，它封装了某个节点上的多维度资源，如内存、CPU、磁盘和网络等，当 AM 向 RM 申请资源时，RM 为 AM 返回的资源便是用 Container 表示的。YARN 会为每个任务分配一个 Container，且该任务只能使用该 Container 中描述的资源。

需要注意的是，Container 不同于 MRv1 中的 slot，它是一个动态资源划分单位，是根据应用程序的需求动态生成的。截至本书完成时，YARN 仅支持 CPU 和内存两种资源，且使用了轻量级资源隔离机制 Cgroups 进行资源隔离。

7.2.2 YARN 的工作流程

YARN 的工作原理是：首先将数据上传到集群中；然后将写好的程序打成架包通过命令提交 MR 作业，提交到集群后由集群管理者 MR 开始调度分配资源；最后 HDFS 读取数据执行 MapReduce 相关进程对数据进行计算。

当用户向 YARN 中提交一个应用程序后，YARN 将分两个阶段运行该应用程序：第一个阶段是启动 ApplicationMaster；第二个阶段是由 ApplicationMaster 创建的应用程序，为它申请资源，并监控它的整个运行过程，直到运行完成。

总体来说，YARN 的工作流程分为以下几个步骤。
（1）用户向 YARN 中提交应用程序，其中包括 ApplicationMaster 程序、启动 ApplicationMaster 的命令、用户程序等。

（2）ResourceManager 为该应用程序分配第一个 Container，并与对应的 NodeManager 通信，要求它在这个 Container 中启动应用程序的 ApplicationMaster。

（3）ApplicationMaster 首先向 ResourceManager 注册，这样用户可以直接通过 ResourceManage 查看应用程序的运行状态，然后它将为各个任务申请资源，并监控它的运行状态，直到运行结束，即重复步骤（4）～（7）。

（4）ApplicationMaster 采用轮询的方式通过 RPC 协议向 ResourceManager 申请和领取资源。

（5）一旦 ApplicationMaster 申请到资源后，便与对应的 NodeManager 通信，要求它启动任务。

（6）NodeManager 为任务设置好运行环境（包括环境变量、jar 包、二进制程序等）后，将任务启动命令写到一个脚本中，并通过运行该脚本启动任务。

（7）各个任务通过某个 RPC 协议向 ApplicationMaster 汇报自己的状态和进度，以让 ApplicationMaster 随时掌握各个任务的运行状态，从而可以在任务失败时重新启动任务。在应用程序运行过程中，用户可随时通过 RPC 向 ApplicationMaster 查询应用程序的当前运行状态。

（8）应用程序运行完成后，ApplicationMaster 向 ResourceManager 注销并关闭自己。

YARN 的工作流程如图 7-3 所示。

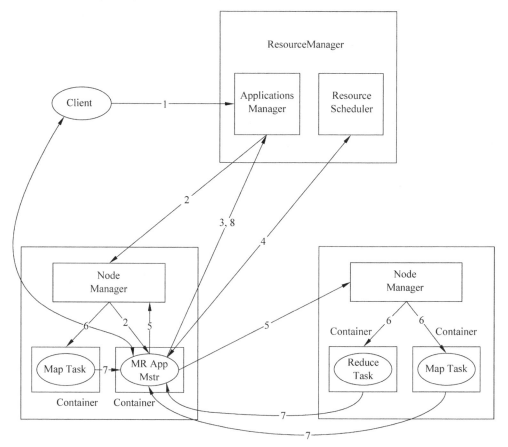

图 7-3　YARN 的工作流程图

7.2.3 YARN 的优势

与 MapReduce 相比，YARN 具有如下优势。

(1) YARN 的设计大大减少了 JobTracker(也就是现在的 ResourceManager)的资源消耗，并且使监测每一个 Job 子任务 (tasks) 状态的程序分布式化，某种程度上更安全、更优美。

(2) 在新的 YARN 中，ApplicationMaster 是一个可变更的部分，用户可以对不同的编程模型编写自己的 AppMst，让更多类型的编程模型能够跑在 Hadoop 集群中，可以参考 Hadoop YARN 官方配置模板中的 mapred-site.xml 配置。

(3) 对于资源的表示以内存为单位（在目前版本的 YARN 中，没有考虑 CPU 的占用），比之前所剩余 slot 数目更合理。

(4) 老的框架中，JobTracker 一个很大的负担就是监控 Job 下的 tasks 的运行状况；现在，这个部分就由 ApplicationMaster 进行操作。ResourceManager 中有一个模块称为 ApplicationsMasters(注意，不是 ApplicationMaster)，它是监测 Application Master 的运行状况，如果出问题，则会将其在其他机器上重启。

(5) Container 是 YARN 为了将来做资源隔离而提出的一个框架。这一点借鉴了 Mesos 的工作，目前是一个框架，仅仅提供 Java 虚拟机内存的隔离，Hadoop 团队的设计思路是后续能支持更多的资源调度和控制，既然资源表示成内存量，那就没有了之前的 map slot/reduce slot 分开造成集群资源闲置的尴尬情况。

7.3 操作实践：YARN Shell 实例

YARN Shell 是 YARN 命令脚本。如果 YARN 命令已经在 Path 环境变量中，那么只需在控制台上输入"yarn"(不需要输入任何参数)即可查看 YARN 命令行，如图 7-4 所示。

```
resourcemanager      run the ResourceManager
nodemanager          run a nodemanager on each slave
timelineserver       run the timeline server
rmadmin              admin tools
version              print the version
jar <jar>            run a jar file
application          prints application(s) report/kill application
applicationattempt   prints applicationattempt(s) report
container            prints container(s) report
node                 prints node report(s)
logs                 dump container logs
classpath            prints the class path needed to get the
                     Hadoop jar and the required libraries
daemonlog            get/set the log level for each daemon
```

图 7-4 YARN 命令行

下面对每个命令进行解释。

(1) yarn resourcemanager：启动 ResourceManager。

(2) yarn nodemanager：在每个节点上启动 nodemanager 进程。

(3) yarn timelineserver：启动 YARN 时间线服务器(Timeline Server)是一个通用

作业历史服务器(Job History Server),不只是针对 MapReduce,而是针对更多的计算框架,包括 Spark 等。

(4) yarn rmadmin:这个命令主要是管理功能的呈现。其中,rmadmin 后面还可以加上更多的参数选项,部分参数选项如图 7-5 所示。

以下是对 rmadmin 参数项功能的讲解。

-refreshQueues:重载队列的 ACL、状态和调度器特定的属性,ResourceManager 将重新加载 mapred-queues 配置文件。

-refreshNodes:无须重启 Namenode 就能刷新 dfs. hosts 和 dfs. hosts. exclude 配置文件。

```
Usage: yarn rmadmin
   -refreshQueues
   -refreshNodes
   -refreshSuperUserGroupsConfiguration
   -refreshUserToGroupsMappings
   -refreshAdminAcls
   -refreshServiceAcl
   -getGroups [username]
   -help [cmd]
```

图 7-5　yarn rmadmin 部分参数选项示例

-refreshUserToGroupsMappings:刷新用户到组的映射。

-resfreshSuperUserGroupsConfiguration:刷新用户到组的配置。

-refreshAdminAcls:刷新 ResourceManager 的 ACL 管理。

-refreshServiceAcl:是 ResourceManager 重载服务级别的授权文件。

-getGroups[username]:获取指定用户所属的组。

-help[cmd]:显示指定命令的帮助。

(5) yarn version:打印 Hadoop 详细的版本。

(6) yarn jar <jar>:运行一个 jar 文件。

(7) yarn application:查看应用程序的状态,以及参数选项,如图 7-6 所示。

```
File Edit View Search Terminal Help
usage: application
 -appStates <States>        Works with -list to filter applications
                            based on input comma-separated list of
                            application states. The valid application
                            state can be one of the following:
                            ALL,NEW,NEW_SAVING,SUBMITTED,ACCEPTED,RUN
                            NING,FINISHED,FAILED,KILLED
 -appTypes <Types>          Works with -list to filter applications
                            based on input comma-separated list of
                            application types.
 -help                      Displays help for all commands.
 -kill <Application ID>     Kills the application.
 -list                      List applications. Supports optional use
                            of -appTypes to filter applications based
                            on application type, and -appStates to
                            filter applications based on application
                            state.
 -movetoqueue <Application ID>  Moves the application to a different
                            queue.
 -queue <Queue Name>        Works with the movetoqueue command to
                            specify which queue to move an
                            application to.
 -status <Application ID>   Prints the status of the application.
```

图 7-6　yarn application 参数选项示例

其中,参数选项-appStates < States >与-appTypes < Types >均要与-lists 配合使用,基于应用程序的状态来过滤应用程序。如果应用程序的状态有多个,则用逗号分隔。而-list 是指从 ResourceManager 返回应用程序列表。另外,需要注意的是,-queue < Queue Name >要与-movetoqueue < Application ID >配合使用,将应用程序移到指定队列。

(8) yarn applicationattempt[options]：打印应用程序尝试报告，有-list < ApplicationId > 和-status < Application Attempt Id >两个参数选项，功能分别为获取应用程序尝试的列表和打印状态。

(9) yarn container[options]：打印 Container 尝试的报告，有-list < Application Attempt Id >和-status < ContainerId >两个参数选项。

(10) yarn node[options]：打印节点的报告，参数选项有-all、-list、-states < States >及-status < NodeId >。

(11) yarn queue[options]：打印队列信息，只有一个参数选项：-status < QueueName >。

(12) yarn logs -applicationId < application ID >[options]：主要用于收集作业日志。注意：应用程序没有完成，日志没有收集完成，是不能打印日志的。此命令在调试作业中起到重要作用，参数选项有-applicationId < application ID >、-appOwner < AppOwner >、-containerId < ContainerId >及-nodeAddress < NodeAddress >。

(13) yarn classpath：打印 Hadoop 相关库和其他所需的库。

(14) yarn daemonlog：指定的守护进程获取或设置日志优先级。

综上所述，YARN 命令行功能丰富，对查看日志、程序顺利运行等方面有较大帮助。

小结

本章首先阐述了统一资源管理和调度平台引例，然后介绍了 YARN 技术，其中包括 YARN 的架构和工作流程，最后介绍了 YARN Shell 实例。

习题

1. 简述统一资源管理和调度平台的优点。
2. 简述 YARN 架构及工作流程。
3. YARN Shell 有哪些命令，如何使用？

第 8 章

Spark集群计算

科学技术及互联网的发展,推动着大数据时代的来临,各行各业每天都在产生数量巨大的数据碎片,众多的信息纷繁复杂,需要搜索、处理、分析、归纳、总结其深层次的规律。大数据是对大量、动态、能持续的数据,通过运用新系统、新工具、新模型的挖掘,从而获得具有洞察力和新价值的东西。Apache Spark 是一个围绕速度、易用性和复杂分析构建的大数据处理框架。

8.1 Spark 简介

Spark 是一个基于内存的开源计算框架,于 2009 年诞生于加州大学伯克利分校,它最初是该大学的研究性项目,后来在 2010 年正式开源,并于 2013 年成为 Apache 基金项目,到 2014 年已成为 Apache 基金的顶级项目。该项目整个发展历程刚过 6 年,但其发展速度非常惊人。正由于 Spark 来自于大学,其整个发展过程都充满了学术研究的标记,是学术带动 Spark 核心架构的发展,如弹性分布式数据集(Resilient Distributed Datasets,RDD)、流处理(Spark Streaming)、机器学习(MLlib)、SQL 分析(Spark SQL)和图计算(GraphX)。Apache Spark 是一个快速通用集群系统,提供 Java、Scala、Python 和 R 语言的高阶 API 接口和优化引擎,支持通用图形计算,也提供丰富的工具集,包括 Spark SQL 和结构化数据处理、机器学习、图像处理和流计算。

Apache Spark 是大规模数据处理的快速通用引擎,其特点如下。

(1) 速度快。Spark 比 Hadoop MapReduce 在内存中的运行速度提升 100 倍,甚至能够将应用在磁盘上的运行速度提升 10 倍。

(2) 易使用。Spark 让开发者可以快速地使用 Java、Scala、Python 和 R 语言编写程序。它本身自带了一个超过 80 个高阶操作符的集合,而且还可以用它在 Shell 中交互式地查询数据。

(3) 通用性。支持 SQL 查询、流数据、机器学习和图表数据处理。Spark 提供强大的库(包括 SQL、DataFrames、MLlib、GraphX 和 Spark Streaming),可以在同一应用程序中无缝地组合这些库。

(4) 在任何地方运行。Spark 可以运行在 Hadoop、Mesos、standalone 或云中。它可以访问各种数据源,包括 HDFS、Cassandra、HBase 和 S3。

8.1.1 Spark 生态系统

Spark 的生态系统不同于 Hadoop 的 MapReduce 和 HDFS,Spark 主要包括 Spark Core,以及在 Spark Core 基础之上建立的应用框架 Spark Streaming、Spark SQL、MLlib 和 GraphX,如图 8-1 所示。

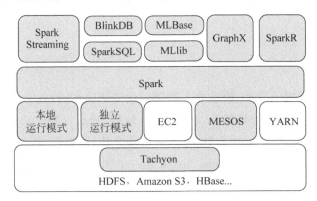

图 8-1 Spark 的生态系统

1. Spark Streaming

Spark Streaming 基于微批量方式的计算和处理,可以用于处理实时的流数据。它使用 DStream,简单来说,它就是一个弹性分布式数据集(RDD)系列,进行实时数据处理。

2. Spark SQL

Spark SQL 可以通过 JDBC API 将 Spark 数据集暴露出去,而且还可以用传统的 BI 和可视化工具在 Spark 数据上执行类似 SQL 的查询。用户还可以用 Spark SQL 对不同格式的数据(如 JSON、Parquet 及数据库等)执行 ETL,将其转化,然后暴露给特定的查询。

3. MLlib

MLlib 是一个可扩展的 Spark 机器学习库,由通用的学习算法和工具组成,包括二元分类、线性回归、聚类、协同过滤、梯度下降及底层的优化原语。

4. GraphX

GraphX 是用于图计算和并行图计算的新的（alpha）Spark API。通过引入弹性分布式属性图（resilient distributed property graph），一种顶点和边都带有属性的有向多重图，扩展了 Spark RDD。为了支持图计算，GraphX 暴露了一个基础操作符集合（如 subgraph、joinVertices 和 aggregateMessages）和一个经过优化的 Pregel API 变体。此外，GraphX 还包括一个持续增长的用于简化图分析任务的图算法和构建器集合。

除了这些库以外，还有一些其他的库，如 BlinkDB 和 Tachyon。

BlinkDB 是一个近似查询引擎，用于在海量数据上执行交互式 SQL 查询。BlinkDB 可以通过牺牲数据精度来提升查询响应时间。通过在数据样本上执行查询并展示包含有意义的错误线注解的结果，操作大数据集合。

Tachyon 是一个以内存为中心的分布式文件系统，能够提供内存级别速度的跨集群框架（如 Spark 和 MapReduce）的可信文件共享。它将工作集文件缓存在内存中，从而避免到磁盘中加载需要经常读取的数据集。通过这一机制，不同的作业/查询和框架可以以内存级的速度访问缓存的文件。

此外，还有一些用于与其他产品集成的适配器，如 Cassandra（Spark Cassandra 连接器）和 R（SparkR）。Cassandra Connector 可用于访问存储在 Cassandra 数据库中的数据并在这些数据上执行数据分析。

8.1.2 Spark 架构

Spark 架构采用了分布式计算中的 Master-Slave 模型。Master 是对应集群中的含有 Master 进程的节点，Slave 是集群中含有 Worker 进程的节点。Master 作为整个集群的控制器，负责整个集群的正常运行；Worker 相当于计算节点，接收主节点命令并进行状态汇报；Executor 负责任务的执行；Client 作为用户的客户端负责提交应用；Driver 负责控制一个应用的执行，如图 8-2 所示。

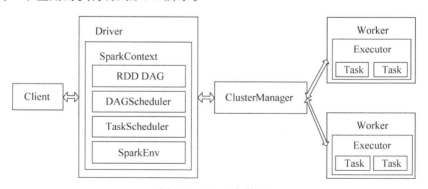

图 8-2　Spark 架构图

Spark 集群部署后，需要在主节点和从节点分别启动 Master 进程和 Worker 进程，对整个集群进行控制。在一个 Spark 应用的执行过程中，Driver 和 Worker 是两个重要角色。Driver 程序是应用逻辑执行的起点，负责作业的调度，即 Task 任务的分发，而多个

Worker 用来管理计算节点和创建 Executor 并行处理任务。在执行阶段，Driver 会将 Task 和 Task 所依赖的 file 和 jar 序列化后传递给对应的 Worker 机器，同时 Executor 对相应数据分区的任务进行处理。

Spark 的整体流程是：Client 提交应用，Master 找到一个 Worker 启动 Driver，Driver 向 Master 或资源管理器申请资源，之后将应用转化为 RDD Graph，再由 DAGScheduler 将 RDD Graph 转化为 Stage 的有向无环图提交给 TaskScheduler，由 TaskScheduler 提交任务给 Executor 执行，在任务执行的过程中，其他组件协同工作，确保整个应用顺利执行。

弹性分布式数据集（Resilient Distributed Datasets，RDD）是 Spark 框架中的核心概念，可以将 RDD 视作数据库中的一张表，其中可以保存任何类型的数据，Spark 将数据存储在不同分区上的 RDD 中，RDD 可以帮助重新安排计算并优化数据处理过程。

此外，它还具有容错性，因为 RDD 知道如何重新创建和重新计算数据集，RDD 是不可变的，可以用变换（transformation）修改 RDD，但是这个变换所返回的是一个全新的 RDD，而原有的 RDD 仍然保持不变。

8.2 Spark RDD

传统的 MapReduce 虽然具有自动容错、平衡负载和可拓展性的优点，但是其最大缺点是采用非循环式的数据流模型，使得在迭代计算时要进行大量的磁盘 IO 操作。RDD（resilientdistributed dataset）正是解决这一缺点的抽象方法，它是一个只读的、可分区的分布式数据集，这个数据集的全部或部分可以缓存在内存中，在多次计算间重用。RDD 也是一种有容错机制的特殊集合，可以分布在集群的节点上，以函数式编程操作集合的方式，进行各种并行操作。它是弹性的，计算过程中内存不够时它会和磁盘进行数据交换。

8.2.1 RDD 的依赖关系

RDD 作为数据结构，本质上是一个只读的分区记录集合。一个 RDD 可以包含多个分区，每个分区就是一个 Dataset 片段。RDD 可以相互依赖。如果 RDD 的每个分区最多只能被一个 Child RDD 的一个分区使用，则称为窄依赖（narrow dependency）；如果多个 Child RDD 分区都可以依赖，则称为宽依赖（wide Dependency）。依据其特性，不同的操作可能会产生不同的依赖。例如，map 操作会产生窄依赖，而 join 操作则产生宽依赖。

Spark 之所以将依赖分为窄依赖与宽依赖，基于以下两点原因。

首先，窄依赖可以支持在同一个 Cluster Node 上以管道形式执行多条命令，如在执行了 map 后，紧接着执行 filter。相反，宽依赖需要所有的父分区都是可用的，可能还需要调用类似 MapReduce 之类的操作进行跨节点传递。

其次,从失败恢复的角度考虑,窄依赖的失败恢复更有效,因为它只需要重新计算丢失的父分区即可,而且可以并行地在不同节点进行重计算;而宽依赖牵涉到 RDD 各级的多个父分区。如图 8-3 所示,说明了窄依赖与宽依赖之间的区别,每一个方框表示一个 RDD,其中的阴影矩形表示 RDD 的分区。

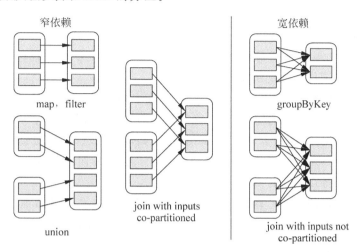

图 8-3　窄依赖与宽依赖

8.2.2　作业调度

当用户对一个 RDD 执行 Action(如 count 或 save)操作时,调度器会根据该 RDD 的 Lineage,来构建一个由若干阶段(stage)组成的一个 DAG(有向无环图)来执行程序,如图 8-4 所示。

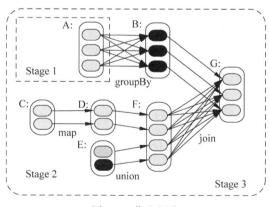

图 8-4　作业调度

实线椭圆标识的是 RDD,阴影背景的椭圆是分区,若已存于内存中则用黑色背景标识。RDD G 上一个 Action 的执行将会以宽依赖为分区来构建各个 Stage,对各 Stage 内部的窄依赖则前后连接构成流水线。在本例中,Stage 1 的输出已经存在于 RAM 中,所以直接执行 Stage 2,然后执行 Stage 3。

每个 Stage 都包含尽可能多的、连续的窄依赖型的 RDD。各个阶段之间的分界则是宽依赖所需的 Shuffle 操作，或者是 DAG 中一个经由该分区能更快到达父 RDD 的已计算分区。然后，调度器运行多个任务来计算各个阶段所缺失的分区，直到最终得出目标 RDD。

调度器向各机器的任务分配采用延时调度机制并根据数据存储位置（本地性）来确定。若一个任务需要处理的某个分区刚好存储在某个节点的内存中，则该任务会分配给那个节点；否则，如果一个任务处理的某个分区含有的 RDD 提供较佳的位置（如一个 HDFS 文件），则把该任务分配到这些位置。

对应宽依赖类的操作（如 Shuffle 依赖），会将中间记录"物化"到父分区的节点上。这和 MapReduce 物化 map 的输出类似，能简化数据的故障恢复过程。

对于执行失败的任务，只要它对应 Stage 的父类信息仍然可用，便会在其他节点上重新执行。如果某些 Stage 变为不可用（例如，Shuffle 在 map 阶段的某个输出丢失了），则重新提交相应的任务以并行计算丢失的分区。

若某个任务执行缓慢（即"落后者"Straggler），则系统会在其他节点上执行该任务的复制，这与 MapReduce 做法类似，并取最先得到的结果作为最终的结果。

8.2.3 内存管理

Spark 提供了 3 种对持久化 RDD 的存储策略：未序列化 Java 对象存于内存中、序列化后的数据存于内存以及磁盘存储。第一种存储策略的性能表现是 3 种存储策略中最优秀的，因为可以直接访问在 Java 虚拟机内存里的 RDD 对象。在空间有限的情况下，第二种存储策略可以让用户采用比 Java 对象图更有效的内存组织方式，代价是降低了性能。第三种存储策略适用于 RDD 太大，难以存储在内存的情形，但每次重新计算该 RDD 会带来额外的资源开销。

对于有限可用内存，Spark 使用以 RDD 为对象的 LRU 回收算法来进行管理。当计算得到一个新的 RDD 分区，但却没有足够空间来存储它时，系统会从最近最少使用的 RDD 中回收一个分区的空间。除非该 RDD 是新分区对应的 RDD，这种情况下，Spark 会将旧的分区继续保留在内存，防止同一个 RDD 的分区被循环调入和调出。因为大部分的操作会在一个 RDD 的所有分区上进行，所以很有可能已经存在内存中的分区将会被再次使用。

8.2.4 检查点支持

虽然 lineage 可用于错误后 RDD 的恢复，但对于很长的 lineage 的 RDD 来说，这样的恢复耗时较长。因此，对某些 RDD 进行检查点操作（dheckpoint）保存到稳定存储上，是有帮助的。

通常情况下，对于包含宽依赖的长"血统"的 RDD 设置检查点操作是非常有用的，在这种情况下，集群中某个节点的故障会使得从各个父 RDD 得出某些数据丢失，这时就需要完全重算。相反，对于那些窄依赖于稳定存储上数据的 RDD 来说，对其进行检查点操作就不是必需的了。如果一个节点发生故障，RDD 在该节点中丢失的分区数据可以通过并行的方式从其他节点中重新计算出来,计算成本只是复制整个 RDD 的很小一部分。

Spark 当前提供了为 RDD 设置检查点(用一个 REPLICATE 标志来持久化)操作的 API,让用户自行决定需要为哪些数据设置检查点操作。

最后,由于 RDD 的只读特性使得比常用的共享内存更容易做 checkpoint,因为不需要关心一致性的问题,RDD 的写出可在后台进行,而不需要程序暂停或进行分布式快照。

8.3　Spark 集群部署及应用案例

视频讲解

Spark 集群部署需要一台 Master 节点和多台 Slave 节点,Spark 使用 Hadoop 的数据源。Spark 集群环境需要部署 Hadoop,部署 Hadoop 参考第 3 章。

Spark 集群部署需要 scala-2.12.7.tgz 和 spark-2.4.0-bin-hadoop2.7.tgz 两个软件包,下载网址为 www.scala-lang.org 和 mirrors.aliyun.com。先下载软件包,然后解压到/opt 目录下,解压后子目录名带有版本号,对其重命名,使其简洁。

8.3.1　Spark 参数配置

Spark 需要添加环境变量和修改 Spark 参数,修改的配置文件有 spark-env.sh 和 slaves,3 台节点机安装配置方法一样。

(1) 修改/etc/hosts 文件,添加内容如下。

```
172.30.0.10    master
172.30.0.11    slave1
172.30.0.12    slave2
```

(2) 修改/root/.bash_profile 文件,添加内容如下。

```
export SCALA_HOME = /opt/scala
export SPARK_HOME = /opt/spark
export PATH = $PATH:$SCALA_HOME/bin:$SPARK_HOME/bin
```

(3) 在 Spark 的配置目录下新建 spark-env.sh 文件,内容如下。

```
export JAVA_HOME = /opt/jdk1.8
export HADOOP_HOME = /opt/hadoop
export HADOOP_CONF_DIR = $HADOOP_HOME/etc/hadoop
export SCALA_HOME = /opt/scala
export SPARK_HOME = /opt/spark
export SPARK_MASTER_IP = master
export SPARK_WORKER_MEMORY = 2g
```

(4) 在 Spark 的配置目录下新建 slaves 文件,内容如下。

```
master
slave1
slave2
```

8.3.2 Spark 集群运行

(1) 启动 Hadoop,操作如下。

```
[root@master ~]# start-dfs.sh
[root@master ~]# start-yarn.sh
[root@master ~]# mr-jobhistory-daemon.sh start historyserver
```

(2) 启动 Spark,操作如下。

```
[root@master ~]# /opt/spark/sbin/start-all.sh
```

(3) 查看 Spark 进程,操作如下。

```
[root@master ~]# jps
13250 SecondaryNamenode
13027 Namenode
17574 Master
17688 Worker
13435 ResourceManager
13725 JobHistoryServer
18239 Jps
[root@slave1 ~]# jps
3172 Jps
1511 Datanode
1626 NodeManager
3035 Worker
[root@slave2 ~]# jps
3409 Worker
1432 Datanode
1547 NodeManager
3549 Jps
```

(4) 打开浏览器输入"http://master:8080",查看 Spark 集群运行信息,如图 8-5 所示。

(5) 打开浏览器输入"http://slave1:8081",查看 Worker 运行信息,如图 8-6 所示。

8.3.3 Spark 交互

Spark 交互主要有 Spark-Shell 和 PYSpark。Spark-Shell 支持 Scala 语言;PYSpark 支持 Python 语言。脚本可以直接在交互窗口运行,非常方便程序的调试。

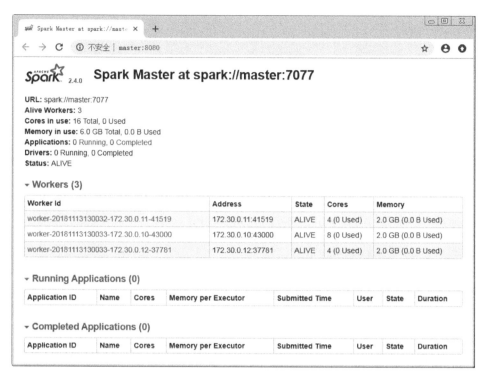

图 8-5　Spark 集群运行信息

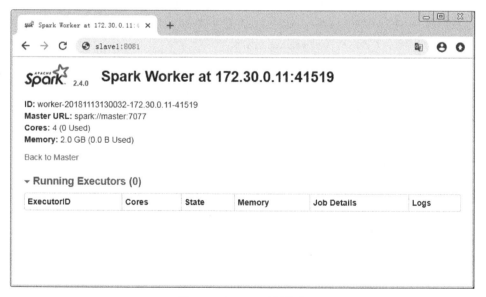

图 8-6　Worker 运行信息

(1) 启动 Spark-Shell,操作如下。

```
# spark-shell
…
Spark context Web UI available at http://master:4040
Spark context available as 'sc' (master = local[*], app id = local-1542085724102).
Spark session available as 'spark'.
Welcome to
      ____              __
     / __/__  ___ _____/ /__
    _\ \/ _ \/ _ `/ __/  '_/
   /___/ .__/\_,_/_/ /_/\_\   version 2.4.0
      /_/

Using Scala version 2.11.12 (Java HotSpot(TM) 64-Bit Server VM, Java 1.8.0_131)
Type in expressions to have them evaluated.
Type :help for more information.

scala>
```

(2) 启动 PYSpark,操作如下。

```
# pyspark
Python 2.7.5 (default, Jul 13 2018, 13:06:57)
…
Welcome to
      ____              __
     / __/__  ___ _____/ /__
    _\ \/ _ \/ _ `/ __/  '_/
   /__ / .__/\_,_/_/ /_/\_\   version 2.4.0
      /_/

Using Python version 2.7.5 (default, Jul 13 2018 13:06:57)
SparkSession available as 'spark'.
>>>
```

8.3.4　Spark 算子

Spark 算子大致可以分为转化(Transformation)和行动(Action)两类。

Transformation 并不触发提交作业,完成作业中间过程处理。Transformation 操作是延迟计算的。也就是说,从一个 RDD 转换生成另一个 RDD 的转换操作不是马上执行的,而是需要等到有 Action 操作时才会真正触发运算。Transformation 函数包括 map、filter、flatMap、groupByKey、reduceByKey、aggregateByKey、pipe 和 coalesce 等。

Action 算子会触发 SparkContext 提交作业(Job);还会触发 Spark 提交作业(Job),并将数据输出 Spark 系统。Action 操作包括 reduce、collect、count、first、take、countByKey 及 foreach 等。

1. 基本 RDD 的转化操作

(1) map(func): 将 func 函数作用到 RDD 的每个元素,生成一个新的 RDD 返回。

例 8.1　计算 rdd1 中各元素值的平方。

Scala 语言：

```
scala> var rdd1 = sc.parallelize(List(1, 2, 3, 4, 5))
scala> var rdd2 = rdd1.map(x => x * x)
scala> println(rdd2.collect().toList)
List(1, 4, 9, 16, 25)
```

Python 语言：

```
>>> rdd1 = sc.parallelize([1, 2, 3, 4, 5])
>>> rdd2 = rdd1.map(lambda x : x * x).collect()
>>> print(rdd2)
[1, 4, 9, 16, 25]
```

(2) flatMap(func): 与 map 相似,但是每个输入的 item 能够被 map 到 0 个或更多的 items 输出。

例 8.2　将 rdd1 中元素数据切分为单词。

Scala 语言：

```
scala> var lst = List("I am spark","She is python","He is scala")
scala> var rdd1 = sc.parallelize(lst, 3)
scala> var rdd2 = rdd1.flatMap(x => x.split(" "))
scala> println(rdd2.collect().toList)
List(I, am, spark, She, is, python, He, is, scala)
```

Python 语言：

```
>>> lst = ['I am spark','She is python','He is scala']
>>> rdd1 = sc.parallelize(lst, 3)
>>> rdd2 = rdd1.flatMap(lambda x: x.split()).collect()
>>> print(rdd2)
['I', 'am', 'spark', 'She', 'is', 'python', 'He', 'is', 'scala']
```

(3) filter(func): 选出所有 func 返回值为 true 的元素,作为一个新的 RDD 返回。

例 8.3　过滤 rdd1 中的值不是 2 的元素。

Scala 语言：

```
scala> var rdd1 = sc.parallelize(List(1, 2, 3, 4, 5))
scala> var rdd2 = rdd1.filter(x => x!= 2)
scala> println(rdd2.collect().toList)
List(1, 3, 4, 5)
```

Python 语言：

```
>>> rdd1 = sc.parallelize([1, 2, 3, 4, 5])
>>> rdd2 = rdd1.filter(lambda x: x!= 2).collect()
>>> print(rdd2)
[1, 3, 4, 5]
```

(4) mapPartitions(func)：与 map 相似，但是 mapPartitions 的输入函数单独作用于 RDD 的每个分区(block)上，因此 func 的输入和返回值都必须是迭代器 iterator。

例 8.4 假设 rdd1 有 0~9 共 10 个元素，分成 3 个区，使用 mapPartitions 返回每个元素的平方。如果使用 map 方法，map 中的输入函数会被调用 10 次；而使用 mapPartitions 方法，输入函数只会被调用 3 次，每个分区被调用 1 次。

Scala 语言：

```
scala> def squareFunc(iter: Iterator[Int]) : Iterator[(Int)] = {
     |     var res = List[(Int)]()
     |     while (iter.hasNext) {
     |         val cur = iter.next;
     |         res .::= (cur * cur)
     |     }
     |     res.iterator
     | }
scala> val rdd1 = sc.parallelize(0 to 9, 3)
scala> val rdd2 = rdd1.mapPartitions(squareFunc)
scala> println(rdd2.collect().toList)
List(4, 1, 0, 25, 16, 9, 81, 64, 49, 36)
```

Python 语言：

```
>>> def squareFunc(a):
...     for i in a:
...         yield i * i
>>> rdd1 = sc.parallelize(range(10), 3)
>>> rdd2 = rdd1.mapPartitions(squareFunc).collect()
>>> print(rdd2)
[0, 1, 4, 9, 16, 25, 36, 49, 64, 81]
```

(5) mapPartitionsWithIndex(func)：与 mapPartitions 相似，但是输入函数 func 提供了一个正式的参数，可以用来表示分区的编号。

例 8.5 对 rdd1 的分区进行求和。

Scala 语言：

```
scala> def func(index:Int,iter: Iterator[Int]): Iterator[(Int,Int)] = {
     |     var res = List[(Int,Int)]()
     |     var sum = 0
     |     iter.foreach (sum += _)
```

```
        |       res .::= (index,sum)
        |       res.iterator
        | }
scala> var rdd1 = sc.parallelize(0 to 9, 3)
scala> var rdd2 = rdd1.mapPartitionsWithIndex(func)
scala> println(rdd1.glom().collect().deep.mkString(" "))
Array(0, 1, 2) Array(3, 4, 5) Array(6, 7, 8, 9)
scala> println(rdd2.glom().collect().deep.mkString(" "))
Array((0,3)) Array((1,12)) Array((2,30))
```

Python 语言:

```
>>> def func(index,iterator):yield (index,sum(iterator))
>>> rdd1 = sc.parallelize(range(10), 3)
>>> rdd2 = rdd1.mapPartitionsWithIndex(func)
>>> print(rdd1.glom().collect())
[[0, 1, 2], [3, 4, 5], [6, 7, 8, 9]]
>>> print(rdd2.glom().collect())
[[(0, 3)], [(1, 12)], [(2, 30)]]
```

(6) distinct([numTasks]): 删除 RDD 中相同的元素。

例 8.6 删除 rdd1 中相同的元素。

Scala 语言:

```
scala> var rdd1 = sc.parallelize(List(1, 1, 2, 5, 2, 9, 6, 1))
scala> var rdd2 = rdd1.distinct()
scala> println(rdd2.collect().toList)
List(1, 9, 2, 5, 6)
```

Python 语言:

```
>>> rdd1 = sc.parallelize([1, 1, 2, 5, 2, 9, 6, 1])
>>> rdd2 = rdd1.distinct()
>>> print(rdd2.collect())
[1, 9, 2, 5, 6]
```

(7) sample(withReplacement, fraction, seed): 以指定的种子随机抽样数量为 fraction 的数据, withReplacement 为抽样数据是否放回, true 为有放回的抽样, false 为无放回的抽样。

例 8.7 从 rdd1 中随机且有放回地抽出 50% 的数据, 随机种子值为 3。

Scala 语言:

```
scala> var rdd1 = sc.parallelize(0 to 20)
scala> var rdd2 = rdd1.sample(true,0.5,3)
scala> println(rdd2.collect().toList)
List(8, 10, 11, 11, 13, 18, 19, 19)
```

Python 语言:

```
>>> rdd1 = sc.parallelize(range(20))
>>> rdd2 = rdd1.sample(True, 0.5, 3)
>>> print(rdd2.collect())
[0, 1, 4, 5, 6, 8, 13, 14]
```

(8) cartesian(otherDataset):对两个 RDD 内的所有元素进行笛卡儿积操作。

例 8.8 对 rdd1 和 rdd2 进行笛卡儿积操作。

Scala 语言:

```
scala> val rdd1 = sc.parallelize(List(('a', 1),('a', 2)))
scala> val rdd2 = sc.parallelize(List(('c', 3),('d', 4)))
scala> val rdd3 = rdd1.cartesian(rdd2)
scala> println(rdd3.collect.toList)
List(((a,1),(c,3)), ((a,1),(d,4)), ((a,2),(c,3)), ((a,2),(d,4)))
```

Python 语言:

```
>>> rdd1 = sc.parallelize([('a', 1),('a', 2)])
>>> rdd2 = sc.parallelize([('c', 3),('d', 4)])
>>> rdd3 = rdd1.cartesian(rdd2)
>>> print(rdd3.collect())
[(('a', 1), ('c', 3)), (('a', 1), ('d', 4)), (('a', 2), ('c', 3)), (('a', 2), ('d', 4))]
```

(9) union(otherDataset):合并两个 RDD 元素,两个 RDD 元素的数据类型要相同。

例 8.9 合并 rdd1 和 rdd2。

Scala 语言:

```
scala> val rdd1 = sc.parallelize(List(('a', 1),('a', 2)))
scala> val rdd2 = sc.parallelize(List(('c',3),('d', 4)))
scala> val rdd3 = rdd1.union(rdd2)
scala> println(rdd3.collect.toList)
List((a,1), (a,2), (c,3), (d,4))
```

Python 语言:

```
>>> rdd1 = sc.parallelize([('a', 1),('a', 2)])
>>> rdd2 = sc.parallelize([('c', 3),('d', 4)])
>>> rdd3 = rdd1.union(rdd2)
>>> print(rdd3.collect())
[('a', 1), ('a', 2), ('c', 3), ('d', 4)]
```

(10) intersection(otherDataset):对两个 RDD 内的所有元素提取交集。

例 8.10 计算 rdd1 和 rdd2 的交集。

Scala 语言:

```
scala> val rdd1 = sc.parallelize(0 to 9)
scala> val rdd2 = sc.parallelize(6 to 15)
scala> val rdd3 = rdd1.intersection(rdd2)
scala> println(rdd1.collect.toList)
List(0, 1, 2, 3, 4, 5, 6, 7, 8, 9)
scala> println(rdd2.collect.toList)
List(6, 7, 8, 9, 10, 11, 12, 13, 14, 15)
scala> println(rdd3.collect.toList)
List(8, 9, 6, 7)
```

Python 语言:

```
>>> rdd1 = sc.parallelize(range(10))
>>> rdd2 = sc.parallelize(range(6,15))
>>> rdd3 = rdd1.intersection(rdd2)
>>> print(rdd1.collect())
[0, 1, 2, 3, 4, 5, 6, 7, 8, 9]
>>> print(rdd2.collect())
[6, 7, 8, 9, 10, 11, 12, 13, 14]
>>> print(rdd3.collect())
[6, 7, 8, 9]
```

(11) subtract(otherDataset):进行集合的差操作。

例 8.11 rdd1 去除 rdd1 和 rdd2 交集中的所有元素。

Scala 语言:

```
scala> val rdd1 = sc.parallelize(0 to 9)
scala> val rdd2 = sc.parallelize(6 to 15)
scala> val rdd3 = rdd1.subtract(rdd2)
scala> println(rdd3.collect.toList)
List(0, 1, 2, 3, 4, 5)
```

Python 语言:

```
>>> rdd1 = sc.parallelize(range(10))
>>> rdd2 = sc.parallelize(range(6,15))
>>> rdd3 = rdd1.subtract(rdd2)
>>> print(rdd3.collect())
[0, 1, 2, 3, 4, 5]
```

2. 基本 RDD 的行动操作

(1) collect():返回 RDD 中的所有元素。

例 8.12 显示 rdd1 中的元素。

Scala 语言：

```
scala> var rdd1 = sc.parallelize( 0 to 9)
scala> println(rdd1.collect().toList)
List(0, 1, 2, 3, 4, 5, 6, 7, 8, 9)
```

Python 语言：

```
>>> rdd1 = sc.parallelize(range(10))
>>> print(rdd1.collect())
[0, 1, 2, 3, 4, 5, 6, 7, 8, 9]
```

(2) count()：返回 RDD 中的元素个数。

例 8.13 计算 rdd1 中的元素个数。

Scala 语言：

```
scala> var rdd1 = sc.parallelize(List(1, 2, 'a', 'b'))
scala> rdd1.count()
res2: Long = 4
```

Python 语言：

```
>>> rdd1 = sc.parallelize([1, 2, 'a', 'b'])
>>> rdd1.count()
4
```

(3) countByValue()：返回 RDD 中的各元素出现次数。

例 8.14 计算 rdd1 中各元素出现的次数。

Scala 语言：

```
scala> var rdd1 = sc.parallelize(List(1, 2,'a',1,2,'a',1))
scala> println(rdd1.countByValue())
Map(97 -> 2, 1 -> 3, 2 -> 2)
```

Python 语言：

```
>>> rdd1 = sc.parallelize([1, 2, 'a', 1, 2, 'a', 1])
>>> print(rdd1.countByValue())
defaultdict(<type 'int'>, {'a': 2, 1: 3, 2: 2})
```

(4) take(num)：返回 RDD 前 num 个元素。

例 8.15 取 rdd1 中前 3 个元素。

Scala 语言：

```
scala> var rdd1 = sc.parallelize(List(2, 3, 'a', 1, 4))
scala> println(rdd1.take(3).toList)
List(2, 3, 97)
```

Python 语言：

```
>>> rdd1 = sc.parallelize([2, 3, 'a', 1, 4])
>>> print(rdd1.take(3))
[2, 3, 'a']
```

（5）top(num)：返回 RDD 最大的 num 个元素。

例 8.16　取 rdd1 中最大的 3 个元素。

Scala 语言：

```
scala> var rdd1 = sc.parallelize(List(2, 3, 'a', 1, 4))
scala> println(rdd1.top(3).toList)
List(97, 4, 3)
```

Python 语言：

```
>>> rdd1 = sc.parallelize([2, 3, 'a', 1, 4])
>>> print(rdd1.top(3))
['a', 4, 3]
```

（6）takeOrdered(num)：返回 RDD 最小的 num 个元素。

例 8.17　取 rdd1 中最小的 3 个元素。

Scala 语言：

```
scala> var rdd1 = sc.parallelize(List(2, 3, 'a', 1, 4))
scala> println(rdd1.takeOrdered(3).toList)
List(1, 2, 3)
```

Python 语言：

```
>>> rdd1 = sc.parallelize([2, 3, 'a', 1, 4])
>>> print(rdd1.takeOrdered(3))
[1, 2, 3]
```

（7）takeSample(withReplacement,num,seed)：takeSample 函数类似于 sample 函数，withReplacement 表示采样是否放回，true 表示有放回的采样，false 表示无放回的采样；num 表示返回的采样数据的个数；seed 表示用于指定的随机数生成器种子。

例 8.18　从 rdd1 中随机且有放回地抽出 10 个数据，随机种子值为 5。

Scala 语言：

```
scala> var rdd1 = sc.parallelize(0 to 20)
scala> var rdd2 = rdd1.takeSample(true,10,3)
scala> println(rdd2.toList)
List(10, 3, 2, 11, 10, 2, 19, 0, 14, 12)
```

Python 语言:

```
>>> rdd1 = sc.parallelize(range(20))
>>> rdd2 = rdd1.takeSample(True,10,3)
>>> print(rdd2)
[8, 12, 7, 9, 9, 18, 16, 19, 3, 14]
```

(8) reduce(func): 将 RDD 中元素两两传递给输入函数,同时产生一个新的值;新产生的值与 RDD 中下一个元素再被传递给输入函数,直到最后只有一个值为止。

例 8.19 对 rdd1 中的元素求和。

Scala 语言:

```
scala> var rdd1 = sc.parallelize(1 to 10)
scala> var sum = rdd1.reduce((x,y) => x + y)
sum: Int = 55
```

Python 语言:

```
>>> rdd1 = sc.parallelize(range(1,11))
>>> sum = rdd1.reduce(lambda x,y : x + y)
>>> print(sum)
55
```

(9) fold(zero)(func): 和 reduce 一样,但是需要提供初始值。

例 8.20 对 rdd1 中的元素加上初始值求和。

Scala 语言:

```
scala> var rdd1 = sc.parallelize(1 to 10)
scala> var sum = rdd1.fold(10)((x,y) => x + y)
sum: Int = 145
```

Python 语言:

```
>>> rdd1 = sc.parallelize(range(1,11))
>>> sum = rdd1.fold(10, lambda x,y : x + y)
>>> print(sum)
145
```

(10) aggregate(zeroValue)(seqOp, combOp): 将每个分区里面的元素进行聚合,然后用 combine 函数对每个分区的结果和初始值(zeroValue)进行 combine 操作。

aggregate 与 fold 和 reduce 的不同之处在于：aggregate 相当于采用归并的方式进行数据聚集，这种聚集是并行化的。

例 8.21 对 rdd1 进行聚集计算。

Scala 语言：

```
scala> var rdd1 = sc.parallelize(List(1, 2, 3, 3))
scala> var rdd2 = rdd1.aggregate((0,0))((x,y) => (x._1 + y, x._2 + 1),
     |      (x,y) => (x._1 + y._1, x._2 + y._2))
rdd2: (Int, Int) = (9,4)
```

Python 语言：

```
>>> rdd1 = sc.parallelize([1, 2, 3, 3])
>>> rdd2 = rdd1.aggregate((0,0), lambda x,y : (x[0] + y, x[1] + 1),
...    lambda x,y : (x[0] + y[0], x[1] + y[1]))
>>> print(rdd2)
(9, 4)
```

（11）foreach(func)：对 RDD 中的每个元素使用给定的函数。

例 8.22 打印 rdd1 中的元素。

Scala 语言：

```
scala> var rdd1 = sc.parallelize(List(1, 2, 3, 3))
scala> rdd1.foreach( x => print(x + " "))
3 3 2 1
```

Python 语言：

```
>>> def prn(x):print(x,;print(' '),
>>> rdd1 = sc.parallelize([1, 2, 3, 3])
>>> rdd1.foreach(prn)
1 3 2 3
```

3. Pair RDD 的转化操作

（1）reduceByKey(func)：对元素为 K-V 对的 RDD 中 Key 相同的元素的 Value 进行 reduce 操作。

例 8.23 对 rdd1 中 Key 相同的元素的值求和。

Scala 语言：

```
scala> var rdd1 = sc.parallelize(List((1, 2),(3, 4),(3, 6)))
scala> var rdd2 = rdd1.reduceByKey((x,y) => x + y)
scala> print(rdd2.collect().toList)
List((1,2), (3,10))
```

Python 语言：

```
>>> rdd1 = sc.parallelize([(1, 2),(3, 4),(3, 6)])
>>> rdd2 = rdd1.reduceByKey(lambda x,y : x + y)
>>> print(rdd2.collect())
[(1, 2), (3, 10)]
```

(2) groupBy(func)：func 返回 Key，传入的 RDD 的各个元素根据这个 Key 进行分组。

例 8.24 对 rdd1 中的元素按奇偶进行分组。

Scala 语言：

```
scala> var rdd1 = sc.parallelize(1 to 10)
scala> var rdd2 = rdd1.groupBy(x => x%2 )
scala> print(rdd2.collect().toList
List((0,CompactBuffer(2, 4, 6, 8, 10)), (1,CompactBuffer(1, 3, 5, 7, 9)))
```

Python 语言：

```
>>> rdd1 = sc.parallelize(range(1,10))
>>> rdd2 = rdd1.groupBy(lambda x : x%2)
>>> print(rdd2.map(lambda x:(x[0],list(x[1]))).collect())
[(0, [2, 4, 6, 8]), (1, [1, 3, 5, 7, 9])]
```

(3) groupByKey()：对元素为 K-V 对的 RDD 中 Key 相同的元素进行分组。

例 8.25 对 rdd1 中相同的 Key 进行分组。

Scala 语言：

```
scala> var rdd1 = sc.parallelize(List(('A',1),('B',2),('A',3),('C',4),('B',0)))
scala> var rdd2 = rdd1.groupByKey()
scala> print(rdd2.collect().toList)
List((A,CompactBuffer(1, 3)), (B,CompactBuffer(2, 0)), (C,CompactBuffer(4)))
```

Python 语言：

```
>>> rdd1 = sc.parallelize([('A',1),('B',2),('A',3),('C',4),('B',0)])
>>> rdd2 = rdd1.groupByKey()
>>> print(rdd2.map(lambda x:(x[0],list(x[1]))).collect())
[('A', [1, 3]), ('C', [4]), ('B', [2, 0])]
```

(4) combineByKey(createCombiner, mergeValue, mergeCombiners, partitioner, mapSideCombine)：在第一次遇到 Key 时创建组合器函数(createCombiner)，将 RDD 数据集中的 V 类型值转换成 C 类型值(V => C)；再次遇到相同的 Key 时，合并值函数(mergeValue)将 C 类型值与这次传入的 V 类型值合并成一个 C 类型值(C,V)=> C；合并组合器函数(mergeCombiners)将 C 类型值两两合并成一个 C 类型值；partitioner 表示

使用已有的或自定义的分区函数,默认是 HashPartitioner;mapSideCombine 表示是否在 map 端进行 Combine 操作,默认为 true。

例 8.26 统计男性和女生的个数,并以(性别,(名字,名字,…),个数)的形式输出。

Scala 语言:

```
scala> val people = List(("male", "Mobin"), ("male", "Kpop"), ("female", "Lucy"),
     |  ("male", "Lufei"), ("female", "Amy"))
scala> val rdd1 = sc.parallelize(people)
scala> def createCombiner = { (x: String) => (List(x), 1)}
scala> def mergeVal = { (peo: (List[String], Int), x : String) => (x :: peo._1, peo._2 + 1)}
scala> def mergeComb = { (sex1: (List[String], Int), sex2: (List[String], Int))
     |     => (sex1._1 ::: sex2._1, sex1._2 + sex2._2) }
mergeComb: ((List[String], Int), (List[String], Int)) => (List[String], Int)
scala> val rdd2 = rdd1.combineByKey(createCombiner, mergeVal, mergeComb)
scala> rdd2.foreach(println)
(female,(List(Lucy, Amy),2))
(male,(List(Mobin, Kpop, Lufei),3))
```

Python 语言:

```
>>> people = [("male", "Mobin"), ("male", "Kpop"), ("female", "Lucy"),
...  ("male", "Lufei"), ("female", "Amy")]
>>> rdd1 = sc.parallelize(people)
>>> createCombiner = (lambda x : ([x], 1))
>>> mergeVal = (lambda peo, x : (x + peo[0], peo[1] + 1))
>>> mergeComb = (lambda sex1, sex2 : (sex1[0] + sex2[0], sex1[1] + sex2[1]))
>>> rdd2 = rdd1.combineByKey(createCombiner, mergeVal, mergeComb)
>>> def prn(x):print(x)
>>> rdd2.foreach(prn)
('female', (['Lucy', 'Amy'], 2))
('male', (['Mobin', 'Kpop', 'Lufei'], 3))
```

(5) mapValues(func):原 RDD 中的 Key 保持不变,与新的 Value 一起组成新的 RDD 中的元素。因此,该函数只适用于元素为 K-V 对的 RDD。

例 8.27 对 rdd1 中的 Value 加 3。

Scala 语言:

```
scala> var rdd1 = sc.parallelize(List(("Pig", 3),("Rabbit", 5),("Python", 6)))
scala> var rdd2 = rdd1.mapValues(_ + 3)
scala> println(rdd2.collect.toList)
List((Pig,6), (Rabbit,8), (Python,9))
```

Python 语言:

```
>>> rdd1 = sc.parallelize([("Pig", 3),("Rabbit", 5),("Python", 6)])
>>> rdd2 = rdd1.mapValues(lambda x: x + 3)
```

```
>>> print(rdd2.collect())
[('Pig', 6), ('Rabbit', 8), ('Python', 9)]
```

(6) flatMapValues(func)：flatMapValues 类似于 mapValues，不同的在于 flatMapValues 应用于元素为 K-V 对的 RDD 中的 V 值。每个元素的 Value 被输入函数映射为一系列的值，然后这些值再与原 RDD 中的 Key 组成一系列新的 K-V 对。

例 8.28 对 rdd1 中的 Value 取平方。

Scala 语言：

```
scala> var rdd1 = sc.parallelize(List(('A',List(1, 2, 3)),('B',List(4, 5, 6))))
scala> var rdd2 = rdd1.flatMapValues(x =>(for (i <- x) yield i * i))
scala> println(rdd2.collect.toList)
List((A,1), (A,4), (A,9), (B,16), (B,25), (B,36))
```

Python 语言：

```
>>> rdd1 = sc.parallelize([('A',(1, 2, 3)),('B',(4, 5, 6))])
>>> rdd2 = rdd1.flatMapValues(lambda x: [i * i for i in x])
>>> print(rdd2.collect())
[('A', 1), ('A', 4), ('A', 9), ('B', 16), ('B', 25), ('B', 36)]
```

(7) keys()：返回 RDD 中的 Key。

例 8.29 列出 rdd1 中的 Key。

Scala 语言：

```
scala> var rdd1 = sc.parallelize(List(("Pig", 3),("Rabbit", 5),("Python", 6)))
scala> var rdd2 = rdd1.keys
scala> print(rdd2.collect().toList)
List(Pig, Rabbit, Python)
```

Python 语言：

```
>>> rdd1 = sc.parallelize([("Pig", 3),("Rabbit", 5),("Python", 6)])
>>> rdd2 = rdd1.keys()
>>> print(rdd2.collect())
['Pig', 'Rabbit', 'Python']
```

(8) values()：返回 RDD 中的 Value。

例 8.30 列出 rdd1 中的 Value。

Scala 语言：

```
scala> var rdd1 = sc.parallelize(List(("Pig", 3),("Rabbit", 5),("Python", 6)))
scala> var rdd2 = rdd1.values
scala> print(rdd2.collect().toList)
List(3, 5, 6)
```

Python 语言：

```
>>> rdd1 = sc.parallelize([("Pig", 3),("Rabbit", 5),("Python", 6)])
>>> rdd2 = rdd1.values()
>>> print(rdd2.collect())
[3, 5, 6]
```

(9) sortBy(func，ascending)：根据给定的 func 函数对 RDD 中的元素进行排序，ascending 默认是 true，也就是升序。

例 8.31 对 rdd1 中的元素进行降序排序。

Scala 语言：

```
scala> var rdd1 = sc.parallelize(List(3, 1, 50, 3, 6, 18))
scala> var rdd2 = rdd1.sortBy(x => x, false)
scala> print(rdd2.collect().toList)
List(50, 18, 6, 3, 3, 1)
```

Python 语言：

```
>>> rdd1 = sc.parallelize([3, 1, 50, 3, 6, 18])
>>> rdd2 = rdd1.sortBy(lambda x : x , False)
>>> print(rdd2.collect())
[50, 18, 6, 3, 3, 1]
```

(10) sortByKey()：将元素通过函数生成相应的 Key，数据转化为 Key-Value 格式，之后将 Key 相同的元素分为一组。

例 8.32 对 rdd1 中的元素按 Key 排序。

Scala 语言：

```
scala> var rdd1 = sc.parallelize(List(("B", 66),("A", 65),("D", 68)))
scala> var rdd2 = rdd1.sortByKey (true)
scala> print(rdd2.collect().toList)
List((A,65), (B,66), (D,68))
```

Python 语言：

```
>>> rdd1 = sc.parallelize([("B", 66),("A", 65),("D", 68)])
>>> rdd2 = rdd1.sortByKey(True)
>>> print(rdd2.collect())
[('A', 65), ('B', 66), ('D', 68)]
```

(11) zip(otherDataset)：将两个 RDD 的元素聚合在一起。

例 8.33 将 rdd1 和 rdd2 聚合在一起。

Scala 语言：

```
scala> var rdd1 = sc.parallelize(List('A','B','C'))
scala> var rdd2 = sc.parallelize(List(65, 66, 67))
scala> var rdd3 = rdd1.zip(rdd2)
scala> print(rdd3.collect.toList)
List((A,65), (B,66), (C,67))
```

Python 语言:

```
>>> rdd1 = sc.parallelize(['A','B','C'])
>>> rdd2 = sc.parallelize([65, 66, 67])
>>> rdd3 = rdd1.zip(rdd2)
>>> print(rdd3.collect())
[('A', 65), ('B', 66), ('C', 67)]
```

(12) subtractByKey(otherDataset):删除 RDD 中 Key 与另一个 RDD 中 Key 相同的元素。

例 8.34 删除 rdd1 中 Key 与 rdd2 相同的元素。

Scala 语言:

```
scala> var rdd1 = sc.parallelize(List((1, 2),(3, 4),(3, 6)))
scala> var rdd2 = sc.parallelize(List((3, 8),(4, 9)))
scala> var rdd3 = rdd1.subtractByKey(rdd2)
scala> print(rdd3.collect.toList)
List((1,2))
```

Python 语言:

```
<<< rdd1 = sc.parallelize([(1, 2),(3, 4),(3, 6)])
<<< rdd2 = sc.parallelize([(3, 8),(4, 9)])
<<< rdd3 = rdd1.subtractByKey(rdd2)
<<< print(rdd3.collect())
[(1, 2)]
```

(13) cogroup(otherDataSet, numPartitions):对两个 RDD,如(K,V)和(K,W),Key 相同的元素分别做聚合,然后返回(K, Iterator < V >, Iterator < W >)形式的 RDD, numPartitions 设置分区数,提高作业并行度。

例 8.35 将 rdd1 与 rdd2 分组聚合。

Scala 语言:

```
scala> var rdd1 = sc.parallelize(List((1, 2),(3, 4),(3, 6)))
scala> var rdd2 = sc.parallelize(List((3, 8),(4, 9)))
scala> var rdd3 = rdd1.cogroup(rdd2)
scala> print(rdd3.collect.toList)
List((1,(CompactBuffer(2),CompactBuffer())), (3,(CompactBuffer(4, 6),CompactBuffer(8))), (4,(CompactBuffer(),CompactBuffer(9))))
```

Python 语言:

```
>>> rdd1 = sc.parallelize([(1, 2),(3, 4),(3, 6)])
>>> rdd2 = sc.parallelize([(3, 8),(4, 9)])
>>> rdd3 = rdd1.cogroup(rdd2)
>>> def prn(x):
...     print('['),; print(x[0]),; print(','),
...     for i in x[1]: print(list(i)),
...     print(']'),
>>> rdd3.foreach(prn)
[ 1 , [2] [ ] ] [ 3 , [4, 6] [8] ] [ 4 , [] [9] ]
```

(14) join(otherDataset,numPartitions): 对两个 RDD 先进行 cogroup 操作形成新的 RDD, 再对每个 Key 下的元素进行笛卡儿积, 运算 numPartitions 设置分区数, 提高作业并行度。

例 8.36 将 rdd1 与 rdd2 内连接。

Scala 语言:

```
scala> var rdd1 = sc.parallelize(List((1, 2),(3, 4),(3, 6)))
scala> var rdd2 = sc.parallelize(List((3, 8),(4, 9)))
scala> var rdd3 = rdd1.join(rdd2)
scala> print(rdd3.collect.toList)
List((3,(4,8)), (3,(6,8)))
```

Python 语言:

```
>>> rdd1 = sc.parallelize([(1,2),(3,4),(3,6)])
>>> rdd2 = sc.parallelize([(3, 8),(4, 9)])
>>> rdd3 = rdd1.join(rdd2)
>>> print(rdd3.collect())
[(3, (4, 8)), (3, (6, 8))]
```

(15) leftOuterJoin(otherDataset,numPartitions): 左外连接, 包含左 RDD 的所有数据, 如果右边没有与之匹配的用 None 表示, numPartitions 设置分区数, 提高作业并行度。

例 8.37 将 rdd1 与 rdd2 左连接。

Scala 语言:

```
scala> var rdd1 = sc.parallelize(List((1, 2),(3, 4),(3, 6)))
scala> var rdd2 = sc.parallelize(List((3, 8),(4, 9)))
scala> var rdd3 = rdd1.leftOuterJoin(rdd2)
scala> print(rdd3.collect.toList)
List((1,(2,None)), (3,(4,Some(8))), (3,(6,Some(8))))
```

Python 语言：

```
<<< rdd1 = sc.parallelize([(1,2),(3,4),(3,6)])
<<< rdd2 = sc.parallelize([(3,8),(4,9)])
<<< rdd3 = rdd1.leftOuterJoin(rdd2)
<<< print(rdd3.collect())
[(1, (2, None)), (3, (4, 8)), (3, (6, 8))]
```

（16）rightOuterJoin(otherDataset,numPartitions)：右外连接，包含右 RDD 的所有数据，如果左边没有与之匹配的用 None 表示，numPartitions 设置分区数，提高作业并行度。

例 8.38 将 rdd1 与 rdd2 右连接。

Scala 语言：

```
scala> var rdd1 = sc.parallelize(List((1, 2),(3, 4),(3, 6)))
scala> var rdd2 = sc.parallelize(List((3, 8),(4, 9)))
scala> var rdd3 = rdd1.rightOuterJoin(rdd2)
scala> print(rdd3.collect.toList)
List((3,(Some(4),8)), (3,(Some(6),8)), (4,(None,9)))
```

Python 语言：

```
>>> rdd1 = sc.parallelize([(1,2),(3,4),(3,6)])
>>> rdd2 = sc.parallelize([(3,8),(4,9)])
>>> rdd3 = rdd1.rightOuterJoin(rdd2)
>>> print(rdd3.collect())
[(3, (4, 8)), (3, (6, 8)), (4, (None, 9))]
```

4. Pair RDD 的行动操作

（1）countByKey()：统计每个 Key 对应的 Value 个数。

例 8.39 统计 rdd1 中每个 Key 对应的 Value 个数。

Scala 语言：

```
scala> var rdd1 = sc.parallelize(List(('B', 1),('B', 2),('A', 3),('A',4),('A', 5)))
scala> var rdd2 = rdd1.countByKey()
scala> print(rdd2)
Map(A -> 3, B -> 2)
```

Python 语言：

```
>>> rdd1 = sc.parallelize([('B', 1),('B', 2),('A', 3),('A', 4),('A', 5)])
>>> rdd2 = rdd1.countByKey()
>>> print(rdd2)
defaultdict(<type 'int'>, {'A': 3, 'B': 2})
```

(2) collectAsMap( )：作用于 K-V 类型的 RDD 上，与 collect 不同的是，collectAsMap 函数不包含重复的 Key。对于重复的 Key，后面的元素覆盖前面的元素。

例 8.40 删除 rdd1 中重复的 Key 值，只保留一个。

Scala 语言：

```
scala> var rdd1 = sc.parallelize(List(('B', 1),('B', 2),('A', 3),('A',4),('A', 5)))
scala> var rdd2 = rdd1.collectAsMap()
scala> print(rdd2)
Map(A -> 5, B -> 2)
```

Python 语言：

```
>>> rdd1 = sc.parallelize([('B', 1),('B', 2),('A', 3),('A', 4),('A', 5)])
>>> rdd2 = rdd1.collectAsMap()
>>> print(rdd2)
{'A': 5, 'B': 2}
```

(3) lookup(Key)：作用于 K-V 类型的 RDD 上，返回指定 Key 所有的 Value。

例 8.41 查找 rdd1 中 Key 对应的 Value。

Scala 语言：

```
scala> var rdd1 = sc.parallelize(List(('B', 1),('B', 2),('A', 3),('A',4),('A', 5)))
scala> var rdd2 = rdd1.lookup('B')
scala> print(rdd2)
WrappedArray(1, 2)
```

Python 语言：

```
>>> rdd1 = sc.parallelize([('B', 1),('B', 2),('A', 3),('A', 4),('A', 5)])
>>> rdd2 = rdd1.lookup('B')
>>> print(rdd2)
[1, 2]
```

5. RDD 的其他操作

(1) glom()：将 RDD 的每个分区中的类型为 T 的元素转换成数组 Array[T]。

例 8.42 读取 rdd1 各分区元素。

Scala 语言：

```
scala> var rdd1 = sc.parallelize(List('A','B','C','D','E'), 2)
scala> var n = rdd1.glom()
scala> print(n.collect().deep.mkString(" "))
Array(A, B) Array(C, D, E)
```

Python 语言：

```
>>> rdd1 = sc.parallelize(['A','B','C','D','E'], 2)
>>> n = rdd1.glom()
>>> print(n.collect())
[['A', 'B'], ['C', 'D', 'E']]
```

(2) getNumPartitions()：读取 RDD 分区数。

例 8.43 读取 rdd1 分区数。

Scala 语言：

```
scala> var rdd1 = sc.parallelize(0 to 9, 2)
scala> var n = rdd1.getNumPartitions
scala> println(rdd1.glom().collect().deep.mkString(" "))
Array(0, 1, 2, 3, 4) Array(5, 6, 7, 8, 9)
scala> println(n)
2
```

Python 语言：

```
<<< rdd1 = sc.parallelize(range(10), 2)
<<< n = rdd1.getNumPartitions()
<<< print(rdd1.glom().collect())
[[0, 1, 2, 3, 4], [5, 6, 7, 8, 9]]
<<< print(n)
2
```

(3) coalesce(numPartitions，shuffle=False)：将 RDD 的分区数减小到 numPartitions 个。当数据集通过过滤规模减小时，使用这个操作可以提升性能。

例 8.44 对 rdd1 重新调整分区。

Scala 语言：

```
scala> var rdd1 = sc.parallelize(0 to 9, 3)
scala> var rdd2 = rdd1.coalesce(2)
scala> print(rdd1.glom().collect().deep.toList)
List(Array(0, 1, 2), Array(3, 4, 5), Array(6, 7, 8, 9))
scala> print(rdd2.glom().collect().deep.toList)
List(Array(0, 1, 2), Array(3, 4, 5, 6, 7, 8, 9))
```

Python 语言：

```
>>> rdd1 = sc.parallelize(range(9), 3)
>>> rdd2 = rdd1.coalesce(2)
>>> print(rdd1.glom().collect())
[[0, 1, 2], [3, 4, 5], [6, 7, 8]]
>>> print(rdd2.glom().collect())
[[0, 1, 2], [3, 4, 5, 6, 7, 8]]
```

（4）repartition(numPartitions)：重组数据，数据被重新随机分区为 numPartitions 个，numPartitions 可以比原来大，也可以比原来小，平衡各个分区。这一操作会将整个数据集在网络中重组。

例 8.45 对 rdd1 重新随机分区。

Scala 语言：

```
scala> var rdd1 = sc.parallelize(0 to 9, 3)
scala> var rdd2 = rdd1.repartition(3)
scala> print(rdd1.glom().collect().deep.toList)
List(Array(0, 1, 2), Array(3, 4, 5), Array(6, 7, 8, 9))
scala> print(rdd2.glom().collect().deep.toList)
List(Array(2, 5, 7), Array(0, 3, 8), Array(1, 4, 6, 9))
```

Python 语言：

```
>>> rdd1 = sc.parallelize(range(9), 3)
>>> rdd2 = rdd1.repartition(3)
>>> print(rdd1.glom().collect())
[[0, 1, 2], [3, 4, 5], [6, 7, 8]]
>>> print(rdd2.glom().collect())
[[], [0, 1, 2, 3, 4, 5], [6, 7, 8]]
```

（5）cache()：将 RDD 元素从磁盘缓存到内存，相当于 persist(MEMORY_ONLY) 函数的功能。

例 8.46 将 rdd1 元素缓存到内存。

Scala 语言：

```
scala> var rdd1 = sc.parallelize(List(("Pig", 3),("Rabbit", 5),("Python", 6)))
scala> rdd1.cache()
```

Python 语言：

```
>>> rdd1 = sc.parallelize([("Pig", 3),("Rabbit", 5),("Python", 6)])
>>> rdd1.cache()
```

（6）persist()：对 RDD 元素进行持久化操作，持久化级别包括 MEMORY_ONLY、MEMORY_ONLY_SER、MEMORY_AND_DISK、MEMORY_AND_DISK_SER 和 DISK_ONLY。RDD 还有一个方法称为 unpersist()，调用该方法可以手动把持久化的 RDD 从缓存中删除。

例 8.47 将 rdd1 元素缓存到磁盘。

Scala 语言：

```
scala> var rdd1 = sc.parallelize(List(("Pig", 3),("Rabbit", 5),("Python", 6)))
scala> rdd1.persist(org.apache.spark.storage.StorageLevel.DISK_ONLY)
```

Python 语言:

```
>>> rdd1 = sc.parallelize([("Pig", 3),("Rabbit", 5),("Python", 6)])
>>> rdd1.persist(pyspark.StorageLevel.DISK_ONLY)
```

(7) pipe(command,[envVars]):将驱动程序中的 RDD 交给 shell 处理(外部进程),RDD 元素作为标准输入传给脚本,脚本处理之后的标准输出会作为新的 RDD 返回给驱动程序。

例 8.48 显示系统时间。

Scala 语言:

```
scala> var rdd1 = sc.parallelize(List("www.swvtc.cn","swpt"),2)
scala> var rdd2 = rdd1.pipe("date")
scala> print(rdd2.collect().toList)
List(2017 年 08 月 27 日 星期日 21:26:05 CST, 2017 年 08 月 27 日 星期日 21:26:05 CST)
```

Python 语言:

```
>>> rdd1 = sc.parallelize(["www.swvtc.cn","swpt"],2)
>>> rdd2 = rdd1.pipe("date")
>>> print(rdd2.collect())
[u'Sun Aug 27 21:27:18 CST 2017', u'Sun Aug 27 21:27:18 CST 2017']
```

(8) wholeTextFiles(path: String,minPartitions: Int):读取整个文件内容,每个文件读取为一个 record,返回的是一个 key-value 的 pairRDD,key 是这个文件的路径名,value 是这个文件的具体内容。

例 8.49 读取整个文件内容。

Scala 语言:

```
scala> var rdd1 = sc.wholeTextFiles("file:///root/swpt.txt")
scala> rdd1.foreach(println)
(file:/root/swpt.txt,hello swpt
www.swvtc.cn
)
```

Python 语言:

```
>>> rdd1 = sc.wholeTextFiles("file:///root/swpt.txt")
>>> print(rdd1.collect())
[(u'file:/root/swpt.txt', u'hello swpt \nwww.swvtc.cn\n')]
```

(9) saveAsTextFile(path,compressionCodecClass=None):把 RDD 保存为文本文件。

例 8.50 把 rdd1 保存为文本文件。

Scala 语言：

```
scala> var rdd1 = sc.parallelize(List(("Pig", 3),("Rabbit", 5),("Python", 6)),2)
scala> rdd1.saveAsTextFile("file:///root/sw1")

# cat /root/sw1/*
(Pig,3)
(Rabbit,5)
(Python,6)
```

Python 语言：

```
>>> rdd1 = sc.parallelize([("Pig", 3),("Rabbit", 5),("Python", 6)],2)
>>> rdd1.saveAsTextFile("file:///root/sw1")

# cat /root/sw1/*
('Pig', 3)
('Rabbit', 5)
('Python', 6)
```

8.3.5 Spark 算法实例 1：词频统计

Spark 支持 Scala、Python、Java 和 R 等编程语言。Scala 作为 Spark 的原生语言，代码简洁而且功能完善，很多开发者都比较认可，它是业界广泛使用的 Spark 程序开发语言。Spark 也提供了 Python 的编程模型 PySpark，使得 Python 可以作为 Spark 开发语言之一。尽管现在 PySpark 还不能支持所有的 Spark API，但是以后的支持度会越来越高。Java 也可以作为 Spark 的开发语言之一，而相对于前两者逊色了很多，但是 Java 8 却很好地适应了 Spark 的开发风格。无论使用 Scala、Python，还是 Java 编程，都需要遵循 Spark 编程模型，考虑到对 Spark 平台支持的有力程度，Spark 对 Scala 语言的支持是最好的，因为它有最丰富的和最易用的编程接口。

1. 词频统计描述

词频统计（word count）的主要功能是统计文本中某单词出现的次数，其形式如图 8-7 所示。

2. 词频统计编程

（1）单词文件 sw.txt 内容如下。

```
Spark Python Scala
Hadoop Spark Python
```

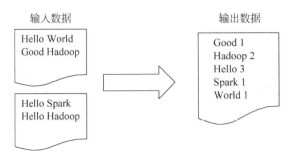

图 8-7 词频统计

(2) 上传 sw.txt 到 HDFS。

```
# hdfs dfs -mkdir /input
# hdfs dfs -put sw.txt /input
```

(3) 运行 Scala-IDE,创建项目 WCPG 和类 WordCount.scala,如图 8-8 所示。

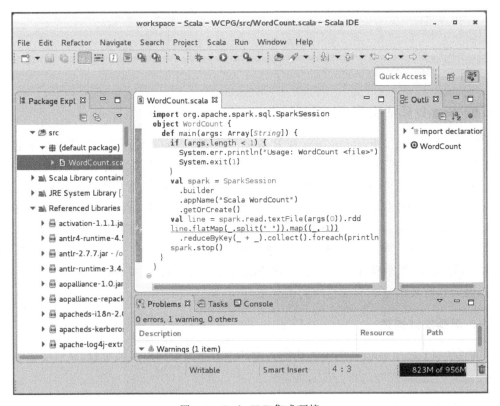

图 8-8 Scala-IDE 集成环境

Scala 语言编写的 WordCount.scala 代码如下。

```
import org.apache.spark.sql.SparkSession
object WordCount {
```

```
  def main(args: Array[String]) {
    if (args.length < 1) {
      System.err.println("Usage: WordCount <file>")
      System.exit(1)
    }
    val spark = SparkSession
      .builder
      .appName("Scala WordCount")
      .getOrCreate()
    val line = spark.read.textFile(args(0)).rdd
    line.flatMap(_.split(" ")).map((_, 1))
      .reduceByKey(_ + _).collect().foreach(println)
    spark.stop()
  }
}
```

(4) 在 Scala-IDE 中导出 WCPG.jar。

(5) 运行 WCPG.jar。

```
# spark-submit -- master spark://master:7077 \
-- class WordCount WCPG.jar /input \
> WordCount.txt
```

(6) 查看 WordCount 运行结果如下。

```
# cat WordCount.txt
(Spark,2)
(Python,2)
(Scala,1)
(Hadoop,1)
```

(7) Python 语言编写的 WordCount.py 代码如下。

```
from __future__ import print_function
import sys
from operator import add
from pyspark.sql import SparkSession
if __name__ == "__main__":
    if len(sys.argv) != 2:
        print("Usage: wordcount <file>", file=sys.stderr)
        exit(-1)
    spark = SparkSession\
        .builder\
        .appName("PythonWordCount")\
        .getOrCreate()
    lines = spark.read.text(sys.argv[1]).rdd.map(lambda r: r[0])
    counts = lines.flatMap(lambda x: x.split(' ')) \
```

```
            .map(lambda x: (x, 1)) \
            .reduceByKey(add)
output = counts.collect()
for (word, count) in output:
    print("% s: % i" % (word, count))
spark.stop()
```

(8) 运行 WordCount.py,操作如下。

```
# spark-submit -- master spark://master:7077 \
-- name PythonWC WordCount.py /input \
> WordCount.txt
```

(9) 查看 WordCount.py,运行结果如下。

```
# cat WordCount.txt
Python: 2
Spark: 2
Hadoop: 1
Scala: 1
```

8.3.6 Spark 算法实例2：相关系数

1. 相关系数概述

相关系数(correlation coefficient)是用以反映变量之间关系密切程度的统计指标。相关系数是按积差方法计算,同样以两变量与各自平均值的离差为基础,通过两个离差相乘来反映两变量之间的相关程度；着重研究线性的单相关系数。

常用的相关系数有 Pearson 相关系数(pearson correlation coefficient)和 Spearman 相关系数(spearman correlation coefficient)。Pearson 相关系数表达的是两个数值变量的线性相关性,它一般适用于正态分布,其取值范围为[-1,1]。当取值为 0 时,表示不相关；当取值为(0～-1]时,表示负相关；当取值为(0,1]时,表示正相关。Spearman 相关系数也用来表达两个变量的相关性,但是它没有 Pearson 相关系数对变量的分布要求那么严格。另外,Spearman 相关系数可以更好地用于测度变量的排序关系。

Pearson 相关系数计算公式为

$$\rho = \frac{\sum_{i=1}^{N}(x_i - \bar{x})(y_i - \bar{y})}{\sqrt{\sum_{i=1}^{N}(x_i - \bar{x})^2 \sum_{i=1}^{N}(y_i - \bar{y})^2}}$$

Spearman 相关系数计算公式为

$$\rho = 1 - \frac{6\sum_{i=1}^{N} d_i^2}{N(N^2 - 1)}$$

式中,ρ 为相关系数值,N 为数据个数。

2. 相关系数示例

例如,9 个小麦品种(分别用 A1、A2、…、A9 表示)的 6 个性状资料如表 8-1 所示,做相关系数计算并检验。

表 8-1　9 个小麦品种的 6 个性状资料

	A1	A2	A3	A4	A5	A6	A7	A8	A9
X1(冬季分蘖)	11.5	9.0	7.5	9.1	11.6	13.0	11.6	10.7	11.1
X2(株高)	95.3	97.7	110.7	89.0	88.0	87.7	79.7	119.3	87.7
X3(每穗粒数)	26.4	30.8	39.7	35.4	29.3	24.6	25.6	29.9	32.3
X4(千粒数)	39.2	46.8	39.1	35.3	37.0	44.8	43.7	38.8	35.6
X5(抽穗期)	4/9	4/17	4/17	4/18	4/20	4/19	4/19	4/19	4/18
X6(成熟期)	6/2	6/6	6/3	6/2	6/7	6/7	6/5	6/5	6/3

由相关系数计算公式可计算出 6 个性状间的相关系数,分析及检验结果如表 8-2 所示。从表 8-2 可以看出,冬季分蘖与每穗粒数之间呈现负相关($\rho = -0.8965$),即小麦冬季分蘖越多,那么每穗的小麦粒数越少;其他性状之间的关系不显著。

表 8-2　6 个性状的相关系数

性状	X1	X2	X3	X4	X5	X6
X1	1.0					
X2	−0.3863	1.0				
X3	−0.8965	0.2465	1.0			
X4	0.1448	0.19749	−0.4732	1.0		
X5	0.0425	−0.29589	0.1040	0.0259	1.0	
X6	0.4397	−0.45072	−0.4938	0.5684	0.5826	1.0

(1) 小麦性状数据文件 wheat.txt 内容如下。

```
1: 11.5, 95.3, 26.4, 39.2, 9, 2
2: 9.0, 97.7, 30.8, 46.8, 17, 6
3: 7.5, 110.7, 39.7, 39.1, 17, 3
4: 9.1, 89.0, 35.4, 35.3, 18, 2
5: 11.6, 88.0, 29.3, 37.0, 20, 7
6: 13.0, 87.7, 24.6, 44.8, 19, 7
7: 11.6, 79.7, 25.6, 43.7, 19, 5
8: 10.7, 119.3, 29.9, 38.8, 19, 5
9: 11.1, 87.7, 32.3, 35.6, 18, 3
```

(2) 上传 wheat.txt 到 HDFS。

```
# hdfs dfs - put wheat.txt /input
```

(3) Scala 语言编写的 WheatCorr.scala 代码如下。

```scala
import org.apache.spark.sql.SparkSession
import org.apache.spark.mllib.linalg.Vectors
import org.apache.spark.mllib.regression.LabeledPoint
import org.apache.spark.mllib.stat.Statistics
object KMeansEx {
  def main(args: Array[String]) {
    val spark = SparkSession.builder.appName("WheatCorrEx").getOrCreate()
    //装载数据
    val data = spark.read.textFile("/input/wheat.txt").rdd
    //将文本文件的内容转化为 LabeledPoint 类型
    val parsedData = data.map { line =>
      val parts = line.split(':')
      LabeledPoint(parts(0).toDouble,Vectors.dense(parts(1).split(',')
      .map(_.toDouble)))
    }.cache()
    val result = Statistics.corr(parsedData.map(_.features),"pearson")
    print(result)
    spark.stop()
  }
}
```

(4) 运行 WheatCorr.scala,结果如下。

```
 1.0                   -0.3863434467644298   -0.8965178785206713  ...  (6 total)
-0.3863434467644298    1.0                    0.24647054787346923 ...
-0.8965178785206713    0.24647054787346923    1.0                 ...
 0.1448076877722131    0.1973970190793534    -0.473183264108617   ...
 0.04249974394276884  -0.29589448248404826    0.10395442668359    ...
 0.43973409594661567  -0.450717829743439     -0.49384408762076265 ...
```

(5) Python 语言编写的 wheatcorr.py 代码如下。

```python
from __future__ import print_function
import sys
from operator import add
from pyspark.sql import SparkSession
import numpy as np
from pyspark.mllib.regression import LabeledPoint
from pyspark.mllib.stat import Statistics
def parsePoint(line):
    parts = line.split(':')
    values = [float(x) for x in parts[1].split(',')]
    return LabeledPoint(float(parts[0]), values)

if __name__ == "__main__":
    spark = SparkSession.builder.appName("PyWheatCorr").getOrCreate()
```

```
lines = spark.read.text("/input/wheat.txt").rdd.map(lambda r: r[0])
parsedData = lines.map(parsePoint).cache()
features = parsedData.map(lambda p : p.features)
result = Statistics.corr( features, method = "pearson")
print(result)
spark.stop()
```

(6) 运行 wheatcorr.py,结果如下。

```
[[  1.          -0.38634345  -0.89651788   0.14480769   0.04249974   0.4397341 ]
 [ -0.38634345   1.           0.24647055   0.19739702  -0.29589448  -0.45071783]
 [ -0.89651788   0.24647055   1.          -0.47318326   0.10395443  -0.49384409]
 [  0.14480769   0.19739702  -0.47318326   1.           0.02586193   0.56837048]
 [  0.04249974  -0.29589448   0.10395443   0.02586193   1.           0.58255858]
 [  0.4397341   -0.45071783  -0.49384409   0.56837048   0.58255858   1.        ]]
```

小结

Spark 是大数据处理的最佳选择,而 MLlib 是 Spark 最核心、最重要的部分。RDD 是 Spark 的基础,需要掌握 RDD 的基本工作原理和特性。本章详细介绍了 RDD 中最基本的使用,包括 RDD 转化(Transformation)和行动(Action)的 API。Spark 算法提供两个实例:词频统计是集群算法中最典型的算法;分析 9 个小麦品种的 6 个性状,并对其进行相关系数的计算。本章代码使用 Scala 和 Python 两种语言编写,学习本章之前需要了解 Scala 和 Python 两种语言的基本语法。

习题

1. 简述 Spark 开源框架及其特点。
2. 简述 Spark 生态系统及其主要组件。
3. 简述 RDD 的基本工作原理和特性。
4. 使用蒙特卡罗方法计算圆周率 π 值。
5. 使用 Spark 实现词频统计和相关系数计算。

第 9 章

Spark机器学习

机器学习(Machine Learning，ML)是人工智能中一个重要的研究领域，一直受到人工智能及认知心理学家的普遍关注。如何从大量数据中提取有价值的信息，成为非常重要的课题，机器学习就是一种能从无序的数据中提取有用信息的工具。

9.1 机器学习概述

视频讲解

机器学习是一门多领域交叉学科，涉及概率论、统计学、逼近论、凸分析和算法复杂度理论等多门学科，专门研究计算机怎样模拟或实现人类的学习行为，以获取新的知识或技能，重新组织已有的知识结构使之不断改善自身的性能。机器学习是用数据或以往的经验优化计算机程序的性能标准。机器学习使计算机能够模拟人的学习行为，自动地通过学习获取知识和技能，不断改善性能，实现自我完善。

机器学习是一个庞大的家族体系，涉及众多算法、任务和学习理论。

(1) 按任务类型分，机器学习模型可以分为回归模型、分类模型和结构化学习模型。回归模型又称预测模型，输出的是一个不能枚举的数值；分类模型又分为二分类模型和多分类模型，常见的二分类问题有垃圾邮件过滤，常见的多分类问题有文档自动归类；结构化学习模型的输出不再是一个固定长度的值，如图片语义分析，输出是图片的文字描述。

(2) 从方法的角度分，机器学可模型可以分为线性模型和非线性模型。线性模型较为简单，但作用不可忽视，线性模型是非线性模型的基础，很多非线性模型都是在线性模型的基础上变换而来的；非线性模型又可以分为传统机器学习模型(SVM、KNN、决策树等)和深度学习模型。

(3) 按照学习理论分，机器学习模型可以分为有监督学习、半监督学习、无监督学习、迁移学习和强化学习。当训练样本带有标签时是有监督学习；训练样本部分有标签，部分无标签时是半监督学习；训练样本全部无标签时是无监督学习。迁移学习是把已经训练好的模型参数迁移到新的模型上以帮助新模型训练；强化学习是一个学习最优策略，可以让本体在特定环境中，根据当前状态做出行动，从而获得最大回报。强化学习和有监督学习最大的不同是，每次的决定没有对与错，而是希望获得最多的累计奖励。

9.1.1 机器学习的发展史

机器学习是人工智能研究较为"年轻"的分支，它的发展过程大体上可分为4个时期。

（1）第一阶段是在20世纪50年代中叶到60年代中叶，属于机器学习的热烈时期。

（2）第二阶段是在20世纪60年代中叶至70年代中叶，称为机器学习的冷静时期。

（3）第三阶段是从20世纪70年代中叶至80年代中叶，称为机器学习的复兴时期。

（4）第四阶段始于1986年。

机器学习进入新阶段的重要表现有以下几个方面。

（1）机器学习已成为新的边缘学科，并在高校形成一门课程。它综合应用心理学、生物学和神经生理学，以及数学、自动化和计算机科学形成机器学习理论基础。

（2）结合各种学习方法，取长补短的多种形式的集成学习系统研究正在兴起。特别是连接学习符号学习的耦合，因其可以更好地解决连续性信号处理中知识与技能的获取与求精问题而受到重视。

（3）机器学习与人工智能各种基础问题的统一性观点正在形成。例如，学习与问题求解结合进行，知识表达与学习的观点产生了通用智能系统SOAR的组块学习。类比学习与问题求解结合的基于案例方法已成为经验学习的重要方向。

（4）各种学习方法的应用范围不断扩大，一部分已形成商品。归纳学习的知识获取工具已在诊断分类型专家系统中广泛使用；连接学习在声、图、文识别中占优势；分析学习已用于设计综合型专家系统；遗传算法与强化学习在工程控制中有较好的应用前景；与符号系统耦合的神经网络连接学习将在企业的智能管理与智能机器人运动规划中发挥作用。

（5）与机器学习有关的学术活动空前活跃。国际上除每年一次的机器学习研讨会外，还有计算机学习理论会议及遗传算法会议等。

9.1.2 机器学习步骤

机器学习是一类算法的总称，这些算法企图从大量历史数据中挖掘出隐含的规律，并用于预测或分类。更具体地说，机器学习可以看作是寻找一个函数，输入是样本数据，输出是期望的结果，只是这个函数过于

视频讲解

复杂，以至于不太方便形式化表达。需要注意的是，机器学习的目标是使学到的函数很好地适用于"新样本"，而不仅是在训练样本上表现很好。学到的函数适用于新样本的能力，称为泛化（generalization）能力。

学习一个好的函数，过程分为以下3步。

（1）选择一个合适的模型。这通常需要依据实际问题而定，针对不同的问题和任务

需要选取恰当的模型,模型就是一组函数的集合。

(2) 判断一个函数的好坏。这需要确定一个衡量标准,也就是通常说的损失函数(loss function),损失函数的确定也需要依据具体问题而定,如回归问题一般采用欧几里得距离,分类问题一般采用交叉熵代价函数。

(3) 找出最好的函数。如何从众多函数中最快地找出最好的那一个,这一步是最大的难点,做到又快又准往往不是一件容易的事情。常用的方法有梯度下降法、最小二乘法等。机器学习得到最好的函数后,需要在新样本上进行测试,只有在新样本上表现很好,才算是一个好的函数。

9.2 Spark MLlib 概述

MLlib 是 Spark 的机器学习(ML)库,其目标是使实用的机器学习具有可扩展性并且变得容易。它提供了以下工具。

(1) ML 算法(ML Algorithms):常用的学习算法,如分类、回归、聚类和协同过滤。

(2) 特征(Featurization):特征提取、变换、降维和选择。

(3) 管道(Pipelines):用于构建、评估和调整 ML Pipelines 的工具。

(4) 持久性(Persistence):保存和加载算法、模型和 Pipelines。

(5) 实用(Utilities):线性代数、统计学、数据处理等。

MLlib 目前分为两个代码包。

(1) spark.mllib 包含基于 RDD 的原始算法 API。

(2) spark.ml 提供了基于 DataFrames 高层次的 API,可以用来构建机器学习管道。

从 Spark 2.0 版本开始,spark.mllib 包中基于 RDD 的 API 已进入维护模式。Spark 的主要机器学习 API 现在是 spark.ml 包中基于 DataFrame 的 API。基于 DataFrame 的 API 有如下优点。

(1) DataFrames 提供比 RDD 更加用户友好的 API。DataFrames 的许多好处包括 Spark 数据源、SQL/DataFrames 查询、Tungsten 和 Catalyst 优化及跨语言的统一 API。

(2) 基于 DataFrames 的 MLlib API 跨 ML 算法和多种语言提供统一的 API。

(3) DataFrames 有助于实用的 ML 管道,特别是功能转换。

MLlib 使用线性代数包 Breeze,这取决于 netlib-java 进行优化数字处理。如果 native libraries1(本地库)在运行时不可用,那么将会看到一条警告消息。

9.2.1 数据类型

MLlib 支持单机上的本地向量存储和矩阵,也支持由一个或多个 RDD 支持的分布式矩阵。本地向量和本地矩阵是用作公共接口的简单数据模型,由 Breeze 提供基本的线性代数运算。在监督学习中使用的训练示例在 MLlib 中称为 Labeled Point。

1. Local Vector

Local Vector(本地向量)具有整数类型的行,从 0 开始索引、double 类型值,存储在单台机器上。MLlib 支持两种类型的本地向量：密集向量和稀疏向量。密集向量由一个 double 类型数组组成，而一个稀疏向量必须由索引和一个 double 类型向量组成。例如，矢量(1.0,0.0,3.0)，密集向量表示为[1.0,0.0,3.0]，稀疏向量表示为(3,[0,2],[1.0,3.0])，其中，3 是向量的大小。

例 9.1 本地向量。

Scala 语言：

```
# spark - shell
scala> import org.apache.spark.mllib.linalg.{Vector, Vectors}
scala> val dv: Vector = Vectors.dense(1.0, 0.0, 3.0)
scala> val sv1: Vector = Vectors.sparse(3, Array(0, 2), Array(1.0, 3.0))
scala> val sv2: Vector = Vectors.sparse(3, Seq((0, 1.0), (2, 3.0)))
scala> println(sv1)
(3,[0,2],[1.0,3.0])
scala> println(sv2)
(3,[0,2],[1.0,3.0])
```

Python 语言：

```
# pyspark
>>> import numpy as np
>>> import scipy.sparse as sps
>>> from pyspark.mllib.linalg import Vectors
>>> dv1 = np.array([1.0, 0.0, 3.0])
>>> dv2 = [1.0, 0.0, 3.0]
>>> sv1 = Vectors.sparse(3, [0, 2], [1.0, 3.0])
>>> sv2 = sps.csc_matrix((np.array([1.0, 3.0]), np.array([0, 2]),
    np.array([0, 2])), shape = (3, 1))
>>> print(sv1)
(3,[0,2],[1.0,3.0])
>>> print(sv2)
  (0, 0)    1.0
  (2, 0)    3.0
```

2. Labeled Point

Labeled Point(标签向量)是一个本地向量、密集向量或者稀疏向量，并且带有一个标签。在 MLlib 中，标签向量用于监督学习算法。使用 double 值来存储标签，因此可以在回归和分类中使用标签向量。对于二进制分类，标签应为 0(－)或 1(＋)。对于多类分类，标签应该是从 0 开始的类索引：0,1,2,…。

例 9.2 标签向量。

Scala 语言：

```
# spark - shell
scala> import org.apache.spark.mllib.linalg.Vectors
scala> import org.apache.spark.mllib.regression.LabeledPoint
```

```
scala> val pos = LabeledPoint(1.0, Vectors.dense(1.0, 0.0, 3.0))
scala> val neg = LabeledPoint(0.0, Vectors.sparse(3, Array(0, 2), Array(1.0, 3.0)))
scala> print(pos)
(1.0,[1.0,0.0,3.0])
scala> print(neg)
(0.0,(3,[0,2],[1.0,3.0]))
```

Python 语言：

```
# pyspark
>>> from pyspark.mllib.linalg import SparseVector
>>> from pyspark.mllib.regression import LabeledPoint
>>> pos = LabeledPoint(1.0, [1.0, 0.0, 3.0])
>>> neg = LabeledPoint(0.0, SparseVector(3, [0, 2], [1.0, 3.0]))
>>> print(pos)
(1.0,[1.0,0.0,3.0])
>>> print(neg)
(0.0,(3,[0,2],[1.0,3.0]))
```

3. Local Matrix

Local Matrix(本地矩阵)具有整数类型的行和列索引及 double 类型值，存储在单台机器上。MLlib 支持密集矩阵，其输入值按照列 column-major 顺序存储在单个 double 数组中，稀疏矩阵是其非零值按照主要列顺序以压缩稀疏列(CSC)格式存储。例如，以下密集矩阵

$$\begin{pmatrix} 1.0 & 2.0 \\ 3.0 & 4.0 \\ 5.0 & 6.0 \end{pmatrix}$$

存储在具有矩阵大小(3,2)的一维数组[1.0,3.0,5.0,2.0,4.0,6.0]中。

例 9.3 本地矩阵。

Scala 语言：

```
# spark-shell
scala> import org.apache.spark.mllib.linalg.{Matrix, Matrices}
scala> val dm: Matrix = Matrices.dense(3, 2, Array(1.0, 3.0, 5.0, 2.0, 4.0, 6.0))
scala> val sm: Matrix = Matrices.sparse(3, 2, Array(0, 1, 3), Array(0, 2, 1),
         Array(9, 6, 8))
scala> print(dm)
1.0  2.0
3.0  4.0
5.0  6.0
scala> print(sm)
3 x 2 CSCMatrix
(0,0) 9.0
(2,1) 6.0
(1,1) 8.0
```

Python 语言：

```
# pyspark
>>> from pyspark.mllib.linalg import Matrix, Matrices
>>> dm2 = Matrices.dense(3, 2, [1, 3, 5, 2, 4, 6])
>>> sm = Matrices.sparse(3, 2, [0, 1, 3], [0, 2, 1], [9, 6, 8])
>>> print(dm2)
DenseMatrix([[ 1.,  2.],
             [ 3.,  4.],
             [ 5.,  6.]])
>>> print(sm)
3 X 2 CSCMatrix
(0,0) 9.0
(2,1) 6.0
(1,1) 8.0
```

4. RowMatrix

RowMatrix 是面向行的分布式矩阵，没有有意义的行索引，由其行的 RDD 支持，其中每行是本地向量。由于每一行都由本地向量表示，因此列数受整数范围的限制，但实际上它应该小得多。

例 9.4 RowMatrix。

Scala 语言：

```
# spark-shell
scala> import org.apache.spark.rdd.RDD
scala> import org.apache.spark.mllib.linalg.{Vector,Vectors}
scala> import org.apache.spark.mllib.linalg.distributed.RowMatrix
scala> val dv1 : Vector = Vectors.dense(1.0,2.0,3.0)
scala> val dv2 : Vector = Vectors.dense(2.0,3.0,4.0)
scala> val dv3 : Vector = Vectors.dense(3.0,4.0,5.0)
scala> val rows : RDD[Vector] = sc.parallelize(Array(dv1,dv2,dv3))
scala> val mat: RowMatrix = new RowMatrix(rows)
scala> val m = mat.numRows()
m: Long = 3
scala> val n = mat.numCols()
n: Long = 3
scala> mat.rows.foreach(println)
[3.0,4.0,5.0]
[2.0,3.0,4.0]
[1.0,2.0,3.0]
scala> val qrResult = mat.tallSkinnyQR(true)
```

Python 语言：

```
# pyspark
>>> from pyspark.mllib.linalg.distributed import RowMatrix
>>> rows = sc.parallelize([[1, 2, 3], [4, 5, 6], [7, 8, 9], [10, 11, 12]])
```

```
>>> mat = RowMatrix(rows)
>>> m = mat.numRows()
>>> n = mat.numCols()
>>> rowsRDD = mat.rows
>>> m
4L
>>> n
3L
>>> print(rowsRDD.collect())
[DenseVector([1.0, 2.0, 3.0]), DenseVector([4.0, 5.0, 6.0]),
DenseVector([7.0, 8.0, 9.0]), DenseVector([10.0, 11.0, 12.0])]
```

5. IndexedRowMatrix

IndexedRowMatrix 类似于 RowMatrix,但具有有意义的行索引。它由索引行的 RDD 支持,因此每行由其长整类型索引和本地向量表示。

例 9.5 IndexedRowMatrix。

Scala 语言:

```
# spark-shell
scala> import org.apache.spark.mllib.linalg.distributed.{IndexedRow,
    IndexedRowMatrix, RowMatrix}
scala> val idxr1 = IndexedRow(1,dv1)
scala> val idxr2 = IndexedRow(2,dv2)
scala> val rows: RDD[IndexedRow] = sc.parallelize(Array(idxr1,idxr2))
scala> val mat: IndexedRowMatrix = new IndexedRowMatrix(rows)
scala> val m = mat.numRows()
m: Long = 3
scala> val n = mat.numCols()
n: Long = 3
scala> val rowMat: RowMatrix = mat.toRowMatrix()
scala> mat.rows.foreach(println)
IndexedRow(2,[2.0,3.0,4.0])
IndexedRow(1,[1.0,2.0,3.0])
scala> rowMat.rows.foreach(println)
[1.0,2.0,3.0]
[2.0,3.0,4.0]
```

Python 语言:

```
# pyspark
>>> from pyspark.mllib.linalg.distributed import IndexedRow, IndexedRowMatrix
>>> indexedRows = sc.parallelize([IndexedRow(0, [1, 2, 3]),
...                               IndexedRow(1, [4, 5, 6]),
...                               IndexedRow(2, [7, 8, 9]),
...                               IndexedRow(3, [10, 11, 12])])
>>> indexedRows = sc.parallelize([(0, [1, 2, 3]), (1, [4, 5, 6]),
...                               (2, [7, 8, 9]), (3, [10, 11, 12])])
```

```
>>> mat = IndexedRowMatrix(indexedRows)
>>> m = mat.numRows()
>>> n = mat.numCols()
>>> rowsRDD = mat.rows
>>> rowMat = mat.toRowMatrix()
>>> m
4L
>>> n
3L
>>> print(rowsRDD.collect())
[IndexedRow(0, [1.0,2.0,3.0]), IndexedRow(1, [4.0,5.0,6.0]),
IndexedRow(2, [7.0,8.0,9.0]), IndexedRow(3, [10.0,11.0,12.0])]
>>> print(rowMat.rows.collect())
[DenseVector([1.0, 2.0, 3.0]), DenseVector([4.0, 5.0, 6.0]),
DenseVector([7.0, 8.0, 9.0]), DenseVector([10.0, 11.0, 12.0])]
```

6. CoordinateMatrix

CoordinateMatrix 是由 RDD 支持的分布式矩阵。每个条目都是一个元组(i：Long，j：Long，value：Double)，其中 i 是行索引，j 是列索引，value 是条目值。只有当矩阵的两个维度都很大且矩阵非常稀疏时，才应使用 CoordinateMatrix。

例 9.6 CoordinateMatrix

Scala 语言：

```
# spark-shell
scala> import org.apache.spark.mllib.linalg.distributed.{CoordinateMatrix,
       MatrixEntry}
scala> val ent1 = new MatrixEntry(0,1,0.5)
scala> val ent2 = new MatrixEntry(2,2,1.8)
scala> val entries: RDD[MatrixEntry] = sc.parallelize(Array(ent1,ent2))
scala> val mat: CoordinateMatrix = new CoordinateMatrix(entries)
scala> val m = mat.numRows()
m: Long = 3
scala> val n = mat.numCols()
n: Long = 3
scala> val indexedRowMatrix = mat.toIndexedRowMatrix()
scala> mat.entries.foreach(println)
MatrixEntry(2,2,1.8)
MatrixEntry(0,1,0.5)
scala> indexedRowMatrix.rows.foreach(println)
IndexedRow(2,(3,[2],[1.8]))
IndexedRow(0,(3,[1],[0.5]))
```

Python 语言：

```
# pyspark
>>> from pyspark.mllib.linalg.distributed import CoordinateMatrix, MatrixEntry
>>> entries = sc.parallelize([MatrixEntry(0, 0, 1.2), MatrixEntry(1, 0, 2.1),
```

```
            MatrixEntry(2, 1, 3.7)])
>>> entries = sc.parallelize([(0, 0, 1.2), (1, 0, 2.1), (2, 1, 3.7)])
>>> mat = CoordinateMatrix(entries)
>>> m = mat.numRows()
>>> n = mat.numCols()
>>> entriesRDD = mat.entries
>>> rowMat = mat.toRowMatrix()
>>> indexedRowMat = mat.toIndexedRowMatrix()
>>> blockMat = mat.toBlockMatrix()
>>> m
3L
>>> n
2L
>>> print(entriesRDD.collect())
[MatrixEntry(0, 0, 1.2), MatrixEntry(1, 0, 2.1), MatrixEntry(2, 1, 3.7)]
>>> print(rowMat.rows.collect())
[SparseVector(2, {0: 1.2}), SparseVector(2, {0: 2.1}),
SparseVector(2, {1: 3.7})]
>>> print(indexedRowMat.rows.collect())
[IndexedRow(0, (2,[0],[1.2])), IndexedRow(1, (2,[0],[2.1])),
IndexedRow(2, (2,[1],[3.7]))]
>>> print(blockMat.toLocalMatrix())
DenseMatrix([[ 1.2,  0. ],
             [ 2.1,  0. ],
             [ 0. ,  3.7]])
```

7. BlockMatrix

BlockMatrix 是由一个 MatrixBlocks 类型的 RDD 支持的分布式矩阵，其中，MatrixBlock 是一个元组((Int,Int),Matrix)，(Int,Int)是块索引，而 Matrix 索引指定的子矩阵，其大小为 rowsPerBlock * colsPerBlock。BlockMatrix 支持的方法有 add 和 multiply。BlockMatrix 还有一个辅助方法 validate，可用于检查 BlockMatrix 是否设置正确。

例 9.7 BlockMatrix。

Scala 语言：

```
# spark-shell
scala> import org.apache.spark.mllib.linalg.distributed.{BlockMatrix,
       CoordinateMatrix, MatrixEntry}
scala> val ent1 = new MatrixEntry(0,0,1)
scala> val ent2 = new MatrixEntry(1,1,1)
scala> val ent3 = new MatrixEntry(2,0,-1)
scala> val ent4 = new MatrixEntry(2,1,2)
scala> val ent5 = new MatrixEntry(2,2,1)
scala> val ent6 = new MatrixEntry(3,0,1)
scala> val ent7 = new MatrixEntry(3,1,1)
scala> val ent8 = new MatrixEntry(3,3,1)
scala> val entries : RDD[MatrixEntry] = sc.parallelize(Array(ent1,ent2,
       ent3,ent4,ent5,ent6,ent7,ent8))
```

```
scala> val coordMat: CoordinateMatrix = new CoordinateMatrix(entries)
scala> val matA: BlockMatrix = coordMat.toBlockMatrix().cache()
scala> matA.validate()
scala> val ata = matA.transpose.multiply(matA)
scala> entries.foreach(println)
MatrixEntry(0,0,1.0)
MatrixEntry(1,1,1.0)
MatrixEntry(3,1,1.0)
MatrixEntry(3,3,1.0)
MatrixEntry(2,0,-1.0)
MatrixEntry(2,1,2.0)
MatrixEntry(2,2,1.0)
MatrixEntry(3,0,1.0)
scala> coordMat.entries.foreach(print)
MatrixEntry(2,0,-1.0)MatrixEntry(0,0,1.0)MatrixEntry(1,1,1.0)
MatrixEntry(2,1,2.0)MatrixEntry(2,2,1.0)MatrixEntry(3,0,1.0)
MatrixEntry(3,1,1.0)MatrixEntry(3,3,1.0)
scala> print(matA.toLocalMatrix)
  1.0   0.0   0.0   0.0
  0.0   1.0   0.0   0.0
 -1.0   2.0   1.0   0.0
  1.0   1.0   0.0   1.0
scala> matA.numColBlocks
res40: Int = 1
scala>  matA.numRowBlocks
res41: Int = 1
scala> print(ata.toLocalMatrix)
  3.0   -1.0   -1.0   1.0
 -1.0    6.0    2.0   1.0
 -1.0    2.0    1.0   0.0
  1.0    1.0    0.0   1.0
```

Python 语言：

```
# pyspark
>>> from pyspark.mllib.linalg import Matrices
>>> from pyspark.mllib.linalg.distributed import BlockMatrix
>>> blocks = sc.parallelize([((0, 0), Matrices.dense(3, 2, [1, 2, 3, 4, 5, 6])),
...                          ((1, 0), Matrices.dense(3, 2, [7, 8, 9, 10, 11, 12]))])
>>> mat = BlockMatrix(blocks, 3, 2)
>>> m = mat.numRows()
>>> n = mat.numCols()
>>> blocksRDD = mat.blocks
>>> ocalMat = mat.toLocalMatrix()
>>> indexedRowMat = mat.toIndexedRowMatrix()
>>> coordinateMat = mat.toCoordinateMatrix()
>>> m
6L
```

```
>>> n
2L
>>> print(blocksRDD.collect())
[((0, 0), DenseMatrix(3, 2, [1.0, 2.0, 3.0, 4.0, 5.0, 6.0], 0)),
((1, 0), DenseMatrix(3, 2, [7.0, 8.0, 9.0, 10.0, 11.0, 12.0], 0))]
>>> print(ocalMat)
DenseMatrix([[ 1.,  4.],
             [ 2.,  5.],
             [ 3.,  6.],
             [ 7., 10.],
             [ 8., 11.],
             [ 9., 12.]])
>>> print(indexedRowMat.rows.collect())
[IndexedRow(0, [1.0,4.0]), IndexedRow(1, [2.0,5.0]), IndexedRow(2, [3.0,6.0]),
IndexedRow(3, [7.0,10.0]), IndexedRow(4, [8.0,11.0]), IndexedRow(5, [9.0,12.0])]
>>> print(coordinateMat.entries.collect())
[MatrixEntry(0, 0, 1.0), MatrixEntry(1, 0, 2.0), MatrixEntry(2, 0, 3.0),
MatrixEntry(0, 1, 4.0), MatrixEntry(1, 1, 5.0), MatrixEntry(2, 1, 6.0),
MatrixEntry(3, 0, 7.0), MatrixEntry(4, 0, 8.0), MatrixEntry(5, 0, 9.0),
MatrixEntry(3, 1, 10.0), MatrixEntry(4, 1, 11.0), MatrixEntry(5, 1, 12.0)]
```

9.2.2 基本统计——基于 DataFrame 的 API

视频讲解

基于 DataFrame 的 API 使用的是 spark.ml,这里给出 3 个示例:相关性(correlation)、假设检验(hypothesis testing)和累积器(summarizer)。

1. 相关性

计算两个数据系列之间的相关性是统计学中的常见操作。在 spark.ml 中,提供了计算多个系列之间的相关性的方法。支持的相关方法目前是 Pearson 和 Spearman 的相关性。

Correlation 使用指定的方法计算 Vector 的输入数据集的相关矩阵。输出将是一个 DataFrame,它包含向量列的相关矩阵。

例 9.8 相关性。

Scala 语言:

```
# spark - shell
scala> import org.apache.spark.ml.linalg.{Matrix, Vectors}
scala> import org.apache.spark.ml.stat.Correlation
scala> import org.apache.spark.sql.Row
scala> val data = Seq(
     |   Vectors.sparse(4, Seq((0, 1.0), (3, -2.0))),
     |   Vectors.dense(4.0, 5.0, 0.0, 3.0),
     |   Vectors.dense(6.0, 7.0, 0.0, 8.0),
     |   Vectors.sparse(4, Seq((0, 9.0), (3, 1.0)))
     | )
scala> val df = data.map(Tuple1.apply).toDF("features")
```

```
scala> val Row(coeff1: Matrix) = Correlation.corr(df, "features").head
scala> println(s"Pearson correlation matrix:\n $ coeff1")
Pearson correlation matrix:
1.0                    0.055641488407465814   NaN   0.4004714203168137
0.055641488407465814   1.0                    NaN   0.9135958615342522
NaN                    NaN                    1.0   NaN
0.4004714203168137     0.9135958615342522     NaN   1.0
scala> val Row(coeff2: Matrix) = Correlation.corr(df, "features", "spearman").head
scala> println(s"Spearman correlation matrix:\n $ coeff2")
Spearman correlation matrix:
1.0                    0.10540925533894532    NaN   0.40000000000000174
0.10540925533894532    1.0                    NaN   0.9486832980505141
NaN                    NaN                    1.0   NaN
0.40000000000000174    0.9486832980505141     NaN   1.0
```

Python 语言：

```
# pyspark
>>> from pyspark.ml.linalg import Vectors
>>> from pyspark.ml.stat import Correlation
>>> data = [(Vectors.sparse(4, [(0, 1.0), (3, -2.0)]),),
...         (Vectors.dense([4.0, 5.0, 0.0, 3.0]),),
...         (Vectors.dense([6.0, 7.0, 0.0, 8.0]),),
...         (Vectors.sparse(4, [(0, 9.0), (3, 1.0)]),)]
>>> df = spark.createDataFrame(data, ["features"])
>>> r1 = Correlation.corr(df, "features").head()
>>> print("Pearson correlation matrix:\n" + str(r1[0]))
Pearson correlation matrix:
DenseMatrix([[ 1.        ,  0.05564149,         nan,  0.40047142],
             [ 0.05564149,  1.        ,         nan,  0.91359586],
             [        nan,         nan,  1.        ,         nan],
             [ 0.40047142,  0.91359586,         nan,  1.        ]])
>>> r2 = Correlation.corr(df, "features", "spearman").head()
>>> print("Spearman correlation matrix:\n" + str(r2[0]))
Spearman correlation matrix:
DenseMatrix([[ 1.        ,  0.10540926,         nan,  0.4       ],
             [ 0.10540926,  1.        ,         nan,  0.9486833 ],
             [        nan,         nan,  1.        ,         nan],
             [ 0.4       ,  0.9486833 ,         nan,  1.        ]])
```

2. 假设检验

假设检验是统计学中一种强有力的工具，用于确定结果是否具有统计显著性，无论该结果是否偶然发生。spark.ml 目前支持 Pearson 的 Chi-squared(x^2) 独立测试。

ChiSquareTest 针对标签对每个功能进行 Pearson 独立测试。对于每个特征，将"特征,标签"对转换为应急矩阵，对其计算卡方统计量。所有标签和特征值必须是分类的。

例 9.9 假设检验。

Scala 语言：

```
# spark-shell
scala> import org.apache.spark.ml.linalg.{Vector, Vectors}
scala> import org.apache.spark.ml.stat.ChiSquareTest
scala> val data = Seq(
     |     (0.0, Vectors.dense(0.5, 10.0)),
     |     (0.0, Vectors.dense(1.5, 20.0)),
     |     (1.0, Vectors.dense(1.5, 30.0)),
     |     (0.0, Vectors.dense(3.5, 30.0)),
     |     (0.0, Vectors.dense(3.5, 40.0)),
     |     (1.0, Vectors.dense(3.5, 40.0))
     | )
scala> val df = data.toDF("label", "features")
scala> val chi = ChiSquareTest.test(df, "features", "label").head
scala> println(s"pValues = ${chi.getAs[Vector](0)}")
pValues = [0.6872892787909721,0.6822703303362126]
scala> println(s"degreesOfFreedom ${chi.getSeq[Int](1).mkString("[", ",", "]")}")
degreesOfFreedom [2,3]
scala> println(s"statistics ${chi.getAs[Vector](2)}")
statistics [0.75,1.5]
```

Python 语言：

```
# pyspark
>>> from pyspark.ml.linalg import Vectors
>>> from pyspark.ml.stat import ChiSquareTest
>>> data = [(0.0, Vectors.dense(0.5, 10.0)),
...         (0.0, Vectors.dense(1.5, 20.0)),
...         (1.0, Vectors.dense(1.5, 30.0)),
...         (0.0, Vectors.dense(3.5, 30.0)),
...         (0.0, Vectors.dense(3.5, 40.0)),
...         (1.0, Vectors.dense(3.5, 40.0))]
>>> df = spark.createDataFrame(data, ["label", "features"])
>>> r = ChiSquareTest.test(df, "features", "label").head()
>>> print("pValues: " + str(r.pValues))
pValues: [0.687289278791,0.682270330336]
>>> print("degreesOfFreedom: " + str(r.degreesOfFreedom))
degreesOfFreedom: [2, 3]
>>> print("statistics: " + str(r.statistics))
statistics: [0.75,1.5]
```

3. 累积器

通过 Summarizer 为 Dataframe 提供矢量列摘要统计，可用指标是列的最大值、最小值、平均值、方差、非零数以及总计数。

例 9.10 累积器。

Scala 语言：

```
# spark-shell
scala> import org.apache.spark.ml.linalg.{Vector, Vectors}
scala> import org.apache.spark.ml.stat.Summarizer
scala> val data = Seq(
     |    (Vectors.dense(2.0, 3.0, 5.0), 1.0),
     |    (Vectors.dense(4.0, 6.0, 7.0), 2.0)
     | )
scala> val df = data.toDF("features", "weight")
scala> val (meanVal, varianceVal) =
         df.select(Summarizer.metrics("mean", "variance")
         .summary($"features", $"weight")
         .as("summary")).select("summary.mean", "summary.variance")
         .as[(Vector, Vector)].first()
scala> println(s"with weight: mean = ${meanVal}, variance = ${varianceVal}")
with weight: mean = [3.333333333333333,5.0,6.333333333333333],
        variance = [2.0,4.5,2.0]
scala> val (meanVal2, varianceVal2) = df.select(Summarizer.mean($"features"),
        Summarizer.variance($"features"))
        .as[(Vector, Vector)].first()
scala> println(s"without weight: mean = ${meanVal2}, sum = ${varianceVal2}")
without weight: mean = [3.0,4.5,6.0], sum = [2.0,4.5,2.0]
```

Python 语言：

```
# pyspark
>>> from pyspark.ml.stat import Summarizer
>>> from pyspark.sql import Row
>>> from pyspark.ml.linalg import Vectors
>>> df = sc.parallelize([Row(weight=1.0, features=Vectors.dense(1.0, 1.0, 1.0)),
...         Row(weight=0.0, features=Vectors.dense(1.0, 2.0, 3.0))]).toDF()
>>> summarizer = Summarizer.metrics("mean", "count")
>>> df.select(summarizer.summary(df.features, df.weight)).show(truncate=False)
+-----------------------------+
|aggregate_metrics(features, weight)|
+-----------------------------+
|[[1.0,1.0,1.0], 1]           |
+-----------------------------+
>>> df.select(summarizer.summary(df.features)).show(truncate=False)
+-----------------------------+
|aggregate_metrics(features, 1.0) |
+-----------------------------+
|[[1.0,1.5,2.0], 2]           |
+-----------------------------+
>>> df.select(Summarizer.mean(df.features, df.weight)).show(truncate=False)
```

```
+----------+
|mean(features) |
+----------+
|[1.0,1.0,1.0] |
+----------+
>>> df.select(Summarizer.mean(df.features)).show(truncate = False)
+----------+
|mean(features) |
+----------+
|[1.0,1.5,2.0] |
+----------+
```

9.2.3 基本统计——基于 RDD 的 API

视频讲解

基于 RDD 的 API 使用的是 spark.mllib，这里给出 6 个示例：概要统计(summary statistics)、相关性(correlations)、分层抽样(stratified sampling)、假设检验(hypothesis testing)、随机数据生成(random data generation)和核密度估计(kernel density estimation)。

1. 概要统计

通过 Statistics 中提供的函数 colStats 为 RDD[Vector]提供列摘要统计信息。

colStats()返回 MultivariateStatisticalSummary 的一个实例，其中包含列的 max、min、mean、variance、非零数以及总计数。

例 9.11 概要统计。

Scala 语言：

```
# spark - shell
scala > import org.apache.spark.mllib.linalg.Vectors
scala > import org.apache.spark.mllib.stat.{MultivariateStatisticalSummary,
        Statistics}
scala > val observations = sc.parallelize(
        |   Seq(
        |     Vectors.dense(1.0, 10.0, 100.0),
        |     Vectors.dense(2.0, 20.0, 200.0),
        |     Vectors.dense(3.0, 30.0, 300.0)
        |   )
        | )
scala > val summary: MultivariateStatisticalSummary =
        Statistics.colStats(observations)
scala > println(summary.mean)
[2.0,20.0,200.0]
scala > println(summary.variance)
[1.0,100.0,10000.0]
scala > println(summary.numNonzeros)
[3.0,3.0,3.0]
```

Python 语言：

```
# pyspark
>>> import numpy as np
>>> from pyspark.mllib.stat import Statistics
>>> mat = sc.parallelize(
...     [np.array([1.0, 10.0, 100.0]), np.array([2.0, 20.0, 200.0]),
     np.array([3.0, 30.0, 300.0])]
... )
>>> summary = Statistics.colStats(mat)
>>> print(summary.mean())
[  2.  20.  200.]
>>> print(summary.variance())
[ 1.00000000e+00   1.00000000e+02   1.00000000e+04]
>>> print(summary.numNonzeros())
[ 3.  3.  3.]
```

2. 相关性

Statistics 提供了计算序列之间相关性的方法。根据输入类型，有两个 RDD[Double] 或 RDD[Vector]，输出将分别为 Double 或相关矩阵。

例 9.12 相关性。

Scala 语言：

```
# spark-shell
scala> import org.apache.spark.mllib.linalg._
scala> import org.apache.spark.mllib.stat.Statistics
scala> import org.apache.spark.rdd.RDD
scala> val seriesX: RDD[Double] = sc.parallelize(Array(1, 2, 3, 3, 5))
scala> val seriesY: RDD[Double] = sc.parallelize(Array(11, 22, 33, 33, 555))
scala> val correlation: Double = Statistics.corr(seriesX, seriesY, "pearson")
scala> println(s"Correlation is: $correlation")
Correlation is: 0.8500286768773001

scala> val data: RDD[Vector] = sc.parallelize(
    |   Seq(
    |     Vectors.dense(1.0, 10.0, 100.0),
    |     Vectors.dense(2.0, 20.0, 200.0),
    |     Vectors.dense(5.0, 33.0, 366.0))
    | )
scala> val correlMatrix: Matrix = Statistics.corr(data, "pearson")
scala> println(correlMatrix.toString)
1.0                  0.9788834658894731  0.9903895695275673
0.9788834658894731   1.0                 0.9977483233986101
0.9903895695275673   0.9977483233986101  1.0
```

Python 语言：

```
# pyspark
>>> from pyspark.mllib.stat import Statistics
>>> seriesX = sc.parallelize([1.0, 2.0, 3.0, 3.0, 5.0])
>>> seriesY = sc.parallelize([11.0, 22.0, 33.0, 33.0, 555.0])
>>> print("Correlation is: " + str(Statistics.corr(seriesX,
      seriesY, method = "pearson")))
Correlation is: 0.850028676877

>>> data = sc.parallelize(
...     [np.array([1.0, 10.0, 100.0]), np.array([2.0, 20.0, 200.0]),
        np.array([5.0, 33.0, 366.0])]
... )
>>> print(Statistics.corr(data, method = "pearson"))
[[ 1.          0.97888347  0.99038957]
 [ 0.97888347  1.          0.99774832]
 [ 0.99038957  0.99774832  1.        ]]
```

3. 分层抽样

与驻留在 spark.mllib 中的其他统计函数不同，可以对 RDD 的键值对执行分层抽样 sampleByKey 和 sampleByKeyExact 的方法。对于分层抽样，可以将键视为标签，将值视为特定属性。例如，键可以是人或文档 ID，并且相应的值可以是人的年龄列表或文档中的单词列表。sampleByKey 方法将翻转硬币以决定是否对样本进行采样，因此需要对数据进行一次传递，并提供预期的样本大小。sampleByKeyExact 比 sampleByKey 中使用的每层简单随机抽样需要更多的资源，但是会提供 99.99% 置信度的精确抽样大小。Python 语言目前不支持 sampleByKeyExact。

sampleByKeyExact() 允许用户准确地采样 $\lceil f_k \cdot n_k \rceil \forall k \in K$ 项，其中 f_k 是密钥 k 的期望分数，n_k 是密钥 k 的键值对的数量，K 是一组键。无须更换的采样需要在 RDD 上额外通过一次以确保样本大小，而更换采样则需要两次额外通过。

例 9.13 分层抽样。

Scala 语言：

```
# spark - shell
scala> val data = sc.parallelize(
    | Seq((1, 'a'), (1, 'b'), (2, 'c'), (2, 'd'), (2, 'e'), (3, 'f')))
scala> val fractions = Map(1 -> 0.1, 2 -> 0.6, 3 -> 0.3)
scala> val approxSample = data.sampleByKey(withReplacement = false,
        fractions = fractions)
scala> val exactSample = data.sampleByKeyExact(withReplacement = false,
        fractions = fractions)
scala> print(fractions)
Map(1 -> 0.1, 2 -> 0.6, 3 -> 0.3)
scala> approxSample.foreach(println)
(2,c)
```

```
(2,e)
scala> exactSample.foreach(println)
(1,a)
(3,f)
(2,c)
(2,d)
```

Python 语言：

```
# pyspark
>>> data = sc.parallelize([(1, 'a'), (1, 'b'), (2, 'c'), (2, 'd'), (2, 'e'),
    (3, 'f')])
>>> fractions = {1: 0.1, 2: 0.6, 3: 0.3}
>>> approxSample = data.sampleByKey(False, fractions)
>>> print(fractions)
{1: 0.1, 2: 0.6, 3: 0.3}
>>> print(approxSample.collect())
[(1, 'b'), (2, 'c'), (2, 'e')]
```

4. 假设检验

Statistics 提供了运行 Pearson 卡方检验的方法。以下示例演示了如何运用这种方法运行和解释假设检验。

例 9.14 运用 Pearson 卡方检验方法运行和解释假设检验。

Scala 语言：

```
# spark-shell
scala> import org.apache.spark.mllib.linalg._
scala> import org.apache.spark.mllib.regression.LabeledPoint
scala> import org.apache.spark.mllib.stat.Statistics
scala> import org.apache.spark.mllib.stat.test.ChiSqTestResult
scala> import org.apache.spark.rdd.RDD
scala> val vec: Vector = Vectors.dense(0.1, 0.15, 0.2, 0.3, 0.25)
scala> val goodnessOfFitTestResult = Statistics.chiSqTest(vec)
scala> println(s" $ goodnessOfFitTestResult\n")
Chi squared test summary:
method: pearson
degrees of freedom = 4
statistic = 0.12499999999999999
pValue = 0.998126379239318
No presumption against null hypothesis: observed follows the same distribution
as expected..

scala> val mat: Matrix = Matrices.dense(3, 2, Array(1.0, 3.0, 5.0, 2.0, 4.0, 6.0))
scala> val independenceTestResult = Statistics.chiSqTest(mat)
scala> println(s" $ independenceTestResult\n")
Chi squared test summary:
method: pearson
```

```
degrees of freedom = 2
statistic = 0.14141414141414144
pValue = 0.931734784568187
No presumption against null hypothesis: the occurrence of the outcomes
is statistically independent..

scala> val obs: RDD[LabeledPoint] =
     |   sc.parallelize(
     |     Seq(
     |       LabeledPoint(1.0, Vectors.dense(1.0, 0.0, 3.0)),
     |       LabeledPoint(1.0, Vectors.dense(1.0, 2.0, 0.0)),
     |       LabeledPoint(-1.0, Vectors.dense(-1.0, 0.0, -0.5))
     |     )
     |   )
     | )
scala> val featureTestResults: Array[ChiSqTestResult] = Statistics.chiSqTest(obs)
scala> featureTestResults.zipWithIndex.foreach { case (k, v) =>
     |   println(s"Column ${(v + 1)} :")
     |   println(k)
     | }
Column 1 :
Chi squared test summary:
method: pearson
degrees of freedom = 1
statistic = 3.0000000000000004
pValue = 0.08326451666354884
Low presumption against null hypothesis: the occurrence of the outcomes
is statistically independent..
Column 2 :
Chi squared test summary:
method: pearson
degrees of freedom = 1
statistic = 0.75
pValue = 0.3864762307712326
No presumption against null hypothesis: the occurrence of the outcomes
is statistically independent..
Column 3 :
Chi squared test summary:
method: pearson
degrees of freedom = 2
statistic = 3.0
pValue = 0.22313016014843035
No presumption against null hypothesis: the occurrence of the outcomes
is statistically independent..
```

Python 语言：

```
# pyspark
>>> from pyspark.mllib.linalg import Matrices, Vectors
```

```
>>> from pyspark.mllib.regression import LabeledPoint
>>> from pyspark.mllib.stat import Statistics
>>> vec = Vectors.dense(0.1, 0.15, 0.2, 0.3, 0.25)
>>> goodnessOfFitTestResult = Statistics.chiSqTest(vec)
>>> print("%s\n" % goodnessOfFitTestResult)
Chi squared test summary:
method: pearson
degrees of freedom = 4
statistic = 0.12499999999999999
pValue = 0.998126379239318
No presumption against null hypothesis: observed follows the same distribution
as expected..

>>> mat = Matrices.dense(3, 2, [1.0, 3.0, 5.0, 2.0, 4.0, 6.0])
>>> independenceTestResult = Statistics.chiSqTest(mat)
>>> print("%s\n" % independenceTestResult)
Chi squared test summary:
method: pearson
degrees of freedom = 2
statistic = 0.14141414141414144
pValue = 0.931734784568187
No presumption against null hypothesis: the occurrence of the outcomes
is statistically independent..

>>> obs = sc.parallelize(
...     [LabeledPoint(1.0, [1.0, 0.0, 3.0]),
...      LabeledPoint(1.0, [1.0, 2.0, 0.0]),
...      LabeledPoint(1.0, [-1.0, 0.0, -0.5])]
... )
>>> featureTestResults = Statistics.chiSqTest(obs)
>>> for i, result in enumerate(featureTestResults):
...     print("Column %d:\n%s" % (i + 1, result))
...
Column 1:
Chi squared test summary:
method: pearson
degrees of freedom = 0
statistic = 0.0
pValue = 1.0
No presumption against null hypothesis: the occurrence of the outcomes
is statistically independent..
Column 2:
Chi squared test summary:
method: pearson
degrees of freedom = 0
statistic = 0.0
pValue = 1.0
No presumption against null hypothesis: the occurrence of the outcomes
is statistically independent..
Column 3:
Chi squared test summary:
method: pearson
```

```
degrees of freedom = 0
statistic = 0.0
pValue = 1.0
No presumption against null hypothesis: the occurrence of the outcomes
is statistically independent..
```

此外，spark.mllib 还提供了 Kolmogorov-Smirnov(KS)检验的单样本，双侧实现，用于概率分布的相等性。通过提供理论分布的名称（目前仅支持正态分布）及其参数，或者根据给定的理论分布计算累积分布的函数，用户可以测试其假设，即样本是从 null 分配。在用户针对正态分布(distName = "norm")进行测试但不提供分发参数的情况下，测试初始化为标准正态分布并记录适当的消息。

Statistics 提供了运行单样本、双侧 Kolmogorov-Smirnov 检验的方法。以下示例演示了如何运用这种方法运行和解释假设检验。

例 9.15 运用运行单样本、双侧 Kolmogorov-Smirnov 检验方法运行和解释假设检验。

Scala 语言：

```
# spark-shell
scala> import org.apache.spark.mllib.stat.Statistics
scala> import org.apache.spark.rdd.RDD
scala> val data: RDD[Double] = sc.parallelize(Seq(0.1, 0.15, 0.2, 0.3, 0.25))
scala> val testResult = Statistics.kolmogorovSmirnovTest(data, "norm", 0, 1)
scala> println(testResult)
Kolmogorov-Smirnov test summary:
degrees of freedom = 0
statistic = 0.539827837277029
pValue = 0.06821463111921133
Low presumption against null hypothesis: Sample follows theoretical distribution.
scala> println()
scala> val myCDF = Map(0.1 -> 0.2, 0.15 -> 0.6, 0.2 -> 0.05, 0.3 -> 0.05,
         0.25 -> 0.1)
scala> val testResult2 = Statistics.kolmogorovSmirnovTest(data, myCDF)
scala> println(testResult2)
Kolmogorov-Smirnov test summary:
degrees of freedom = 0
statistic = 0.9500000000000001
pValue = 6.249999999763389E-7
Very strong presumption against null hypothesis: Sample follows theoretical
distribution.
```

Python 语言：

```
# pyspark
>>> from pyspark.mllib.stat import Statistics
>>> parallelData = sc.parallelize([0.1, 0.15, 0.2, 0.3, 0.25])
>>> testResult = Statistics.kolmogorovSmirnovTest(parallelData, "norm", 0, 1)
```

```
>>> print(testResult)
Kolmogorov-Smirnov test summary:
degrees of freedom = 0
statistic = 0.539827837277029
pValue = 0.06821463111921133
Low presumption against null hypothesis: Sample follows theoretical distribution.
```

5. 随机数据生成

随机数据生成对于随机算法、原型设计和性能测试非常有用。spark.mllib 支持使用独立同分布(independent indentically distributed, i.i.d)从给定分布绘制的值：均匀、标准正态或泊松分布随机 RDD。

RandomRDDs 提供工厂方法来生成随机 doubleRDD 或 vecors RDD。以下示例生成随机 doubleRDD，其值遵循标准正态分布 N(0,1)，然后将其映射到 N(1,4)。

例 9.16 生成随机 doubleRDD。

Scala 语言：

```
# spark-shell
scala> import org.apache.spark.SparkContext
scala> import org.apache.spark.mllib.random.RandomRDDs._
scala> val sc: SparkContext = spark.sparkContext
scala> val u = normalRDD(sc, 1000000L, 10)
scala> val v = u.map(x => 1.0 + 2.0 * x)
```

Python 语言：

```
# pyspark
>>> from pyspark.mllib.random import RandomRDDs
>>> sc = spark.sparkContext
>>> u = RandomRDDs.normalRDD(sc, 1000000L, 10)
>>> v = u.map(lambda x: 1.0 + 2.0 * x)
```

6. 核密度估计

核密度估计是用于可视化经验概率分布的技术，不需要从中抽取观察样本的特定分布的假设。它计算随机变量的概率密度函数的估计值，在给定的一组点处进行评估，它通过将特定点的经验分布的 PDF 表示为以每个样本为中心的正态分布的 PDF 的平均值来实现该估计。

KernelDensity 提供了从样本的 RDD 计算核密度估计的方法。以下示例演示了如何执行此操作。

例 9.17 从样本的 RDD 计算核密度估计。

Scala 语言：

```
# spark-shell
scala> import org.apache.spark.mllib.stat.KernelDensity
```

```
scala> import org.apache.spark.rdd.RDD
scala> val data: RDD[Double] = sc.parallelize(Seq(1, 1, 1, 2, 3, 4, 5, 5, 6,7, 8, 9, 9))
scala> val kd = new KernelDensity().setSample(data).setBandwidth(3.0)
scala> val densities = kd.estimate(Array(-1.0, 2.0, 5.0))
scala> densities.foreach(println)
0.04145944023341912
0.07902016933085627
0.08962920127312338
```

Python 语言:

```
# pyspark
>>> from pyspark.mllib.stat import KernelDensity
>>> data = sc.parallelize([1.0, 1.0, 1.0, 2.0, 3.0, 4.0, 5.0, 5.0, 6.0, 7.0, 8.0, 9.0, 9.0])
>>> kd = KernelDensity()
>>> kd.setSample(data)
>>> kd.setBandwidth(3.0)
>>> densities = kd.estimate([-1.0, 2.0, 5.0])
>>> print(densities)
[ 0.04145944  0.07902017  0.0896292 ]
```

9.3 Spark 实例

分类算法属于监督式学习,使用类标签已知的样本建立一个分类函数或分类模型,应用分类模型,能对数据库中的类标签未知的数据进行归类,K-Means、决策树、随机森林都是常见的分类算法。相关系数是考查两个变量之间的线性关系的一种统计方法,用于衡量两个变量因数的相关程度。

9.3.1 聚类问题

1. 聚类问题概述

聚类问题是给定一个元素集合 D,其中每个元素具有 n 个可观察属性,使用某种算法将 D 划分成 k 个子集,要求每个子集内部的元素之间相异度尽可能低,而不同子集的元素相异度尽可能高,其中每个子集称为一个簇(cluster)。

2. K-Means 聚类算法

K-Means 算法是最简单的一种聚类算法。K-Means 算法过程如下。
① 给定集合 D,有 n 个样本点。
② 随机指定 k 个点,作为 k 个子集的质心。
③ 根据样本点与 k 个质心的距离远近,将每个样本点划归最近质心所在的子集。
④ 对两个子集重新计算质心。
⑤ 根据新的质心,重复操作③。

⑥ 重复操作④和⑤，直至结果足够收敛或者不再变化。

3. K-Means 编程

学生成绩如表 9-1 所示，使用 K-Means 算法对其进行分类，计算聚类质心。

表 9-1 学生成绩表

姓 名	语 文	数 学	英 语	物 理
李一飞	88.0	74.0	96.0	85.0
谢晓梅	92.0	99.0	95.0	94.0
关义东	91.0	87.0	99.0	95.0
董晓青	78.0	99.0	97.0	81.0
狄汕尾	88.0	78.0	98.0	84.0
龚品清	100.0	95.0	100.0	92.0

（1）学生成绩 score.txt 内容如下。

```
88.0    74.0    96.0    85.0
92.0    99.0    95.0    94.0
91.0    87.0    99.0    95.0
78.0    99.0    97.0    81.0
88.0    78.0    98.0    84.0
100.0   95.0    100.0   92.0
```

（2）上传 score.txt 到 HDFS。

```
# hdfs dfs -put score.txt /input
```

（3）Scala 语言编写的 KMeansEx.scala 代码如下。

```scala
import org.apache.spark.sql.SparkSession
import org.apache.spark.mllib.clustering.{KMeans, KMeansModel}
import org.apache.spark.mllib.linalg.Vectors
object KMeansEx {
  def main(args: Array[String]) {
    val spark = SparkSession.builder.appName("KMeansEx").getOrCreate()
    //装载数据
    val data = spark.read.textFile("/input/score.txt").rdd
    //将文本文件的内容转化为 Double 类型的 Vector 集合
    val parsedData = data.map(s => Vectors.dense(s.split(' ')
        .map(_.toDouble))).cache()
    //分为 3 个子集，最多 20 次迭代
    val numClusters = 3
    val numIterations = 20
    val clusters = KMeans.train(parsedData, numClusters, numIterations)
    //输出每个子集的质心
```

```
    clusters.clusterCenters.foreach { println }
    //评估聚类计算
    val WSSSE = clusters.computeCost(parsedData)
    println("Within Set Sum of Squared Errors = " + WSSSE)
    //输出每组所属的子集索引
    val predict_target = clusters.predict(parsedData)
    print(predict_target.collect().toList)
    spark.stop()
  }
}
```

(4) 运行 KMeansEx.scala,结果如下。

```
[78.0,99.0,97.0,81.0]
[88.0,76.0,97.0,84.5]
[94.33333333333333,93.66666666666666,98.0,93.66666666666666]
Within Set Sum of Squared Errors = 152.5
List(1, 2, 2, 0, 1, 2)
```

(5) Python 语言编写的 KMeansEx.py 代码如下。

```
from __future__ import print_function
import sys
from operator import import add
from pyspark.sql import SparkSession
import numpy as np
from pyspark.mllib.clustering import KMeans
def parseVector(line):
    return np.array([float(x) for x in line.split(' ')])
if __name__ == "__main__":
    spark = SparkSession.builder.appName("PyKMeans").getOrCreate()
    lines = spark.read.text("/input/score.txt").rdd.map(lambda r: r[0])
    data = lines.map(parseVector)
    k = 3
    model = KMeans.train(data, k)
    print("Final centers: " + str(model.clusterCenters))
    print("Total Cost: " + str(model.computeCost(data)))
    predict_target = model.predict(data)
    print(predict_target.collect())
    spark.stop()
```

(6) 运行 KMeansEx.py,结果如下。

```
Final centers: [array([ 94.33333333,  93.66666667, 98. ,  93.66666667]),
array([ 88. ,  76. ,  97. ,  84.5]), array([ 78.,  99.,  97.,  81.])]
Total Cost: 152.5
[2, 0, 0, 1, 2, 0]
```

9.3.2 随机森林

1. 决策树和随机森林概述

决策树(decision tree)是一种基本的分类器,是在已知各种情况发生概率的基础上,通过构成决策树来求取净现值的期望值大于等于零的概率,评价项目风险,判断其可行性的决策分析方法,是直观运用概率分析的一种图解法。由于这种决策分支画成图形很像一棵树的枝干,因此称为决策树。决策树是一种树状结构,其中每个内部节点表示一个属性上的测试,每个分支代表一个测试输出,每个叶节点代表一种类别。

随机森林(random forest)是指利用多棵树对样本进行训练并预测的一种分类器。该分类器最早由 Leo Breiman 和 Adele Cutler 提出,并被注册成商标。在机器学习中,随机森林是一个包含多个决策树的分类器,并且其输出的类别是由个别树输出的类别的众数而定的。

2. 随机森林示例

鸢尾花(Iris)数据中 Setosa、Versicolor、Virginica 是 3 种有名的鸢尾花品种,根据萼片长度、宽度和花瓣长度、宽度进行分析,3 种鸢尾花各取 50 个值,部分数据如表 9-2 所示。取 80% 的 Iris 数据作为训练数据集决策树,取 20% 的 Iris 数据作为交叉验证集,评估训练模型,任选一组数据(如 5.8,2.7,5.1,1.9)进行预测。

表 9-2 鸢尾花(Iris)部分数据

Species(种类)	Sepal(萼片)		Petal(花瓣)	
	Length(长)	Width(宽)	Length(长)	Width(宽)
Setosa	5.1	3.5	1.4	0.2
Setosa	4.9	3.0	1.4	0.2
Setosa	4.7	3.2	1.3	0.2
Setosa	4.6	3.1	1.5	0.2
Setosa	5.0	3.6	1.4	0.2
Versicolor	7.0	3.2	4.7	1.4
Versicolor	6.4	3.2	4.5	1.5
Versicolor	6.9	3.1	4.9	1.5
Versicolor	5.5	2.3	4.0	1.3
Versicolor	6.5	2.8	4.6	1.5
Virginica	6.3	3.3	6.0	2.5
Virginica	5.8	2.7	5.1	1.9
Virginica	7.1	3.0	5.9	2.1
Virginica	6.3	2.9	5.6	1.8
Virginica	6.5	3.0	5.8	2.2

(1) 分割鸢尾花(Iris)数据。

```
训练集:IrisTrain.txt,占80%,行:01~40    51~90    101~140
验证集:IrisCVS.txt, 占20%,行:41~50    91~100   141~150
```

IrisTrain.txt 部分内容如下。

```
0, 5.1  3.5  1.4  0.2
0, 4.9  3.0  1.4  0.2
…
1, 7.0  3.2  4.7  1.4
1, 6.4  3.2  4.5  1.5
…
2, 6.3  3.3  6.0  2.5
2, 5.8  2.7  5.1  1.9
…
```

IrisCVS.txt 部分内容如下。

```
0, 5.1  3.5  1.4  0.2
0, 4.9  3.0  1.4  0.2
…
1, 7.0  3.2  4.7  1.4
1, 6.4  3.2  4.5  1.5
…
2, 6.3  3.3  6.0  2.5
2, 5.8  2.7  5.1  1.9
…
```

(2) 上传 IrisTrain.txt 和 IrisCVS.txt 到 HDFS。

```
# hdfs dfs -put IrisTrain.txt /input
# hdfs dfs -put IrisCVS.txt /input
```

(3) Scala 语言编写的 IrisRandomForest.scala 代码如下。

```scala
import org.apache.spark.sql.SparkSession
import org.apache.spark.rdd.RDD
import org.apache.spark.mllib.regression.LabeledPoint
import org.apache.spark.mllib.linalg.Vectors
import org.apache.spark.mllib.tree
import org.apache.spark.mllib.tree.model.RandomForestModel
import org.apache.spark.mllib.evaluation.MulticlassMetrics

object IrisRandomForest {
  def main(args: Array[String]) {
    val spark = SparkSession.builder.appName("IrisRandomForest").getOrCreate()
    //读取训练数据集
```

```scala
val data1 = spark.read.textFile("/input/IrisTrain.txt").rdd
//读取交叉验证集
val data2 = spark.read.textFile("/input/IrisCVS.txt").rdd
//转换数据集
val trainData = data1.map { line =>
  val parts = line.split(',')
  LabeledPoint(parts(0).toDouble,Vectors.dense(parts(1).split(' ')
  .map(_.toDouble)))
}.cache()
val cvsData = data2.map { line =>
  val parts = line.split(',')
  LabeledPoint(parts(0).toDouble,Vectors.dense(parts(1).split(' ')
  .map(_.toDouble)))
}.cache()
//分类的数目
val numClasses = 3
//设置输入数据格式
val categoricalFeaturesInfo = Map[Int, Int]()
//设置随机森林中决策树数目
val numTrees = 4
//特征子集采样策略,auto 表示算法自主选取
val featureSubsetStrategy = "auto"
//设定信息增益计算方式
val impurity = "entropy"
//树的最大层次
val maxDepth = 4
//特征最大装箱数
val maxBins = 32
//训练随机森林分类器
val model = tree.RandomForest.trainClassifier(
  trainData, numClasses, categoricalFeaturesInfo,
  numTrees, featureSubsetStrategy, impurity, maxDepth, maxBins)
//测试预测效果
val predictionsAndLables = cvsData.map { d =>
    (model.predict(d.features), d.label)
}
//计算精确度
val multiMetrics = new MulticlassMetrics(predictionsAndLables)
println("Precision = " + multiMetrics.precision)
//计算每个样本的准确度(召回率)
val recall = multiMetrics.labels.map(target =>
  (multiMetrics.precision(target), multiMetrics.recall(target)))
//打印准确度(召回率)
println(" --- (precision,recall) --- ")
recall.foreach(println)
//打印交叉验证集验证结果与原来的种类
println(predictionsAndLables.collect().toList)
```

```
    //任选一组数据预测其类型
    val test = Vectors.dense(Array(5.8,2.7,5.1,1.9))
    val result = model.predict(test)
    println("Species is: " + result)
    spark.stop()
  }
}
```

(4) 运行 IrisRandomForest.scala,结果如下。

```
Precision = 0.9666666666666667
--- (precision,recall) ---
(1.0,0.9)
(0.9090909090909091,1.0)
(1.0,1.0)
List((0.0,0.0), (0.0,0.0), (0.0,0.0), (1.0,0.0), (0.0,0.0),(0.0,0.0),(0.0,0.0),
(0.0,0.0), (0.0,0.0), (0.0,0.0), (1.0,1.0), (1.0,1.0), (1.0,1.0), (1.0,1.0),
(1.0,1.0), (1.0,1.0), (1.0,1.0), (1.0,1.0), (1.0,1.0), (1.0,1.0), (2.0,2.0),
(2.0,2.0), (2.0,2.0), (2.0,2.0), (2.0,2.0), (2.0,2.0), (2.0,2.0), (2.0,2.0),
(2.0,2.0), (2.0,2.0))
Species is: 2.0
```

(5) Python 语言编写的 IrisRandomForest.py 代码如下。

```
from __future__ import print_function
import sys
from operator import add
from pyspark.sql import SparkSession
import numpy as np
from pyspark.mllib.regression import LabbeledPoint
from pyspark.mllib.tree import RandomForest
from pyspark.mllib.evaluation import MulticlassMetrics
from pyspark.mllib.linalg import Vectors
def parsePoint(line):
    values = [float(x) for x in line.replace(',', ' ').split(' ')]
    return LabeledPoint(values[0], values[1:])

if __name__ == "__main__":
    spark = SparkSession.builder.appName("PyIrisRandomForest").getOrCreate()
    # 读取数据集
    data1 = spark.read.text("/input/IrisTrain.txt").rdd.map(lambda r: r[0])
    data2 = spark.read.text("/input/IrisCVS.txt").rdd.map(lambda r: r[0])
    # 转换数据集
    trainData = data1.map(parsePoint).cache()
    cvsData = data2.map(parsePoint).cache()
    # 随机森林训练参数
    numClasses = 3
    categoricalFeaturesInfo = {}
```

```python
numTrees = 4
featureSubsetStrategy = "auto"
impurity = "entropy"
maxDepth = 4
maxBins = 32
# 训练随机森林分类器
model = RandomForest.trainClassifier(
    trainData, numClasses, categoricalFeaturesInfo,
    numTrees, featureSubsetStrategy, impurity, maxDepth, maxBins)
# 输出模型内容(树状结构)
print(model.toDebugString())
# 测试预测效果
preds = model.predict(cvsData.map(lambda p: p.features))
actual = cvsData.map(lambda p: p.label)
predictionsAndLables = preds.zip(actual)
print("predictionsAndLables:" + str(predictionsAndLables.collect()))
# 计算准确率
multiMetrics = MulticlassMetrics(predictionsAndLables)
print("Precision = " + str(multiMetrics.precision()))
# 计算每个样本的准确度(召回率)
print(" --- (precision,recall) --- ")
for i in range(3):
    print((multiMetrics.precision(i),multiMetrics.recall(i)))
# 任选一组数据预测其类型
test = Vectors.dense([5.8,2.7,5.1,1.9])
result = model.predict(test)
print("Species is: " + str(result))
spark.stop()
```

(6) 运行 IrisRandomForest.py,结果如下。

```
TreeEnsembleModel classifier with 4 trees

  Tree 0:
  ...
  Tree 3:
    If (feature 3 <= 0.5)
     Predict: 0.0
    Else (feature 3 > 0.5)
     If (feature 3 <= 1.6)
      If (feature 2 <= 4.9)
       Predict: 1.0
      Else (feature 2 > 4.9)
       Predict: 2.0
     Else (feature 3 > 1.6)
      If (feature 0 <= 5.9)
       If (feature 3 <= 1.8)
        Predict: 1.0
```

```
            Else (feature 3 > 1.8)
              Predict: 2.0
          Else (feature 0 > 5.9)
            Predict: 2.0
predictionsAndLables:[(0.0, 0.0), (0.0, 0.0), (0.0, 0.0), (1.0, 0.0), (0.0, 0.0),
(0.0, 0.0), (0.0, 0.0), (0.0, 0.0), (0.0, 0.0), (0.0, 0.0), (1.0, 1.0), (1.0, 1.0),
(1.0, 1.0), (1.0, 1.0), (1.0, 1.0), (1.0, 1.0), (1.0, 1.0), (1.0, 1.0), (1.0, 1.0),
(1.0, 1.0), (2.0, 2.0), (2.0, 2.0), (2.0, 2.0), (2.0, 2.0), (2.0, 2.0), (2.0, 2.0),
(2.0, 2.0), (2.0, 2.0), (2.0, 2.0), (2.0, 2.0)]
Precision = 0.966666666667
--- (precision,recall) ---
(1.0, 0.9)
(0.9090909090909091, 1.0)
(1.0, 1.0)
Species is: 2.0
```

小结

Spark MLlib 提供了丰富的数据类型（Local Vector、Labeled Point、Local Matrix、RowMatrix、IndexedRowMatrix、CoordinateMatrix、BlockMatrix）和 API。基于 DataFrames 的 API 的基本统计有相关性、假设检验和累积器等，基于 RDD 的 API 的基本统计有概要统计、相关性、分层抽样、假设检验、随机数据生成、核密度估计等，这些函数和 API 为大数据处理提供了很大便利。本章还介绍了机器学习和 3 个典型案例，使用 K-Means 聚类算法对学生成绩进行分类，通过分析鸢尾花数据，学习了随机森林算法；掌握数据集的训练、评估与预测方法。

习题

1. 简述机器学习的基本概念。
2. 简述机器学习的分类。
3. MLlib 的数据类型和 API 函数有哪些？如何使用？
4. 3 种鸢尾花数据（萼片宽度、萼片长度、花瓣宽度、花瓣长度）为（4.9,3.0,1.4,0.2）、（5.0,3.6,1.4,0.2）、（5.2,2.7,3.9,1.4）、（6.1,2.9,4.7,1.4）、（7.7,2.6,6.9,2.3）、（6.6,2.9,4.6,1.3）、（4.4,3.2,1.3,0.2）、（5.7,2.8,4.1,1.3），计算其聚类中心并对数据进行分类。
5. 使用决策树算法对鸢尾花数据进行模型训练和预测。

第10章 Hive数据仓库应用

数据仓库是决策支持系统和联机分析应用数据源的结构化数据环境。数据仓库研究和解决从数据库中获取信息的问题。Hive提供了一个熟悉SQL的用户所能熟悉的编程模型,Hive最适合于不需要实时响应查询和不需要记录级别的插入、更新、删除的数据仓库程序。

10.1 Hive简介

Hive是基于Hadoop的一个数据仓库工具,可以将结构化的数据文件映射为一张数据库表,并提供简单的SQL查询功能,可以将SQL语句转换为MapReduce任务进行运行。其优点是学习成本低,可以通过类SQL语句快速实现简单的MapReduce统计,不必开发专门的MapReduce应用,十分适合数据仓库的统计分析。

Hive由Facebook开发,在某种程度上可以看成是用户编程接口,本身并不存储和处理数据,依赖HDFS存储数据,依赖MapReduce处理数据。有类SQL语言HiveQL,不完全支持SQL标准,如不支持更新操作、索引和事务,其子查询和连接操作也存在很多限制。

Hive把HQL语句转换成MapReduce任务后,采用批处理的方式对海量数据进行处理。数据仓库存储的是静态数据,很适合采用MapReduce进行批处理。Hive还提供了一系列对数据进行提取、转换、加载的工具,可以存储、查询和分析存储在HDFS上的数据。

10.1.1 Hive组成模块

图10-1所示为Hive的主要模块及与Hadoop的交互工作。Karmasphere和Hue是图形用户界面的交互工具,Hive发行版中附带的模块有CLI、一个Hive网页界面(HWI),以及可通过JDBC、ODBC和一个Thrift Server进行编程访问的几个模块。

Qubole 也是一款与大数据交互的工具,可针对存储在 AWS、谷歌或 Azure 云上的数据,简化、加快和扩展大数据分析工作负载。

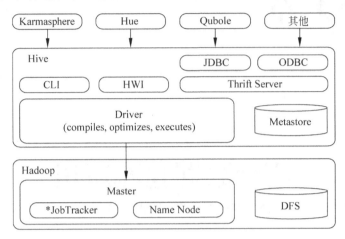

图 10-1　Hive 组成模块

所有的命令和查询都会进入 Driver,通过该模块对输入进行解析编译,对需求的计算进行优化,然后按照指定的步骤执行。当需要启动 MapReduce 任务(job)时,Hive 本身是不会生成 Java MapReduce 算法程序的;相反,Hive 通过一个表示"job 执行计划"的 XML 文件驱动执行内置的、原生的 Mapper 和 Reducer 模块。Hive 通过和 JobTracker 通信来初始化 MapReduce 任务(job),而不必部署在 JobTracker 所在的管理节点上执行。通常,要处理的数据文件是存储在 HDFS 中的,而 HDFS 是由 Namenode 进行管理的。Metastore 是一个独立的关系型数据库,Hive 会在其中保存表模式和其他系统元数据。

10.1.2　Hive 执行流程

Hive 与 Hadoop 交互的主要组件如图 10-2 所示。这些组件的具体说明如下。

(1) UI:包括 Shell 命令、JDBC/ODBC 和 WebUi,其中最常用的是 shell 这个客户端方式对 Hive 进行相应操作。

(2) Driver:Hive 解析器的核心功能是根据用户编写的 SQL 语法匹配出相应的 MapReduce 模板,形成对应的 MapReduce job 进行执行。

(3) Compiler:将 HiveQL 编译成有向无环图(Directed Acyclic Graph,DAG)形式的 MapReduce 任务。

(4) Metastore:Hive 将表中的元数据信息存储在数据库中,如 derby、MySQL,Hive 中的元数据信息包括表的名字、表的列和分区、表的属性(是否为外部表等)、表的数据所在的目录等。编译器 Compiler 根据用户任务去 Metastore 中获取需要的 Hive 的元数据信息。

(5) Execution Engine:执行编译器产生的执行计划,该计划是一个有向无环图,执行引擎管理这些计划的不同阶段之间的依赖关系,并在相关组件上执行这些阶段。

执行流程步骤如下。

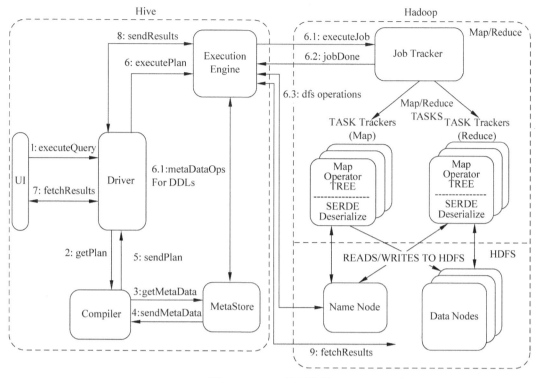

图 10-2　Hive 执行流程

（1）用户提交查询等任务给 Driver。

（2）Compiler 获得该用户的任务 Plan。

（3）Compiler 根据用户任务去 MetaStore 中获取需要的 Hive 的元数据信息。

（4）Compiler 得到元数据信息，对任务进行编译，先将 HiveQL 转换为抽象语法树；然后将抽象语法树转换为查询块，将查询块转化为逻辑的查询计划，重写逻辑查询计划，并将逻辑计划转化为物理的计划（MapReduce）；最后选择最佳的策略。

（5）将最终的计划提交给 Driver。

（6）Driver 将计划（Plan）转交给 Execution Engine 去执行，获取元数据信息，提交给 ResourceManager 或 SourceManager 执行该任务，任务会直接读取 HDFS 中文件进行相应的操作。

（7）获取执行的结果。

（8）取得并返回执行结果。

10.1.3　MetaStore 存储模式

Hive 的安装包括服务端和客户端，服务端可以装在任何节点上。Hive 的 MetaStore 有 3 种存储模式，如图 10-3 所示。

（1）内嵌模式：元数据保存在内嵌 Derby 中，只允许一个会话链接。hive-site.xml 配置如下。

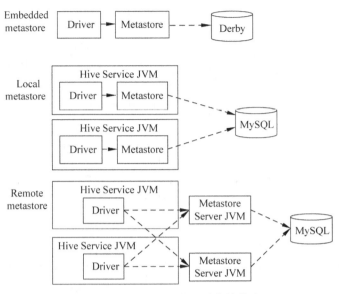

图 10-3 MetaStore 的 3 种存储模式

```
<property>
  <name>hive.metastore.local</name>
  <value>true</value>
  <description>Use false if a production metastore server is used</description>
</property>
<property>
  <name>javax.jdo.option.ConnectionURL</name>
  <value>jdbc:derby:;databaseName = metastore_db;create = true</value>
  <description>JDBC connect string for a JDBC metastore.</description>
</property>
```

(2) 本地模式：本地安装 MySQL 替代 Derby 存储元数据，hive-site.xml 配置如下。

```
<property>
  <name>hive.metastore.local</name>
  <value>true</value>
  <description>Use false if a production metastore server is used</description>
</property>
<property>
  <name>javax.jdo.option.ConnectionURL</name>
  <value>jdbc:mysql://localhost:3306/hive?createDatabaseIfNotExist = true</value>
  <description>JDBC connect string for a JDBC metastore.</description>
</property>
```

(3) 远程模式：远程安装 MySQL 替代 Derby 存储元数据，hive-site.xml 配置如下。

```
<property>
  <name>hive.metastore.local</name>
```

```xml
<value>false</value>
<description>Use false if a production metastore server is used</description>
</property>
<property>
<name>javax.jdo.option.ConnectionURL</name>
<value>jdbc:mysql://server:3306/hive?createDatabaseIfNotExist=true</value>
<description>JDBC connect string for a JDBC metastore.</description>
</property>
```

用户如果不想使用默认的路径,那么可以配置一个不同目录来存储表数据。对于远程模式,默认存储路径是 hdfs://namenode_server/user/hive/warehouse;对于其他模式,默认存储路径是 file:///user/hive/warehouse,代码如下。

```xml
<property>
<name>hive.metastore.warehouse.dir</name>
<value>/user/hive/warehouse</value>
<description>location of default database for the warehouse</description>
</property>
```

对数据量比较小的操作,就可以在本地执行,这样要比提交任务到集群执行效率快很多。配置如下参数,可以开启 Hive 的本地模式。

```
hive> set hive.exec.mode.local.auto = true;(默认为false)
```

10.2 Hive 安装与配置

视频讲解

Hive 分为服务端和客户端,可以选择一台节点或多台节点作为服务端,其他作为客户端。Hive 需要 Hadoop 环境,部署 Hadoop 参照第 3 章。

安装 Hive 需要 apache-hive-2.1.1-bin.tar.gz 软件包,下载网址为 mirrors.aliyun.com。首先下载软件包;然后解压到/opt 目录下,解压后子目录名带有版本号;最后对其改名,使其简洁。

10.2.1 Hive 参数配置

Hive 需要添加环境变量和修改 Hive 参数,修改的配置文件有 hive-env.sh 和 hive-site.xml,3 台节点机安装方法一样。

(1) 修改/etc/hosts 文件,添加内容如下。

```
172.30.0.10     master
172.30.0.11     slave1
172.30.0.12     slave2
```

(2) 修改/root/.bash_profile 文件,添加内容如下。

```
export HIVE_HOME = /opt/hive
export PATH = $PATH:$HIVE_HOME/bin
```

(3) 在 Hive 的配置目录下给文件改名,操作如下。

```
# cd /opt/hive/conf/
# mv beeline-log4j2.properties.template beeline-log4j2.properties
# mv hive-env.sh.template hive-env.sh
# mv hive-exec-log4j2.properties.template hive-exec-log4j2.properties
# mv hive-log4j2.properties.template hive-log4j2.properties
# mv llap-cli-log4j2.properties.template llap-cli-log4j2.properties
# mv llap-daemon-log4j2.properties.template llap-daemon-log4j2.properties
```

(4) 修改配置文件 hive-env.sh,添加内容如下。

```
export HIVE_CONF_DIR = /opt/hive/conf/
export HIVE_AUX_JARS_PATH = /opt/hive/lib/
```

(5) 在 Hive 的配置目录下新建配置文件 hive-site.xml,操作如下。

```
# vi /opt/hive/conf/hive-site.xml
```

服务器端的内容如下。

```xml
<?xml version = "1.0"?>
<?xml-stylesheet type = "text/xsl" href = "configuration.xsl"?>
<configuration>
  <property>
    <name>hive.metastore.warehouse.dir</name>
    <value>/hive/warehouse</value>
    <description>location of default database for the warehouse</description>
  </property>
  <property>
    <name>javax.jdo.option.ConnectionURL</name>
    <value>jdbc:mysql://master:3306/hive?createDatabaseIfNotExist = true</value>
    <description>JDBC connect string for a JDBC metastore.</description>
  </property>
  <property>
    <name>javax.jdo.option.ConnectionDriverName</name>
    <value>com.mysql.jdbc.Driver</value>
    <description>Driver class name for a JDBC metastore</description>
  </property>
  <property>
    <name>javax.jdo.option.ConnectionUserName</name>
    <value>hive</value>
```

```xml
    <description>Username to use against metastore database</description>
  </property>
  <property>
    <name>javax.jdo.option.ConnectionPassword</name>
    <value>123456</value>
    <description>password to use against metastore database</description>
  </property>
  <property>
    <name>hive.querylog.location</name>
    <value>/opt/hive/logs</value>
    <description>Location of Hive run time structured log file</description>
  </property>
  <property>
    <name>hive.metastore.uris</name>
    <value>thrift://master:9083</value>
    <description>Thrift URI for the remote metastore. Used by metastore client to connect to remote metastore.</description>
  </property>
  <property>
    <name>hive.server2.webui.host</name>
    <value>0.0.0.0</value>
  </property>
  <property>
    <name>hive.server2.webui.port</name>
    <value>10002</value>
  </property>
</configuration>
```

客户端的内容如下。

```xml
<?xml version="1.0"?>
<?xml-stylesheet type="text/xsl" href="configuration.xsl"?>
<configuration>
  <property>
    <name>hive.metastore.warehouse.dir</name>
    <value>/hive/warehouse</value>
    <description>location of default database for the warehouse</description>
  </property>
  <property>
    <name>hive.querylog.location</name>
    <value>/opt/apache-hive-2.1.1-bin/logs</value>
    <description>Location of Hive run time structured log file</description>
  </property>
  <property>
    <name>hive.metastore.uris</name>
    <value>thrift://master:9083</value>
    <description>Thrift URI for the remote metastore. Used by metastore client to connect to remote metastore.</description>
```

```
        </property>
</configuration>
```

10.2.2　Hive 运行与测试

运行 Hive 需要先启动 Hadoop 服务,再启动 MetaStore 服务和 HiveServer2 服务。
(1) 启动 Hadoop,操作如下。

```
[root@master ~]# start-dfs.sh
[root@master ~]# start-yarn.sh
[root@master ~]# mr-jobhistory-daemon.sh start historyserver
```

(2) 启动 MetaStore 服务,操作如下。

```
[root@master ~]# hive --service metastore &
```

(3) 启动 HiveServer2 服务,操作如下。

```
[root@master ~]# hiveserver2 &
```

(4) 进行 Hive CLI 测试,操作如下。

```
# hive
hive> CREATE TABLE tbl (i INT);
OK
Time taken: 0.157 seconds
hive> SELECT * FROM tbl;
OK
Time taken: 0.219 seconds
hive> DROP TABLE tbl;
OK
Time taken: 0.146 seconds
hive> EXIT;
```

(5) 设置 Hive 的 CLI 选项,操作如下。

```
# hive --help --service cli
usage: hive
 -d, --define <key=value>          Variable subsitution to apply to hive
                                   commands. e.g. -d A=B or --define A=B
    --database <databasename>      Specify the database to use
 -e <quoted-query-string>          SQL from command line
 -f <filename>                     SQL from files
 -H, --help                        Print help information
    --hiveconf <property=value>    Use value for given property
    --hivevar <key=value>          Variable subsitution to apply to hive
                                   commands. e.g. --hivevar A=B
```

```
-i <filename>              Initialization SQL file
-S, --silent               Silent mode in interactive shell
-v, --verbose              Verbose mode (echo executed SQL to the
                           console)
```

（6）打开浏览器，输入网址"http://master:10002"，查看 HiveServer2 服务，如图 10-4 所示。

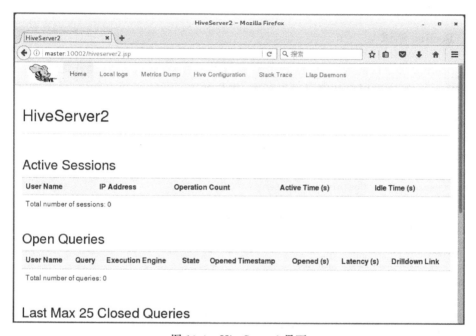

图 10-4　HiveServer2 界面

10.2.3　Hive Beeline

HiveServer2 提供了一个新的命令行工具 Beeline，它基于 SQLLine CLI 的 JDBC 客户端。Beeline 工作模式有两种，即本地嵌入模式和远程模式。嵌入模式下，它返回一个嵌入式的 Hive（类似于 Hive CLI）；而远程模式则是通过 Thrift 协议与某个单独的 HiveServer2 进程进行连接通信。

1. 权限设置

Beeline 使用 JDBC 连接到 HiveServer2 时，如果显示 User：root is not allowed to impersonate hive，是因为远程访问 Hadoop 授权异常导致的，则需要修改 Hadoop 系统的配置文件 core-site.xml，添加如下属性。

```
<property>
    <name>hadoop.proxyuser.root.hosts</name>
    <value>*</value>
```

```
</property>
<property>
   <name>hadoop.proxyuser.root.groups</name>
   <value>*</value>
</property>
```

注意：这里登录的是 root 用户。如果登录的是 hadoop 用户,则配置文件中的"root"改为"hadoop"就可以了。

2. Beeline 测试

在 Beeline 命令行输入"?"可以获得命令帮助。在对 Hive 进行操作之前,需要连接到 HiveServer 2,这里用户和密码可以任意,操作如下。

```
# beeline
Beeline version 2.1.1 by Apache Hive
beeline> ?
beeline> !connect jdbc:hive2://master:10000
Enter username for jdbc:hive2://master:10000: h
Enter password for jdbc:hive2://master:10000: p
0: jdbc:hive2://master:10000> SHOW DATABASES;
+---------------+--+
| database_name |
+---------------+--+
| default       |
+---------------+--+
2 rows selected (1.166 seconds)
0: jdbc:hive2://master:10000> CREATE TABLE tbl (i INT);
No rows affected (1.444 seconds)
0: jdbc:hive2://master:10000> SELECT * FROM tbl;
+--------+--+
| tbl.i  |
+--------+--+
+--------+--+
No rows selected (1.099 seconds)
0: jdbc:hive2://master:10000> DROP TABLE tbl;
No rows affected (1.633 seconds)
0: jdbc:hive2://master:10000> !closeall
Closing: 0: jdbc:hive2://master:10000
beeline> !quit
```

10.3 数据类型和文件格式

Hive 支持关系型数据库中的大多数基本数据类型,同时支持 3 种复杂数据类型。Hive 对其数据在文件中的编码方式具有很大的灵活性,用户可以使用各种工具来管理和处理数据。

10.3.1 数据类型

1. 基本数据类型

Hive 支持多种不同长度的整型和浮点型数据类型,也支持布尔类型,还支持无长度限制的字符类型。从 Hive 0.8.0 版本开始支持时间戳数据类型和二进制数据类型等,如表 10-1 所示。

表 10-1 基本数据类型

数据类型	所占字节	例子
TINYINT	1b 有符号数	10
SMALLINT	2b 有符号数	10
INT	4b 有符号数	10
BIGINT	8b 有符号数	10
BOOLEAN	true 或 false	TRUE
FLOAT	4b 单精度	3.14159
DOUBLE	8b 双精度	3.14159
STRING	字符序列,使用单引号和双引号设置	'hello welcome',"Apache Hive"
BINARY	字节数组	
TIMESTAMP	整型、浮点型或字符串型	123456(UNIX 新纪元秒) '2017-10-01 12:00:56.123456'
DECIMAL	DECIMAL(precision,scale)	Decimal(10,0)
CHAR	255	CHAR(10)
VARCHAR	1~65 535	VARCHAR(50)
DATE	整型、浮点型或字符串型	'2017-10-01 12:00:56'

2. 复杂数据类型

复杂数据类型包括 ARRAY、MAP、STRUCT,这些复杂数据类型是由基础类型组成的,如表 10-2 所示。

表 10-2 复杂数据类型

数据类型	描述	例子
ARRAY	一组有序字段。字段的类型必须相同。例如,数组 A 的值为[1,2],则第 2 个元素为 A[1]	ARRAY(1,2)
MAP	一组无序的键/值对。键的类型必须是原子的,值可以是任何类型,同一个映射的键的类型必须相同,值的类型也必须相同。例如,MAP 数据 M 的键-值对 'a'—>1,'b'—>2,则'b'的值为 M['b']	MAP('a',1,'b',2)
STRUCT	一组命名的字段。字段类型可以不同。例如,name 为 STRUCT{ first STRING, last STRING },则使用 name.last 引用 name 中 last 的值	STRUCT('a',1,2)

例如，创建一个包含复杂数据类型的表，代码如下。

```
CREATE TABLE salesperson (
    name STRING,
    salary FLOAT,
    degree ARRAY < STRING >,
    dues MAP < STRING, FLOAT >,
    address STRUCT < province:STRING, city:STRING, street:STRING, zip:INT >
);
```

10.3.2 文件格式

Hive 支持逗号","分隔符和制表符"\t"分隔符，Hive 使用 field 来表示替换默认分隔符的字符。Hive 中默认的记录和字段分隔符如表 10-3 所示。

表 10-3 Hive 记录和字段默认分隔符

分隔符	描述
\n	对于文本文件来说，每一行都是一条记录，因此换行符可以分隔记录
^A(Ctrl+A)	用于分隔字段(列)。在 CREATE TABLE 语句中使用八进制编码\001 表示
^B	用于分隔 ARRAY 或 STRUCT 中的元素，或者用于 MAP 中的"键/值"对之间的分隔。在 CREATE TABLE 语句中可以使用八进制编码\002 表示
^C	用于 MAP 中的"键/值"对之间的分隔。在 CREATE TABLE 语句中使用八进制编码\003 表示

(1) 销售人员基本信息。salesperson 表的记录如下，其中使用^A 等字符作为字段分隔符。这里包含 4 条记录，为了表达清晰，中间加入空行。

```
王可塘^A12000.0^A学士^B硕士^A公积金^C.2^B保险金^C.05^B养老金^C.1
^A广东省^B汕尾市^B莲塘街36号.^B516600

陈海峰^A8600.0^A学士^A公积金^C.2^B保险金^C.05^B养老金^C.1
^A广东省^B汕尾市^B马宫盐町村.^B516625

萧红海^A6800.0^A^A公积金^C.15^B保险金^C.03^B养老金^C.1
^A广东省^B深圳市^B梅林路18号.^B518049

李凤苑^A6200.0^A^A公积金^C.15^B保险金^C.03^B养老金^C.1
^A广东省^B广州市^B天河北路.^B510630
```

(2) JSON 表示。如果使用 JSON 格式表示第 1 条记录，增加表结构中的字段名，则表示如下。

```
{
    "name": "王可塘",
    "salary": 12000.0,
    "degree": ["学士", "硕士"],
```

```
    "dues": {
        "公积金": .2,
        "保险金":  .05,
        "养老金":  .1
    },
    "address":{
        "province":"广东省",
        "city":"汕尾市",
        "street":"莲塘街36号.",
        "zip":  516600
    }
}
```

（3）定义表结构。如果要定义表结构，则需要明确指定分隔符。创建 salesperson 表命令如下。

```
CREATE TABLE salesperson (
    name STRING,
    salary FLOAT,
    degree ARRAY < STRING >,
    dues MAP < STRING, FLOAT >,
    address STRUCT< province:STRING, city:STRING, street:STRING, zip:INT >
)
ROW FORMAT DELIMITED
FIELDS TERMINATED BY '\001'
COLLECTION ITEMS TERMINATED BY '\002'
MAP KEYS TERMINATED BY '\003'
LINES TERMINATED BY '\n'
STORED AS TEXTFILE;
```

（4）导入数据，命令如下。

```
LOAD DATA LOCAL INPATH '/root/salesperson'
OVERWRITE INTO TABLE salesperson;
```

（5）查看数据，命令如下。

```
hive > SELECT * FROM salesperson;
OK
王可塘    12000.0    ["学士","硕士"]    {"公积金":0.2,"保险金":0.05,"养老金":0.1}
{"province":"广东省","city":"汕尾市","street":"莲塘街36号.","zip":516600}
陈海峰    8600.0    ["学士"]    {"公积金":0.2,"保险金":0.05,"养老金":0.1}
{"province":"广东省","city":"汕尾市","street":"马宫盐町村.","zip":516625}
萧红海    6800.0    []    {"公积金":0.15,"保险金":0.03,"养老金":0.1}
{"province":"广东省","city":"深圳市","street":"梅林路18号.","zip":518049}
李凤苑    6200.0    []    {"公积金":0.15,"保险金":0.03,"养老金":0.1}
{"province":"广东省","city":"广州市","street":"天河北路.","zip":510630}
Time taken: 0.101 seconds, Fetched: 4 row(s)
```

10.4 Hive 数据定义与数据操作

HiveQL 用于提供数据定义、数据操作、查询等。HiveQL 和 MySQL 的查询语言相近，但是还是存在显著差异，HiveQL 不支持行级插入操作、更新操作和删除操作，也不支持事务。

数据定义用于创建、修改和删除数据库、表、视图、函数和索引；数据操作用于装载数据和从表中抽取数据到表中。

10.4.1 基本概念

Hive 与其他 SQL 数据库一样，也有数据库、表、视图、索引，但与其他 SQL 数据库不同，数据存储在 HDFS 中，数据管理也不一样。

1. 数据库

数据库（DATABASE）：相当于关系数据库中的命名空间（namespace），它的作用是将用户和数据库的应用隔离到不同的数据库或模式中，Hive 0.6.0 之后的版本支持数据库，Hive 提供了 CREATE DATABASE dbname、USE dbname 及 DROP DATABASE dbname 等语句。

2. 表

表（TABLE）：Hive 的表逻辑上是由存储的数据和描述表格中的数据形式的相关元数据组成。表存储的数据存放在分布式文件系统中（如 HDFS），元数据存储在关系数据库中，当创建一张 Hive 的表，还没有为表加载数据时，该表在分布式文件系统（如 HDFS）上就是一个文件目录。Hive 里的表有两种类型：一种是表，这种表的数据文件存储在 Hive 的数据仓库中；另一种是外部表，这种表的数据文件可以存放在 Hive 数据仓库外部的分布式文件系统上，也可以放到 Hive 数据仓库中，Hive 的数据仓库也就是 HDFS 上的一个目录，这个目录是 Hive 数据文件存储的默认路径，它可以在 Hive 的配置文件中进行配置，最终也会存放到元数据库中。创建外部表 TABLE 要加关键字 EXTERNAL，同时还要用 LOCATION 指定文件存储的路径；如果不使用 LOCATION，数据文件也会放置到 Hive 的数据仓库中。DROP TABLE 是 Hive 删除表的命令，这两种表运行结果不一样，内部表执行 DROP TABLE 命令时会删除元数据和存储的数据，而外部表执行 DROP TABLE 命令时只删除元数据库中的数据，而不会删除存储的数据。另外，Hive 加载数据（LOAD DATA）时不会对元数据进行任何检查，只是简单地移动文件的位置，如果源文件格式不正确，也只有在做查询操作时才能发现，这时错误格式的字段会以 NULL 来显示。

3. 分区

分区（PARTITION）：Hive 中分区的概念是根据"分区列"的值对表的数据进行粗略

划分的机制,在 Hive 存储上就体现在表的主目录下的一个子目录,这个文件夹的名称就是用户定义的分区列的名称,分区列不是表中的某个字段,而是独立的列,根据这个列存储表中的数据文件。使用分区是为了加快数据分区的查询速度。

4. 桶

桶(BUCKET):TABLE 和 PARTITION 都是目录级别的拆分数据,BUCKET 则是对数据源数据文件本身拆分数据。使用桶的表会将源数据文件按一定规律拆分成多个文件,每个桶就是表(或分区)目录中的一个文件。

10.4.2 数据库管理

数据库管理主要有创建、修改、删除数据库等。
(1) 创建一个数据库,命令如下。

```
hive> CREATE DATABASE shops;
```

如果有同名数据库存在,那么将会抛出一个错误。可以使用如下语句,避免抛出错误。

```
hive> CREATE DATABASE IF NOT EXISTS shops;
```

(2) 查看 Hive 中包含的数据库,命令如下。

```
hive> SHOW DATABASES;
OK
default
shops
Time taken: 0.031 seconds, Fetched: 2 row(s)
```

查看所有以 s 开头并且以其他字符结尾(即.*部分)的数据库名,命令如下。

```
hive> SHOW DATABASES LIKE 's.*';
OK
shops
Time taken: 0.036 seconds, Fetched: 1 row(s)
```

(3) 修改数据库属性,命令如下。

```
hive> ALTER DATABASE shops SET DBPROPERTIES('edited-by' = 'Jie DBA');
```

(4) 查看数据库属性,命令如下。

```
hive> describe database extended shops;
OK
shops    hdfs://master:9000/hive/warehouse/shops.db    hadoop    USER
```

```
{edited-by=Jie DBA}
Time taken: 0.057 seconds, Fetched: 1 row(s)
```

(5) 删除数据库,命令如下。

```
hive> DROP DATABASE IF EXISTS shops;
```

默认情况下,Hive 是不允许用户删除一个包含表的数据库的。如果想直接删除数据库及其表,则语句如下。

```
hive> DROP DATABASE IF EXISTS shops CASCADE;
```

10.4.3 表的管理

表的管理主要有创建、修改、删除表等。使用 CREATE TABLE 语句创建表,CREATE TABLE 语句遵从 SQL 语法惯例,但 Hive 的这个语句中具有明显的功能扩展,使其可以具有广泛的灵活性。例如,可以定义表的数据文件存储位置、存储格式等。

(1) 在 test 数据库中创建表,命令如下。

```
CREATE DATABASE IF NOT EXISTS test;
CREATE TABLE test.salesperson (
    name STRING COMMENT 'Salesperson name',
    salary FLOAT COMMENT 'Salesperson salary',
    degree ARRAY<STRING> COMMENT 'Degree of salesperson',
    dues MAP<STRING, FLOAT>
        COMMENT 'Key are dues item, values are percentages',
    address STRUCT<province:STRING, city:STRING, street:STRING, zip:INT>
        COMMENT 'Home address'
)
LOCATION '/hive/warehouse/test.db/salesperson'
TBLPROPERTIES('creator'='me','create_at'='2017-10-10 20:58:00');
```

(2) 装载数据,命令如下。

```
LOAD DATA LOCAL INPATH '/root/salesperson'
OVERWRITE INTO TABLE test.salesperson;
```

(3) 查看数据库中的表,命令如下。

```
hive> SHOW TABLES;
OK
salesperson
Time taken: 0.033 seconds, Fetched: 1 row(s)
hive> SHOW TABLES IN test;
OK
salesperson
Time taken: 0.038 seconds, Fetched: 1 row(s)
```

(4）查看表结构详细信息。

可以使用 DESCRIBE EXTENDED mydb.employees 命令查看表结构详细信息，使用 FORMATTED 关键词代替 EXTENDED 可以查看更多信息。命令如下。

```
hive> DESCRIBE EXTENDED test.salesperson;
OK
name        string                      Salesperson name
salary      float                       Salesperson salary
degree      array<string>               Degree of salesperson
dues        map<string,float>           Key are dues item, values are percentages
address     struct<province:string,city:string,street:string,zip:int>  Home address

Detailed Table Information   Table(tableName:salesperson, dbName:test, owner:root,
…
location:hdfs://master:9000/hive/warehouse/test.db/salesperson ,
…
parameters:{transient_lastDdlTime=1507960335, creator=me, totalSize=425,
create_at=2017-10-10 20:58:00, numFiles=1}, viewOriginalText:null,
viewExpandedText:null, tableType:MANAGED_TABLE)
Time taken: 0.096 seconds, Fetched: 7 row(s)
```

如果用户只看某一列的信息，那么只要在表名后面加上字段名即可，命令如下。

```
OK
degree              array<string>           from deserializer
Time taken: 0.069 seconds, Fetched: 1 row(s)
```

（5）表的改名。

Hive 支持数据库改名。例如，表 salesperson 改名为 salesmen，命令如下。

```
ALTER TABLE salesperson RENAME TO salesmen;
```

（6）表的删除。

Hive 也支持 DROP TABLE 命令。例如，删除 salesmen 表，命令如下。

```
DROP TABLEsalesmen;
```

（7）增加列。

Hive 可以在分区字段之前增加新的字段到已有字段之后，命令如下。

```
ALTER TABLE salesperson ADD COLUMNS (
    age INT COMMENT 'Salesperson age',
    lastname STRING COMMENT 'Salesperson lastname'
);
```

查看表结构,命令如下。

```
hive> DESCRIBE salesperson;
```

(8) 修改列。

可以对某个字段进行重新命名,命令如下。

```
ALTER TABLE salesperson
CHANGE COLUMN name sale_name STRING
COMMENT 'Salesman name';
```

(9) 替换列(删除列)。

使用命令 REPLACE COLUMNS 替换或删除列,命令如下。

```
ALTER TABLE salesperson REPLACE COLUMNS (
    name STRING COMMENT 'Salesperson name',
    salary FLOAT COMMENT 'Salesperson salary',
    degree ARRAY < STRING > COMMENT 'Degree of salesperson',
    dues MAP < STRING, FLOAT >
            COMMENT 'Key are dues item, values are percentages',
    address STRUCT < province:STRING, city:STRING, street:STRING, zip:INT >
            COMMENT 'Home address'
);
```

(10) 修改表属性。

用户可以增加表属性或修改已经存在的属性,但无法删除属性,命令如下。

```
ALTER TABLE salesperson
SET TBLPROPERTIES ( 'notes' = 'This is test tables');
```

查看表属性,命令如下。

```
hive> DESCRIBE FORMATTED salesperson;
```

10.4.4 外部表的管理

Hive 创建的表默认情况下会将表的数据存储在 hive.metastore.warehouse.dir 定义的目录下。但一些数据库已经在 HDFS 中,不在 hive.metastore.warehouse.dir 定义的目录下,则可以创建一个外部表指向这个数据库。创建外部表与创建表基本一样,外部表使用关键词 EXTERNAL。

(1) 创建外部表。假如需要分析股票数据,则会定期从财经网站下载股市行情数据,这些数据位于 HDFS 的/data/stocks 目录下。可以创建一个外部表,命令如下。

```
CREATE EXTERNAL TABLE IF NOT EXISTS stocks (
    symbol          STRING,
    stocks_name     STRING,
    price_new       FLOAT,
    chg             FLOAT,
    fluctuation     FLOAT,
    purchase        FLOAT,
    sell            FLOAT,
    volume          INT,
    amount          INT,
    price_open      FLOAT,
    prev_close      FLOAT,
    price_high      FLOAT,
    price_low       FLOAT
)
ROW FORMAT DELIMITED FIELDS TERMINATED BY ','
LOCATION '/data/stocks';
```

关键词 EXTERNAL 说明 Hive 这个表是外部的，而 LOCATION 则说明 Hive 数据所在位置。

（2）查询股票涨跌幅最大前 5 只股票，命令如下。

```
hive> SELECT symbol,stocks_name,chg,price_new FROM stocks ORDER BY chg DESC LIMIT 5;
OK
sh603367    辰欣药业    0.44      16.79
sh601086    国芳集团    0.4399    4.55
sh603363    傲农生物    0.1006    9.19
sh600353    旭光股份    0.1005    8.32
sh600460    士兰微      0.1003    10.42
Time taken: 41.233 seconds, Fetched: 5 row(s)
```

（3）对一张存在的表进行表结构复制（不会复制数据），命令如下。

```
CREATE EXTERNAL TABLE IF NOT EXISTS stocks_sch
LIKE stocks
LOCATION '/data/stocks_sch';
```

查询 stocks_sch 表，数据没有复制过来，命令如下。

```
hive> SELECT * FROM stocks_sch;
OK
Time taken: 1.313 seconds
```

（4）表与外部表的转换。

Hive 创建表时，会将数据移动到数据仓库指向的路径；如果创建外部表，仅记录数据所在的路径，不对数据的位置做任何改变。在删除表的时候，表的元数据和数据会被一起删除,而外部表只删除元数据,不删除数据。这样外部表更加安全,数据组织也更加灵活,方便共享源数据。有时表与外部表需要相互转换,操作如下。

```
ALTER TABLE stocks SET TBLPROPERTIES ('EXTERNAL' = 'FALSE');   //外部表转换为表
ALTER TABLE stocks SET TBLPROPERTIES ('EXTERNAL' = 'TRUE');    //表转换为外部表
```

10.4.5 分区管理

数据分区可以有多种形式,但是通常使用分区将数据从物理上转移到和使用最频繁的用户更近的地方,分散压力。

(1) 创建带分区的表,命令如下。

```
DROP TABLE IF EXISTS salesperson;
CREATE TABLE salesperson (
    name STRING COMMENT 'Salesperson name',
    salary FLOAT COMMENT 'Salesperson salary',
    degree ARRAY < STRING > COMMENT 'Degree of salesperson',
    dues MAP < STRING, FLOAT >
            COMMENT 'Key are dues item, values are percentages',
    address STRUCT < province:STRING, city:STRING, street:STRING, zip:INT >
            COMMENT 'Home address'
)
PARTITIONED BY (country STRING, province STRING);
```

(2) 导入数据。导入国家为 CN 和省份为 GD 的销售员信息,命令如下。

```
LOAD DATA LOCAL INPATH '/root/salesperson'
OVERWRITE INTO TABLE salesperson
PARTITION ( country = 'CN', province = 'GD' );
```

导入国家为 CHN 和省份为 GDS 的销售员信息,命令如下。

```
LOAD DATA LOCAL INPATH '/root/salesperson'
OVERWRITE INTO TABLE salesperson
PARTITION ( country = 'CHN', province = 'GDS' );
```

分区改变了 Hive 对数据存储的组织方式,Hive 将会创建反映分区结构的子目录,目录如下。

```
/hive/warehouse/salesperson/country = CHN/province = GDS
/hive/warehouse/salesperson/country = CN/province = GD
```

(3) 查询国家为 CN 和省份为 GD 的销售员信息,命令如下。

```
SELECT * FROM salesperson WHERE country = 'CN' AND province = 'GD';
```

(4) 通过 SHOW PARTITIONS 命令查看表中分区,命令如下。

```
hive> SHOW PARTITIONS salesperson;
OK
country = CHN/province = GDS
country = CN/province = GD
Time taken: 0.107 seconds, Fetched: 2 row(s)

hive> SHOW PARTITIONS salesperson PARTITION ( country = 'CN', province = 'GD');
OK
country = CN/province = GD
Time taken: 0.204 seconds, Fetched: 1 row(s)
```

(5) 外部表分区。外部表同样可以使用分区。在管理大型生产数据时经常要分区,可以优化查询性能。用户可以自己定义目录结构。

例如,表 stocks_pt 结构如下。

```
CREATE EXTERNAL TABLE IF NOT EXISTS stocks_pt (
    symbol          STRING,
    stocks_name     STRING,
    price_new       FLOAT,
    chg             FLOAT,
    fluctuation     FLOAT,
    purchase        FLOAT,
    sell            FLOAT,
    volume          INT,
    amount          INT,
    price_open      FLOAT,
    prev_close      FLOAT,
    price_high      FLOAT,
    price_low       FLOAT
)
PARTITIONED BY ( year INT, month INT, day INT )
ROW FORMAT DELIMITED FIELDS TERMINATED BY ','
LOCATION '/data/stocks_pt';
```

增加一个 2017 年 9 月 29 日的分区,操作如下。

```
ALTER TABLE stocks_pt ADD PARTITION ( year = 2017, month = 9, day = 29 )
LOCATION '/data/stocks_pt/2017/9/29';
```

上传 2017 年 9 月 29 日股票数据到 HDFS,操作如下。

```
# hdfs dfs - put stocks201709291511 /data/stocks_pt/2017/9/29
```

查询 2017 年 9 月 29 日成交量最大的 5 只股票,操作如下。

```
SELECT symbol,stocks_name,volume FROM stocks_pt
WHERE year = 2017 AND month = 9 AND day = 29
ORDER BY volume DESC
LIMIT 5;
```

查询程序运行结果如下。

```
OK
sh600050    中国联通    2529695
sh600307    酒钢宏兴    1947047
sh600460    士兰微     1869326
sh600490    鹏欣资源    1712598
sh600711    盛屯矿业    1582613
Time taken: 43.163 seconds, Fetched: 5 row(s)
```

另外,用户还可以修改某个分区的路径,示例如下。

```
ALTER TABLE stocks_pt PARTITION ( year = 2017, month = 9, day = 29 )
SET LOCATION '/data/stocks_pt/20170929';
```

删除某个分区,示例如下。

```
ALTER TABLE stocks_pt
DROP IF EXISTS PARTITION ( year = 2017, month = 9, day = 29 );
```

10.4.6 数据操作

Hive 没有行级别的数据插入、数据更新和删除操作,向表中输入数据一般使用数据装载操作 LOAD DATA,或者将文件直接上传到 HDFS 正确目录下,也可以使用查询语句向表插入数据。

(1) 使用查询语句装载数据,命令如下。

```
CREATE TABLE IF NOT EXISTS salesman LIKE salesperson;

INSERT OVERWRITE TABLE salesman
PARTITION ( country = 'CN', province = 'GD' )
SELECT name, salary, degree, dues, address FROM salesperson
WHERE country = 'CN' AND province = 'GD';
```

(2) 使用查询语句装载多分区数据。Hive 提供了一个动态分区功能,其基于查询参数推断需要创建的分区名称,动态分区功能默认情况下没有开启。默认是以 strict 模式执行,在这种模式下至少有一列分区字段是静态的。

```
SET hive.exec.dynamic.partition = true;              //开启动态分区功能
SET hive.exec.dynamic.partition.mode = nonstrict;    //允许所有分区都是动态的
```

```
SET hive.exec.max.dynamic.partitions.pernode = 1000;   //创建的最大动态分区个数

CREATE TABLE IF NOT EXISTS salesmen LIKE salesperson;

INSERT OVERWRITE TABLE salesmen
PARTITION ( country , province)
SELECT * FROM salesperson;
```

(3) 使用查询语句创建表。Hive 可以使用查询语句创建表并插入数据。

例如,查询今天股票价格最低的 5 只股票并插入新生成的表中,命令如下。

```
CREATE TABLE stocks_min_price
AS
SELECT symbol,stocks_name,price_open FROM stocks ORDER BY price_open LIMIT 5;

hiv > DESCRIBE stocks_min_price ;
OK
symbol              string
stocks_name         string
price_open          float
Time taken: 0.053 seconds, Fetched: 3 row(s)

hive > SELECT * FROM stocks_min_price;
OK
sh600664    哈药股份      0.0
sh600666    奥瑞德        0.0
sh603111    康尼机电      0.0
sh603039    泛微网络      0.0
sh603729    龙韵股份      0.0
Time taken: 0.066 seconds, Fetched: 5 row(s)
```

(4) 导出数据。Hive 导出数据一般可以使用 HDFS 命令,也可以使用 INSERT … DIRECTORY …。

例如,导出成交额最大的 5 只股票,命令如下。

```
INSERT OVERWRITE LOCAL DIRECTORY '/root/stocks_max'
SELECT symbol,stocks_name,amount FROM stocks
ORDER BY amount DESC
LIMIT 5;
```

查看导出结果如下。

```
# cat /root/stocks_max/ *
sh600522^A 中天科技^A2002837911
sh600887^A 伊利股份^A1911553914
sh600490^A 鹏欣资源^A1890909166
```

```
sh600050^A中国联通^A1877329673
sh600516^A方大炭素^A1865221540
```

10.4.7 桶的操作

使用 BUCKET,需要先打开 Hive 对桶的控制,然后将数据插入到表中。

(1) 创建临时表。

例如,先建立临时表 student_tmp,并导入数据,命令如下。

```
CREATE TABLE student_tmp (
    id INT,
    age INT,
    name STRING,
    state_date STRING
);

LOAD DATA LOCAL INPATH '/root/student'
OVERWRITE INTO TABLE student_tmp;
```

(2) 查询临时表数据,命令如下。

```
hive> SELECT * FROM student_tmp;
OK
61021501    20    张汕尾    20170915
61021502    18    李海丰    20170915
61021503    19    陈可塘    20170915
61021504    21    朱和顺    20170915
61021505    23    谢莲塘    20170915
61021506    22    萧后径    20170915
61021507    24    郑新楼    20170915
Time taken: 0.091 seconds, Fetched: 7 row(s)
```

(3) 建立表 student,设置为 3 个桶,命令如下。

```
CREATE TABLE student (
    id INT,
    age INT,
    name STRING
)
PARTITIONED BY (state_date STRING)
CLUSTERED BY (id)
SORTED BY(age ASC) INTO 3 BUCKETS
ROW FORMAT DELIMITED FIELDS TERMINATED BY ',';
```

(4) 查询临时表,结果插入 student 表,命令如下。

```
SET hive.enforce.bucketing = TRUE;

FROM student_tmp
INSERT OVERWRITE TABLE student PARTITION ( state_date = '20170915')
SELECT id, age, name WHERE state_date = '20170915' SORT BY age ASC;
```

(5) 查看 HDFS 目录,有 3 个文件,内容如下。

```
# hdfs dfs - cat /hive/warehouse/student/state_date = 20170915/000000_0
61021503,19,陈可塘
61021506,22,萧后径
# hdfs dfs - cat /hive/warehouse/student/state_date = 20170915/000001_0
61021501,20,张汕尾
61021504,21,朱和顺
61021507,24,郑新楼
# hdfs dfs - cat /hive/warehouse/student/state_date = 20170915/000002_0
61021502,18,李海丰
61021505,23,谢莲塘
```

10.4.8 索引

Hive 只有有限的索引功能,没有键的概念,但还是可以对一些字段建立索引来加速查询。索引 key 冗余存储,提供基于 key 的数据视图,存储设计以优化查询和检索性能,对于某些查询减少 IO,从而提高性能。

(1) 对字段 name 建立索引,命令如下。

```
CREATE INDEX salesmen_index
ON TABLE salesmen ( name )
AS 'org.apache.hadoop.hive.ql.index.compact.CompactIndexHandler'
WITH DEFERRED REBUILD
IDXPROPERTIES ( 'creator' = 'me' )
IN TABLE salesmen_index_table
COMMENT 'Salesmen indexed by name ';
```

Hive v0.8.0 版本中新增了一个内置 Bitmap 索引处理器。Bitmap 索引普遍应用于排后值较少的列,命令如下。

```
CREATE INDEX salesmen_index
ON TABLE salesmen ( name )
AS 'BITMAP'
WITH DEFERRED REBUILD
IDXPROPERTIES ( 'creator' = 'me' )
IN TABLE salesmen_index_table
COMMENT 'Salesmen indexed by name ';
```

(2) 重建索引。如果用户指定了 WITH DEFERRED REBUILD,那么新索引将呈现空白状态。在任何时候,都可以使用 ALTER INDEX 对索引进行重建,省略 PARTITION,则对所有分区进行重建索引,示例如下。

```
ALTER INDEX salesmen_index
ON salesmen
PARTITION ( country = 'CN',  province = 'GD' )
REBUILD ;
```

(3) 显示索引,命令如下。

```
hive> SHOW FORMATTED INDEX ON salesmen;
OK
idx_name            tab_name          col_names         idx_tab_name
idx_type            comment

salesmen_index      salesmen          name              salesmen_index_table
compact             Salesmen indexed by name
Time taken: 0.059 seconds, Fetched: 4 row(s)
```

(4) 查询索引表中的数据,命令如下。

```
hive> SELECT * FROM salesmen_index_table ;
OK
李凤苑
hdfs://master:9000/hive/warehouse/salesmen/country = CN/province = GD/000000_0
[334]    CN    GD
王可塘
hdfs://master:9000/hive/warehouse/salesmen/country = CN/province = GD/000000_0
[0]    CN    GD
萧红海
hdfs://master:9000/hive/warehouse/salesmen/country = CN/province = GD/000000_0
[229]    CN    GD
陈海峰
hdfs://master:9000/hive/warehouse/salesmen/country = CN/province = GD/000000_0
[118]    CN    GD
Time taken: 0.16 seconds, Fetched: 4 row(s)
```

(5) 删除索引。如果有索引表,删除一个索引将会删除这个索引表,命令如下。

```
DROP INDEX IF EXISTS salesmen_index ON salesmen;
```

10.5 Hive 数据查询

Hive 数据查询语句类似于 SQL 查询语句,在使用 HiveQL 时,应该注意其语法、特性及查询语句对性能的影响。SELECT … FROM 是查询使用最频繁的语句。FROM 子句标识从哪个表、视图或嵌套查询中选择的记录。

10.5.1 简单查询

（1）查询销售员的姓名、工资，命令如下。

```
hive> SELECT name, salary FROM salesperson;
OK
王可塘      12000.0
陈海峰      8600.0
萧红海      6800.0
李凤苑      6200.0
Time taken: 0.143 seconds, Fetched: 4 row(s)
```

（2）查询销售员的姓名、学位，ARRAY类型的值括在"[…]"内，命令如下。

```
hive> SELECT name, degree FROM salesperson;
OK
王可塘   ["学士","硕士"]
陈海峰   ["学士"]
萧红海   []
李凤苑   []
Time taken: 0.211 seconds, Fetched: 4 row(s)
```

（3）查询销售员的姓名、应扣款，MAP和STRUCT类型的值括在"{…}"内，命令如下。

```
hive> SELECT name, dues FROM salesperson;
OK
王可塘   {"公积金":0.2,"保险金":0.05,"养老金":0.1}
陈海峰   {"公积金":0.2,"保险金":0.05,"养老金":0.1}
萧红海   {"公积金":0.15,"保险金":0.03,"养老金":0.1}
李凤苑   {"公积金":0.15,"保险金":0.03,"养老金":0.1}
Time taken: 0.134 seconds, Fetched: 4 row(s)
```

（4）查询销售员的姓名、第1个学位、公积金、城市，命令如下。

```
hive> SELECT name, degree[0], dues["公积金"], address.city FROM salesperson;
OK
王可塘    学士     0.2      汕尾市
陈海峰    学士     0.2      汕尾市
萧红海    NULL    0.15     深圳市
李凤苑    NULL    0.15     广州市
Time taken: 0.77 seconds, Fetched: 4 row(s)
```

(5) 查询销售员的姓名、实发工资(扣除应扣款),命令如下。

```
hive> SELECT name, salary * (1 - dues["公积金"] - dues["保险金"] - dues["养老金"])
    > FROM salesperson;
OK
王可塘    7799.9995
陈海峰    5590.0
萧红海    4896.0
李凤苑    4464.0
Time taken: 0.123 seconds, Fetched: 4 row(s)
```

(6) 聚合查询。查询销售员人数、平均工资、工资总额、最高工资、最低工资,命令如下。

```
hive> SELECT count(*), sum(salary), avg(salary),max(salary),min(salary)
    > FROM salesperson;
OK
4    33600.0    8400.0    12000.0    6200.0
Time taken: 48.963 seconds, Fetched: 1 row(s)
```

10.5.2 复杂查询

1. 嵌套查询

查询实发工资(扣除应扣款)大于 5000 元的销售员姓名、工资、应扣款和实发工资,命令如下。

```
hive> FROM (
    > SELECT name, salary, dues,
    > salary * (1 - dues["公积金"] - dues["保险金"] - dues["养老金"]) as net_salary
    > FROM salesperson
    > ) s
    > SELECT s.name, s.salary, s.dues, s.net_salary
    > WHERE s.net_salary > 5000;
OK
王可塘    12000.0    {"养老金":0.1,"公积金":0.2,"保险金":0.05}    7799.9995
陈海峰    8600.0     {"养老金":0.1,"公积金":0.2,"保险金":0.05}    5590.0
Time taken: 0.14 seconds, Fetched: 2 row(s)
```

也可以使用下面命令查询。

```
hive> SELECT s.* FROM (
    >   SELECT name, salary, dues,
    >   salary * (1 - dues["公积金"] - dues["保险金"] - dues["养老金"]) as net_salary
    >   FROM salesperson ) s
    > WHERE s.net_salary > 5000 ;
```

2. CASE 查询

查询销售员工资,工资高于 10 000 元为"收入高",低于 8000 元为"收入低",中间为"收入中",命令如下。

```
hive> SELECT name, salary, CASE
    >       WHEN salary >= 10000 THEN '收入高'
    >       WHEN salary >= 8000 AND salary < 10000 THEN '收入中'
    >       ELSE '收入低'
    > END   AS bracket FROM   salesperson;
OK
王可塘    12000.0    收入高
陈海峰    8600.0     收入中
萧红海    6800.0     收入低
李凤苑    6200.0     收入低
Time taken: 0.133 seconds, Fetched: 4 row(s)
```

3. 浮点数比较

查询公积金大于 0.15 的销售员姓名、工资、公积金,命令如下。

```
hive> SELECT name, salary, dues["公积金"] FROM salesperson
    > WHERE dues["公积金"] > 0.15;
OK
王可塘    12000.0    0.2
陈海峰    8600.0     0.2
萧红海    6800.0     0.15
李凤苑    6200.0     0.15
Time taken: 0.11 seconds, Fetched: 4 row(s)
```

注意,两条 dues["公积金"] = 0.15 的记录也被输出。由于 dues 的 map 值为 FLOAT 类型,对于数字 0.15,Hive 会将该值保存为 DOUBLE 类型,值是 0.150000000001;FLOAT 类型的 0.15 值是 0.1500001。因此 FLOAT 类型的 0.15 比 DOUBLE 类型的 0.15 大。可以将 0.15 设为 FLOAT 类型,命令如下。

```
hive> SELECT name, salary, dues["公积金"] FROM salesperson
    > WHERE dues["公积金"] > CAST(0.15 AS FLOAT);
OK
王可塘    12000.0    0.2
陈海峰    8600.0     0.2
Time taken: 0.132 seconds, Fetched: 2 row(s)
```

4. LIKE 和 RLIKE

LIKE 和 RLIKE 是谓词操作符。LIKE 使用 SQL 简单正则表达式,"%"表示任意 0 个或多个字符;"_"表示任意单个字符;"[ ]"表示括号内所有字符中的一个;"[^]"表示

不在括号所列之内的单个字符；查询特殊字符"％""_""["用"[ ]"括起便可查询。RLIKE通过Java的正则表达式进行匹配，"."表示任意单个字符；" * "表示重复"左边字符串"0次到多次；"(X|Y)"表示匹配X或Y；" ^ "表示开头；" $ "表示结尾。

例如，查询代码为"123"结尾的股票的代码、名称和最新价，命令如下。

```
hive> SELECT symbol, stocks_name, price_new FROM stocks
    > WHERE symbol LIKE "%123";
OK
sh603123    翠微股份    8.73
sh600123    兰花科创    9.6
Time taken: 0.134 seconds, Fetched: 2 row(s)
```

例如，查询代码为"sh60368"开头和"123"结尾的股票的代码、名称和最新价，命令如下。

```
ive> SELECT symbol, stocks_name, price_new FROM stocks
    > WHERE symbol RLIKE "(^sh60368.*|.*123$)";
OK
sh603686    龙马环卫    32.75
sh603688    石英股份    17.52
sh603123    翠微股份    8.73
sh600123    兰花科创    9.6
sh603689    皖天然气    15.94
Time taken: 0.064 seconds, Fetched: 5 row(s)
```

5. GROUP BY

GROUP BY语句通常和聚合函数一起使用，按照一个列或多个列分组计算。

例如，查询各城市销售人员的平均工资，命令如下。

```
hive> SELECT address.city, avg(salary) FROM salesperson
    > GROUP BY address.city;
OK
广州市    6200.0
汕尾市    10300.0
深圳市    6800.0
Time taken: 44.926 seconds, Fetched: 3 row(s)
```

HAVING语句允许对GROUP BY语句产生的分组进行条件过滤。

例如，查询平均工资大于6500元的情况，命令如下。

```
hive> SELECT address.city, avg(salary) avg_salary FROM salesperson
    > GROUP BY address.city
    > HAVING avg_salary > 6500;
OK
汕尾市    10300.0
```

```
深圳市    6800.0
Time taken: 34.92 seconds, Fetched: 2 row(s)
```

不使用 HAVING 语句,使用嵌套查询也可以实现,命令如下。

```
hive> SELECT s.city, s.avg FROM (
    >   SELECT address.city AS city, avg(salary) AS avg FROM salesperson
    >   GROUP BY address.city) s
    > WHERE s.avg > 6500;
```

10.5.3　JOIN 连接查询

Hive 支持 SQL JOIN 语句,但是只支持等值连接。

1. INNER JOIN

内连接(INNER JOIN)中,只有进行连接的两个表都存在并且匹配的数据才能查询出来。

例如,查找 stocks 表中最新价(小数点有效位两位)相等的两只股票信息,并且最新价大于 45,命令如下。

```
hive> SELECT a.symbol, a.stocks_name, a.price_new, b.symbol, b.stocks_name
    > FROM stocks a JOIN stocks b
    > ON a.price_new = b.price_new
    > WHERE a.symbol <> b.symbol AND a.price_new > 45 ;
OK
sh603288    海天味业    47.39    sh603337    杰克股份
sh603127    昭衍新药    59.0     sh603579    荣泰健康
sh603579    荣泰健康    59.0     sh603127    昭衍新药
sh603337    杰克股份    47.39    sh603288    海天味业
Time taken: 41.313 seconds, Fetched: 4 row(s)
```

下面使用自连接,先创建两张表:stocks_one 和 stocks_two,导入数据。

```
CREATE TABLE IF NOT EXISTS stocks_one LIKE stocks ;
CREATE TABLE IF NOT EXISTS stocks_two LIKE stocks ;

LOAD DATA LOCAL INPATH '/root/stocks_one'
OVERWRITE INTO TABLE stocks_one;

LOAD DATA LOCAL INPATH '/root/stocks_two'
OVERWRITE INTO TABLE stocks_two;
```

查找 stocks_one 表中最新价(小数点有效位两位)等于 stocks_two 表中某只股票最新价的信息,并且最新价大于 45,命令如下:

```
hive> SELECT a.symbol, a.stocks_name, a.price_new,
    >        b.symbol, b.stocks_name, b.price_new
    > FROM stocks_one a JOIN stocks_two b
    > ON a.price_new = b.price_new
    > WHERE a.price_new > 45 OR b.price_new > 45;
OK
sh603337    杰克股份    47.39    sh603288    海天味业    47.39
sh603579    荣泰健康    59.0     sh603127    昭衍新药    59.0
Time taken: 43.406 seconds, Fetched: 2 row(s)
```

2. OUTER JOIN

外连接(OUTER JOIN)可分为左外连接(LEFT OUTER JOIN)、右外连接(RIGHT OUTER JOIN)和全外连接(FULL OUTER JOIN)。

例如,使用左外连接(LEFT OUTER JOIN)查询如下。

```
hive> SELECT a.symbol, a.stocks_name, a.price_new,
    >        b.symbol, b.stocks_name, b.price_new
    > FROM stocks_one a LEFT OUTER JOIN stocks_two b
    > ON a.price_new = b.price_new
    > WHERE a.price_new > 45 OR b.price_new > 45;
OK
...
sh603833    欧派家居    102.11   NULL        NULL        NULL
sh603337    杰克股份    47.39    sh603288    海天味业    47.39
sh603579    荣泰健康    59.0     sh603127    昭衍新药    59.0
sh603933    睿能科技    52.68    NULL        NULL        NULL
...
Time taken: 40.386 seconds, Fetched: 27 row(s)
```

使用右外连接(RIGHT OUTER JOIN)查询如下。

```
hive> SELECT a.symbol, a.stocks_name, a.price_new,
    >        b.symbol, b.stocks_name, b.price_new
    > FROM stocks_one a RIGHT OUTER JOIN stocks_two b
    > ON a.price_new = b.price_new
    > WHERE a.price_new > 45 OR b.price_new > 45;
OK
...
NULL        NULL        NULL     sh603883    老百姓      48.71
sh603579    荣泰健康    59.0     sh603127    昭衍新药    59.0
NULL        NULL        NULL     sh603986    兆易创新    130.26
NULL        NULL        NULL     sh603387    基蛋生物    53.84
NULL        NULL        NULL     sh603716    塞力斯      84.44
sh603337    杰克股份    47.39    sh603288    海天味业    47.39
NULL        NULL        NULL     sh603444    吉比特      219.11
...
Time taken: 34.676 seconds, Fetched: 38 row(s)
```

使用全外连接(FULL OUTER JOIN)查询如下。

```
hive> SELECT a.symbol, a.stocks_name, a.price_new,
    >        b.symbol, b.stocks_name, b.price_new
    > FROM stocks_one a FULL OUTER JOIN stocks_two b
    > ON a.price_new = b.price_new
    > WHERE a.price_new > 45 OR b.price_new > 45;
OK
sh603179    新泉股份    45.5     NULL       NULL       NULL
NULL        NULL        NULL     sh603358   华达科技    45.58
NULL        NULL        NULL     sh603320   迪贝电气    46.36
sh603305    旭升股份    46.65    NULL       NULL       NULL
sh603658    安图生物    46.98    NULL       NULL       NULL
sh603337    杰克股份    47.39    sh603288   海天味业    47.39
NULL        NULL        NULL     sh603038   华立股份    47.8
NULL        NULL        NULL     sh603881   数据港      47.9
...
sh603579    荣泰健康    59.0     sh603127   昭衍新药    59.0
sh600276    恒瑞医药    59.93    NULL       NULL       NULL
sh603096    新经典      60.08    NULL       NULL       NULL
NULL        NULL        NULL     sh603602   纵横通信    61.82
sh603183    建研院      62.27    NULL       NULL       NULL
...
Time taken: 37.74 seconds, Fetched: 63 row(s)
```

3. LEFT SEMI JOIN

Hive 当前没有实现 IN 或 EXISTS 子查询,所以可以用左半开连接(LEFT SEMI JOIN)重写用户的子查询语句。半开连接的限制是 JOIN 子句中右边的表只能在 ON 子句中设置过滤条件。

例如,查找 stocks_one 表中最新价(小数点有效位两位)等于 stocks_two 表中某只股票最新价的信息,并且最新价为 8.3,命令如下。

```
hive> SELECT a.symbol, a.stocks_name, a.price_new
    > FROM stocks_one a LEFT SEMI JOIN stocks_two b
    > ON a.price_new = b.price_new AND a.price_new = 8.3;
OK
sh603366    日出东方    8.3
sh600919    江苏银行    8.3
Time taken: 47.082 seconds, Fetched: 2 row(s)
```

4. 笛卡儿积

笛卡儿积(JOIN)是一种连接,表示左边表的行数乘右边表的行数等于笛卡儿积结果集的大小。笛卡儿积会产生大量的数据,笛卡儿积不是并行执行的,而且使用 MapReduce 计算架构的话,无法进行优化,因此应该避免使用笛卡儿积查询。

5. Map-Side JOIN

Map-Side JOIN 的使用场景是一个大表和一个小表的连接操作，其中，"小表"是指文件足够小，可以加载到内存中。该算法可以将 JOIN 算子执行在 Map 端，无须经历 shuffle 和 reduce 等阶段，因此效率非常高。

10.5.4 其他语句

1. ORDER BY 和 SORT BY

如果在 HADOOP 上进行 ORDER BY 全排序，则会导致所有的数据集中在一台 reducer 节点上，然后进行排序，这样很可能会超过单个节点的磁盘和内存存储能力而导致任务失败。Hive 增加了一个可选方案 DISTRIBUTE BY ＋ SORT BY，被 DISTRIBUTE BY 设定的字段为 KEY，数据会被 HASH 分发到不同的 Reducer 机器上，然后 SORT BY 会对同一个 Reducer 机器上的每组数据进行局部排序，示例如下。

```
hive> SELECT id, age, name FROM student ORDER BY age;
OK
61021502    18    李海丰
61021503    19    陈可塘
61021501    20    张汕尾
61021504    21    朱和顺
61021506    22    萧后径
61021505    23    谢莲塘
61021507    24    郑新楼
Time taken: 33.74 seconds, Fetched: 7 row(s)

hive> SELECT id, age, name FROM student
    > DISTRIBUTE BY id SORT BY id ASC, age DESC;
OK
61021501    20    张汕尾
61021502    18    李海丰
61021503    19    陈可塘
61021504    21    朱和顺
61021505    23    谢莲塘
61021506    22    萧后径
61021507    24    郑新楼
Time taken: 37.126 seconds, Fetched: 7 row(s)
```

2. CLUSTER BY

CLUSTER BY 的功能就是 DISTRIBUTE BY 和 SORT BY 相结合，CLUSTER BY 指定的列不能指定 ASC 和 DESC，只能升序。

如下两条语句是等价的。

```
SELECT id, age, name FROM student DISTRIBUTE BY id SORT BY age ;
SELECT id, age, name FROM studentCLUSTER BY age ;
```

3．抽样查询

抽样可以从被抽取的数据中估计和推断出整体的特性，是科学实验、质量检验、社会调查普遍采用的一种经济有效的工作和研究方法。Hive 支持桶抽样和块抽样，桶表是指在创建表时使用 CLUSTERED BY 子句创建了桶的表。桶表抽样的语法如下。

```
table_sample: TABLESAMPLE (BUCKET x OUT OF y [ON colname])
```

TABLESAMPLE 子句允许用户编写用于数据抽样而不是整个表的查询，该子句出现在 FROM 子句中，可用于任何表中。桶编号从 1 开始，colname 表明抽取样本的列，可以是非分区列中的任意一列，或者使用 rand() 表明在整个行中抽取样本而不是单个列。在 colname 上分桶的行随机进入 1 到 y 个桶中，返回属于桶 x 的行。

例如，使用 rand() 函数对 student 表进行抽样，返回 3 个桶中的第 2 个桶的随机行，命令如下。

```
hive > SELECT * FROM student TABLESAMPLE(BUCKET 2 OUT OF 3 ON rand()) s;
OK
61021506    22    萧后径    20170915
61021501    20    张汕尾    20170915
61021504    21    朱和顺    20170915
61021507    24    郑新楼    20170915
Time taken: 0.112 seconds, Fetched: 4 row(s)
```

从 Hive-0.8 开始可以使用块抽样，语法如下。

```
block_sample: TABLESAMPLE (n PERCENT)
block_sample: TABLESAMPLE (ByteLengthLiteral)
ByteLengthLiteral : (Digit) + ('b' | 'B' | 'k' | 'K' | 'm' | 'M' | 'g' | 'G')
block_sample: TABLESAMPLE (n ROWS)
```

块抽样查询示例如下。

```
hive > SELECT * FROM student TABLESAMPLE(30 PERCENT) s;
OK
61021503    19    陈可塘    20170915
61021506    22    萧后径    20170915
61021501    20    张汕尾    20170915
Time taken: 0.113 seconds, Fetched: 3 row(s)

hive > SELECT * FROM student TABLESAMPLE(30B) s;
OK
61021503    19    陈可塘    20170915
```

```
61021506   22   萧后径   20170915
Time taken: 0.064 seconds, Fetched: 2 row(s)

hive> SELECT * FROM student TABLESAMPLE(4 ROWS) s;
OK
61021503   19   陈可塘   20170915
61021506   22   萧后径   20170915
61021501   20   张汕尾   20170915
61021504   21   朱和顺   20170915
Time taken: 0.103 seconds, Fetched: 4 row(s)
```

4. UNION ALL

UNION ALL 可以将两个或多个表进行合并,每个 UNION 子查询必须都有相同的列,而且对应的每个字段的字段类型必须是一致的,示例如下。

```
hive> SELECT * FROM (
    >   SELECT "Hello" AS a, ARRAY(1,2,3)
    >   UNION ALL
    >   SELECT "World" , ARRAY(4,5)
    > ) hw ;
Total MapReduce CPU Time Spent: 2 seconds 350 msec
OK
Hello     [1,2,3]
World     [4,5]
Time taken: 28.585 seconds, Fetched: 2 row(s)
```

5. 视图

视图(view)是从一个或多个表(或视图)导出的表,是一个逻辑结构,本身不会存储数据,Hive 中的视图的作用总的来说就是为了简化查询语句。

例如,创建视图,命令如下。

```
CREATE VIEW view_salesperson(name, salary, city) AS
SELECT name,salary,address.city FROM salesperson
WHERE salary > 6500;
```

使用视图查询,命令如下。

```
hive> SELECT * FROM view_salesperson WHERE city = "汕尾市";
OK
王可塘    12000.0    汕尾市
陈海峰    8600.0     汕尾市
Time taken: 0.214 seconds, Fetched: 2 row(s)
```

修改视图属性，命令如下。

```
ALTER VIEW view_salesperson
SET TBLPROPERTIES ( 'create_at' = 'shanwei' );
```

查看视图属性，命令如下。

```
hive > DESCRIBE FORMATTED view_salesperson;
OK
# col_name              data_type           comment
name                    string
salary                  float
city                    string

# Detailed Table Information
...
Table Parameters:
  create_at             shanwei
  last_modified_by      root
  last_modified_time    1508284662
  transient_lastDdlTime 1508284662

# Storage Information
...
# View Information
View Original Text:    SELECT name,salary,address.city FROM salesperson
WHERE salary > 6500
View Expanded Text:    SELECT 'name' AS 'name', 'salary' AS 'salary', 'city' AS 'city' FROM (SELECT
'salesperson'.'name','salesperson'.'salary','salesperson'.'address'.'city' FROM 'default'.
'salesperson'
WHERE 'salesperson'.'salary'> 6500) 'default.view_salesperson'
Time taken: 0.079 seconds, Fetched: 33 row(s)
```

删除视图，命令如下。

```
DROP VIEW IF EXISTS view_salesperson;
```

10.6 Hive 编程

Hive 提供了 JDBC 接口，JDBC 是一种可用于执行 SQL 语句的 Java 应用程序设计接口（Application Programming Interface，API）。它由 Java 类组成，JDBC 给 Hive 开发提供一种标准的应用程序设计接口，使开发人员可以用纯 Java 语言编写 Hive 程序。

10.6.1 JDBC 函数

Hive 提供的 JDBC 函数与其他 SQL 类似，包括加载驱动、创建连接对象、创建命令对象、执行 SQL 语句和关闭连接。

1. 加载驱动

连接 Hive 之前需要加载数据库驱动类，Hive 驱动函数为 Class.forName(driver)，driver 为 org.apache.hive.jdbc.HiveDriver。

例如，加载驱动，命令如下。

```
String driverName = "org.apache.hive.jdbc.HiveDriver";
Class.forName(driverName);
```

2. 创建连接对象

定义连接 Hive 的字符串 URL，URL 包括 HiveServer 2 地址及数据库名。Hive 数据库默认数据库为 default。

例如，创建连接，Hive 默认没有设置用户名和密码，可以任意，命令如下。

```
String url = "jdbc:hive2://master:10000/default";
Connection con = DriverManager.getConnection(url, user, passwd);
```

3. 创建命令对象

通过 Statement 类提供的方法对 Hive 数据进行操作。

例如，创建命令对象，命令如下。

```
Statement stmt = con.createStatement();
```

4. 执行 SQL 语言

执行 select 语句，返回结果集 ResultSet。ResultSet 本质上是指向数据行的游标。最初它位于第一行之前，每调用一次 next()方法，游标向下移动一行。

例如，创建命令对象，命令如下。

```
String sql = "SELECT name, age FROM student";
ResultSet rs = stmt.executeQuery(sql);
rs.next()
String name = rs.getString("name");
int age = rs.getInt("age");
```

如果执行 SQL 语句是数据定义类语句，则可以使用 executeUpdate()。

例如，删除学生表，命令如下。

```
String sql = " DROP TABLE IF EXISTS student_tmp";
stmt.executeUpdate(sql);
```

5. 关闭连接

关闭连接是一种好的编程习惯,命令如下。

```
con.close();
```

10.6.2 完整实例

(1) 编写访问 Hive 程序,完成对 student 表中数据的查询。示例(hive_student.java)代码如下。

```java
import java.sql.SQLException;
import java.sql.Connection;
import java.sql.ResultSet;
import java.sql.Statement;
import java.sql.DriverManager;
public class hive_student {
  public static void main(String[] args) throws SQLException {
    try {
      Class.forName("org.apache.hive.jdbc.HiveDriver");
      String url = "jdbc:hive2://master:10000/default";
      Connection conn = DriverManager.getConnection(url,"h","p");
      Statement stmt = conn.createStatement();
      ResultSet rs = stmt.executeQuery("SELECT id, name, age FROM student");
      while (rs.next()) {
        int id = rs.getInt(1);
        String name = rs.getString("name");
        int age = rs.getInt("age");
        System.out.println( id + "\t" + name + "\t" + age );
      }
      conn.close();
    } catch (Exception e) {
      e.printStackTrace();
    }
  }
}
```

(2) 编译代码,操作如下。

```
# javac hive_student.java
```

(3) 编写脚本 hive_student.sh,内容如下。

```
#!/bin/bash
HADOOP_HOME = /opt/hadoop
HIVE_HOME = /opt/hive
```

```
CLASSPATH = . : $ HIVE_HOME/conf : $ (hadoop classpath)
for i in ${HIVE_HOME}/lib/*.jar ; do
    CLASSPATH = $ CLASSPATH : $ i
done
java -cp $ CLASSPATH hive_student
```

(4) 执行脚本 hive_student.sh,运行结果如下。

```
# . hive_student.sh
61021503  陈可塘  19
61021506  萧后径  22
61021501  张汕尾  20
61021504  朱和顺  21
61021507  郑新楼  24
61021502  李海丰  18
61021505  谢连塘  23
```

小结

Hive 是 Hadoop 生态系统中必不可少的一个工具,它提供了一种 SQL 语言,可查询 HDFS 中的数据。本章介绍了 Hive 组成模块和执行流程、安装配置,以及 Hive CLI、Hive Beeline 命令行窗口的使用,详细分析了 Hive 数据类型和文件格式,重点讲解了股票 stocks 表的创建、数据的导入、数据简单查询、嵌套查询和连接查询,最后讲解了 JDBC 编程,完成了一个完整的实例。Hive 非常适合日志分析、股票信息分析、多维度数据分析和海量结构化数据离线分析。相对于其他工具,Hive 也使得开发者将基于 SQL 的应用程序移植到 Hadoop 变得更加容易。

习题

1. 简述 Hive 的组成及执行流程。
2. 比较 ARRAY、MAP 和 STRUCT 这 3 种数据类型的区别。
3. 举例说明 Hive 文件格式的 JSON 表示。
4. 使用 HSQL 进行词频统计。
5. 从腾讯证券下载股市行情,使用 Hive 进行分析。

第 11 章

ZooKeeper 协调服务

分布式处理(distributed processing)和并行处理(parallel processing)是为了提高并行处理速度采用的两种不同的体系架构。并行处理是利用多个功能部件或多个处理机同时工作来提高系统性能或可靠性的计算机系统,这种系统至少包含指令级或指令级以上的并行。分布式处理则是将不同地点的,或者具有不同功能的,或者拥有不同数据的多台计算机通过通信网络连接起来,在控制系统的统一管理下,协调地完成大规模信息处理任务的计算机系统。在分布式环境中协调和管理服务是一个复杂的过程,分布式协调服务有 ZooKeeper、Paxos、Chubby 和 Fourinone 等。

11.1 ZooKeeper 简介

ZooKeeper 是一个开放源代码的分布式应用程序协调服务,是 Google 的一个开源实现。ZooKeeper 的目标是封装好复杂、易出错的关键服务,将简单易用的接口和性能高效、功能稳定的系统提供给用户。ZooKeeper 包含一个简单的原语集,提供 Java 和 C 的接口。ZooKeeper 是一个典型的分布式数据一致性的解决方案,分布式应用程序可以基于它实现诸如数据发布/订阅、负载均衡、命名服务、分布式协调/通知、集群管理、master 选举、分布式锁和分布式队列等功能。ZooKeeper 可以保证分布式顺序一致性、事务原子性、单一视图、可靠性和实时性。ZooKeeper 集群如图 11-1 所示。

在 ZooKeeper 集群中有领导者、跟随者和观察者 3 种角色。ZooKeeper 集群中的所有节点通过一个 leader 选举过程来选定一台为 leader,它为客户端提供读和写服务;跟随者和观察者提供读服务,观察者不参与领导者选举过程。ZooKeeper 角色如表 11-1 所示。

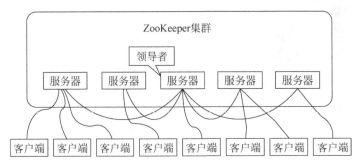

图 11-1 ZooKeeper 集群

表 11-1 ZooKeeper 角色

角 色		描 述
领导者(leader)		领导者负责进行投票的发起和决议,更新系统状态
学习者(learner)	跟随者(follower)	跟随者用于接收客户请求并向客户端返回结果,在选举过程中参与投票
	观察者(observer)	观察者可以接收客户端连接,将写请求转发给领导者节点。但观察者不参与投票过程,只同步领导者的状态,其目的是扩展系统、提高读取速度
客户端(client)		请求发起方

ZooKeeper 客户端会随机连接到 ZooKeeper 集群的一个节点,如果是读请求,就直接从当前节点中读取数据;如果是写请求,节点就会向 Leader 提交事务,Leader 会广播事务,只要有超过半数节点写入成功,该写请求就会被提交。

11.1.1 ZAB 协议

ZooKeeper 的核心是广播,这个机制保证了各个 Server 之间的同步。实现这个机制的协议称为原子广播(ZooKeeper Atomic Broadcast,ZAB)协议,使用 ZAB 协议作为数据一致性的核心算法,ZAB 协议是专为 ZooKeeper 设计的支持崩溃恢复的原子消息广播算法。

1. ZAB 协议核心

所有的事务请求必须由一个全局唯一的服务器(即 Leader)来协调处理,集群其余的服务器称为 Follower 服务器。Leader 服务器负责将一个客户端请求转化为事务提议(proposal),并将该 proposal 分发给集群所有的 Follower 服务器。随后 Leader 服务器需要等待所有的 Follower 服务器的反馈,一旦超过半数的 Follower 服务器进行了正确反馈,Leader 服务器就会再次向所有的 Follower 服务器分发 commit 消息,要求其将前一个 proposal 进行提交。

2. ZAB 协议的 4 个阶段

ZAB 协议分为 4 个阶段:阶段 0 为 Leader 选举,阶段 1 为发现,阶段 2 为同步,阶段

3 为广播。而实际实现时将发现阶级和同步阶段合并为一个恢复阶段。

1）阶段 0：Leader 选举

当集群中没有 Leader 或其他人感受不到 Leader 时会进入这一阶段，这一阶段主要是选出 Zxid（Zxid 是一个 64 位的数字，低 32 代表一个单调递增的计数器，高 32 位代表 Leader 周期）最大的节点作为准 Leader。

2）阶段 1：Discovery 发现

在这个阶段，Followers 与准 Leader 进行通信，同步 Follower 最近接收的事务提议。这个阶段主要是发现当前大多数节点接收的最新提议，并且准 Leader 生成新的 epoch，让 Followers 接受，更新它们的 acceptedEpoch。

一个 Follower 只会连接一个 Leader，如果有一个节点 Follower A 认为另一个节点 Follower B 是 Leader，A 在尝试连接 B 时会被拒绝，A 被拒绝之后，就会进入阶段 0。

3）阶段 2：Synchronization 同步

同步阶段主要是利用 Leader 前一阶段获得的最新提议历史，同步集群中所有的副本。只有当 quorum 都同步完成，准 Leader 才会成为真正的 Leader。Follower 只会接收 Zxid 比自己的 lastZxid 大的提议。

4）阶段 3：Broadcast 广播

到了这个阶段，ZooKeeper 集群才能正式对外提供事务服务，并且 Leader 可以进行消息广播。同时如果有新的节点加入，还需要对新节点进行同步。

值得注意的是，ZAB 提交事务不需要全部 Follower 返回 ACK，只需得到 quorum（超过半数的节点）返回 ACK 就可以了。

3. ZAB 协议内容

ZAB 协议分为两大部分：恢复和广播。

广播（broadcast）：ZAB 协议中，所有的写请求都由 Leader 来处理。正常工作状态下，Leader 接收请求并通过广播协议来处理。

广播的过程实际上是一个简化的二阶段提交过程，如图 11-2 所示。

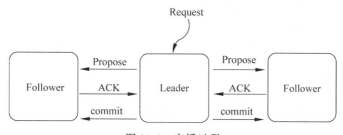

图 11-2　广播过程

① Leader 接收到消息请求后，将消息赋予一个全局唯一的 64 位自增的 Zxid，通过 Zxid 的大小比较，即可实现因果有序特性。

② Leader 通过先进/先出队列实现了全局有序特性，将带有 Zxid 的消息作为一个提案（proposal）分发给所有 Follower。

③ 当 Follower 接收到 Propose，先将 Propose 写到硬盘，写硬盘成功后再向 Leader 返回一个 ACK。

④ 当 Leader 接收到合法数量的 ACK 后，就向所有 Follower 发送 commit 命令，同时会在本地执行该消息。

⑤ 当 Follower 收到消息的 commit 命令时，就会执行该消息。

相较于完整的二阶段提交，ZAB 协议最大的区别是不能终止事务，Follower 要么返回 ACK 给 Leader，要么抛弃 Leader。Leader 不需要所有的 Follower 都响应成功，只要半数的节点返回 ACK 即可。

恢复(recovery)：当服务初次启动或 Leader 挂掉时，系统就会进入恢复模式，直到选出有合法数量 Follower 的新 Leader，然后新 Leader 负责将整个系统同步到最新状态。

由于 ZAB 协议的广播部分不能处理 Leader 挂掉的情况，ZAB 协议引入了恢复模式来处理这一问题。为了使 Leader 挂掉后系统能正常工作，需要解决以下两个问题。

1) 已经被处理的消息不能丢

为了实现已经被处理的消息不能丢这个目的，ZAB 的恢复模式使用了以下策略。

① 选举拥有 proposal 最大值（即 Zxid 最大）的节点作为新的 Leader，由于所有提案被 commit 之前必须有合法数量的 Follower ACK，即必须有合法数量的服务器的事务日志上有该提案的 proposal。因此，只要有合法数量的节点正常工作，就必然有一个节点保存了所有被 commit 消息的 proposal 状态。

② 新的 Leader 将自己事务日志中 proposal 但未 commit 的消息清除。

③ 新的 Leader 与 Follower 建立先进/先出的队列，先将自身有而 Follower 没有的 proposal 发送给 Follower，再将这些 proposal 的 commit 命令发送给 Follower，以保证所有的 Follower 都保存了所有的 proposal，所有的 Follower 都处理了所有的消息。

2) 被丢弃的消息不能再次出现

ZAB 通过巧妙地设计 Zxid 来实现这一目的。一个 Zxid 是 64 位，高 32 位是纪元(epoch)编号，每经过一次 Leader 选举产生一个新的 Leader，新的 Leader 会将 epoch 号加 1；低 32 位是消息计数器，每接收到一条消息这个值加 1，新的 Leader 选举后这个值重置为 0。这样设计的好处是：旧的 Leader 挂掉后重启，它不会被选举为 Leader，因为此时它的 Zxid 肯定小于当前的新的 Leader。当旧的 Leader 作为 Follower 接入新的 Leader 后，新的 Leader 会让它将所有的拥有旧的 epoch 号的未被 commit 的 proposal 清除。

11.1.2 ZooKeeper 数据模型

ZooKeeper 的数据模型是树结构，在内存数据库中，存储了整棵树（ZNode tree）的内容，包括所有的节点路径、节点数据、ACL 信息，ZooKeeper 会定时将这个数据存储到磁盘上。

1. 数据节点

数据节点（ZNode）是一个与 UNIX 文件系统路径相似的节点，可以向这个节点存储或获取数据。通过绝对路径的引用，必须由斜杠"/"开头，如/hbase。

```
[zk: localhost:2181(CONNECTED) 0] ls /
[zookeeper, hbase]
[zk: localhost:2181(CONNECTED) 1] ls /hbase
[replication, meta-region-server, rs, splitWAL, backup-masters, table-lock,
flush-table-proc, region-in-transition, online-snapshot, switch, master,
running, recovering-regions, draining, namespace, hbaseid, table]
```

其中，/zookeeper 是用来保存 ZooKeeper 的配额管理信息，不能轻易删除。ZNode 具有文件和目录两种特点，既像文件一样维护着数据、元信息、ACL、时间戳等数据结构，又像目录一样可以作为路径标识的一部分。每个节点称为一个 ZNode，每个 ZNode 由以下 3 个部分组成。

① stat：此为状态信息，描述该 ZNode 的版本、权限等信息。
② data：与该 ZNode 关联的数据。
③ children：该 ZNode 下的子节点。

ZooKeeper 可以关联一些数据，如分布式应用中的配置文件信息、状态信息、汇集位置等，这些数据的共同特性是数据很小，通常以 KB 为单位。每个 ZNode 的数据至多 1MB，常规使用都小于此值。

2. 状态信息

ZNode 除了存储数据内容之外，还存储了数据节点本身的一些状态信息，通过 get <path> 命令可以获取一个数据节点的内容，代码如下。

```
[zk: localhost:2181(CONNECTED) 2] get /hbase
cZxid = 0x100000002
ctime = Sat Oct 21 15:05:52 CST 2017
mZxid = 0x100000002
mtime = Sat Oct 21 15:05:52 CST 2017
pZxid = 0x800000025
cversion = 57
dataVersion = 0
aclVersion = 0
ephemeralOwner = 0x0
dataLength = 0
numChildren = 17
```

ZooKeeper 上一个数据节点的所有状态信息包括事务 ID、版本信息和子节点个数等，如表 11-2 所示。

表 11-2　ZNode 状态属性

状态属性	说　　明
cZxid	即 Create Zxid，表示该节点创建时的事务 ID
mZxid	即 Modified Zxid，表示该节点最后一次更新时的事务 ID
ctime	即 Create Time，表示该节点创建的时间

续表

状态属性	说明
mtime	即 Modified Time,表示该节点最后一次更新的时间
dataVersion	数据节点的版本号
cversion	子节点的版本号
aclVersion	节点 ACL(授权信息)的版本号
ephemeralOwner	创建 ephemeral 节点的会话 sessionID。如果该节点不是 ephemeral 节点,那么 ephemeralOwner 值为 0
dataLength	数据内容长度
numChildren	当前节点的子节点个数
pZxid	表示该节点的子节点列表最后一次被修改时的事务 ID。注意,只有子节点列表变更了才会变更 pZxid,子节点内容变更不会影响 pZxid

3. 节点类型

在 ZooKeeper 中,每个数据节点都是有生命周期的,其生命周期的长短取决于数据节点的节点类型。在 ZooKeeper 中,节点类型可以分为 persistent nodes(持久化节点)、ephemeral nodes(临时节点)和 sequential nodes(顺序节点)三大类。

① persistent nodes:persistent nodes 不和特定的 session 绑定,不会随着创建该节点的 session 的结束而消失,而是一直存在,除非显式删除该节点。

② ephemeral nodes:ephemeral nodes 的生命周期依赖于创建它们的 session。如果创建该节点的 session 结束了,该节点就会被自动删除,ephemeral nodes 不能拥有子节点。虽然 ephemeral nodes 与创建它的 session 绑定,但只要该节点没有被删除,其他 session 就可以读/写该节点中关联的数据,其创建使用-e 参数指定。

③ sequence nodes:严格地说,sequence nodes 并非节点类型中的一种,它既可以是 ephemeral nodes,也可以是 persistent nodes。创建 sequence nodes 时,ZooKeeper 服务器会在指定的节点名称后加上一个数字序列,该数字序列是递增的。因此,可以多次创建相同的 sequence nodes,而得到不同的节点,其创建使用-s 参数指定。

11.1.3 会话

ZooKeeper 对外的服务端口默认是 2181,客户端启动时,首先会与服务器建立一个 TCP 连接,从第一次连接建立开始,客户端会话(session)的生命周期也就开始了。通过这个连接,客户端能够通过心跳检测与服务器保持有效的会话,也能够向 ZooKeeper 服务器发送请求并接受响应,同时还能够通过连接接收来自服务器的 Watch 事件通知。session 的 sessionTimeout 值用来设置一个客户端会话的超时时间。当服务器压力太大、网络故障或客户端主动断开等各种原因导致客户端连接断开时,只要在 sessionTimeout 规定的时间内能够重新连接上集群中的任意一台服务器,那么之前创建的会话仍然有效。

11.1.4 事件监听器

Watcher(事件监听器)是 ZooKeeper 中的一个很重要的特性。ZooKeeper 允许用户在指定节点上注册一些 Watcher，并且在一些特定事件触发时，ZooKeeper 会将事件通知到设置监控的客户端，该机制是 ZooKeeper 实现分布式协调服务的重要特性。一个 Watch 事件是一个一次性的触发器，即触发一次就会被取消，如果想继续 Watch，需要客户端重新设置 Watcher。

1. Watcher 工作流程

ZooKeeper 的 Watcher 机制主要包括客户端线程、客户端 WatchManager 和 ZooKeeper 服务器三部分，其工作流程为：客户端在向 ZooKeeper 服务器注册 Watcher 的同时，会将 Watcher 对象存储在客户端的 WatchManager 中。当 ZooKeeper 服务器端触发 Watcher 事件后，会向客户端发送通知，客户端线程从 WatchManager 中取出对应的 Watcher 对象执行回调逻辑，如图 11-3 所示。

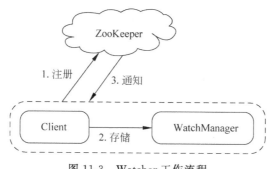

图 11-3　Watcher 工作流程

2. 数据监视和子节点监视

ZooKeeper 设置的不同监视返回不同的数据，getData() 和 exists() 返回 ZNode 的相关信息，而 getChildren() 返回子节点列表。

setData() 数据设置成功会触发设置在某一节点上所设置的数据监视，而一次成功的 create() 操作将会触发当前节点上所设置的数据监视及父节点的子节点监视。一次成功的 delete() 操作将会触发当前节点的数据监视和子节点监视，同时也会触发该节点父节点的 child watch。

可注册 Watcher 的方法有 getData、exists、getChildren，可触发 Watcher 的方法有 create、delete、setData。连接断开的情况下触发的 Watcher 会丢失。一个 Watcher 实例是一个回调函数，被回调一次后就被移除了。如果还需要关注数据的变化，需要再次注册 Watcher。New ZooKeeper 时注册的 Watcher 称为 default watcher，它不是一次性的，只对 Client 的连接状态变化做出反应。

3. Watcher 事件类型

默认 Watcher 事件类型为 Watcher.Event.EventType.None。每次会话事件都会产生相应的状态类型,如表 11-3 所示。

表 11-3 Session 事件

Trigger Event	State Type
AUTH_FAILED	Watcher.Event.KeeperState.AuthFailed
CONNECTED	Watcher.Event.KeeperState.ConnectedReadOnly
CONNECTED	Watcher.Event.KeeperState.SyncConnected
DISCONNECTED	Watcher.Event.KeeperState.Disconnected
SESSION_EXPIRED	Watcher.Event.KeeperState.Expired
SASL-authenticated	Watcher.Event.KeeperState.SaslAuthenticated

每一种 ZNode 操作和子 ZNode 操作都会产生相应的事件类型,使用不同的监视操作,如表 11-4 和表 11-5 所示。

表 11-4 data 事件

Trigger	Event Type	Watches
ZooKeeper.create	Watcher.Event.EventType.NodeCreated	ZooKeeper.exists ZooKeeper.getData
ZooKeeper.setData	Watcher.Event.EventType.NodeDataChanged	ZooKeeper.getData
ZooKeeper.delete	Watcher.Event.EventType.NodeDeleted	ZooKeeper.exists

表 11-5 children 事件

Trigger	Event Type	Watches
ZooKeeper.create	Watcher.EventType.NodeChildrenChanged	ZooKeeper.getChildren
ZooKeeper.delete	Watcher.EventType.NodeChildrenChanged	ZooKeeper.exists ZooKeeper.getChildren

11.1.5 访问权限

ZooKeeper 内部存储了分布式系统运行时状态的元数据,这些元数据会直接影响基于 ZooKeeper 进行构造的分布式系统的运行状态。保障系统中的数据安全,从而避免因误操作带来的数据随意变更而导致的数据库异常十分重要,ZooKeeper 提供了一套完善的 ACL 权限控制机制来保障数据的安全。ZooKeeper 采用 ACL(Access Control Lists)策略来进行权限控制,类似 UNIX 文件系统的权限控制。

ZooKeeper 定义了以下 5 种权限。

(1) CREATE(C):创建子节点的权限。

(2) READ(R):获取节点数据和子节点列表的权限。

(3) WRITE(W):更新节点数据的权限。

（4）DELETE(D)：删除子节点的权限。

（5）ADMIN(A)：设置节点 ACL 的权限。

CREATE 和 DELETE 权限都是针对子节点的权限控制。

11.2 ZooKeeper 集群部署

ZooKeeper 集群由多台节点组成，节点数量一般为奇数个，本集群使用 3 台节点。安装 ZooKeeper 需要 zookeeper-3.4.9.tar.gz 安装包，下载网址参考 mirrors.aliyun.com 或 mirror.hust.edu.cn。首先下载软件包，然后解压到/opt 目录，解压后子目录名带有版本号，对其改名，使其简洁，所有节点的安装配置方法基本一样。

（1）修改/etc/hosts 文件，添加内容如下。

```
172.30.0.10      master
172.30.0.11      slave1
172.30.0.12      slave2
```

（2）修改/root/.bash_profile 文件，添加内容如下。

```
export ZOOKEEPER = /opt/zookeeper
export PATH = $PATH:$ZOOKEEPER/bin
```

（3）在 zookeeper 的配置目录下新建 zoo.cfg 文件，操作如下。

```
# vi /opt/zookeeper/conf/zoo.cfg
内容如下：
# The number of milliseconds of each tick
tickTime = 2000
# The number of ticks that the initial
# synchronization phase can take
initLimit = 10
# The number of ticks that can pass between
# sending a request and getting an acknowledgement
syncLimit = 5
# the port at which the clients will connect
clientPort = 2181
# the maximum number of client connections.
# increase this if you need to handle more clients
#maxClientCnxns = 60
# The number of snapshots to retain in dataDir
autopurge.snapRetainCount = 3
# Purge task interval in hours
# Set to "0" to disable auto purge feature
autopurge.purgeInterval = 1
dataDir = /opt/zookeeper/data
dataLogDir = /opt/zookeeper/logs
```

```
server.0 = master:2888:3888
server.1 = slave1:2888:3888
server.2 = slave2:2888:3888
```

server.id=host:port1:port2 中,id 是每个 ZooKeeper 节点的编号,保存在 dataDir 目录下的 myid 文件中;host 是每个 ZooKeeper 节点的主机名;port1 用于连接 Leader 的端口(2888:集群内通信端口);port2 用于 Leader 选举的端口(3888:集群外通信端口)。

(4) 在 dataDir 指定的目录下创建 myid 文件,并添加相应内容,操作如下。

```
# mkdir /opt/zookeeper/data
# echo 0 > /opt/zookeeper/data/myid
```

注意,不同节点机器的 myid 不同,该值与 server.id=host:port1:port2 中的 id 相同。

11.3 ZooKeeper 基本命令

通过 zkServer.sh 进行 ZooKeeper 相关服务,操作如下。

① 启动 ZooKeeper 服务:zkServer.sh start。
② 查看 ZooKeeper 服务状态:zkServer.sh status。
③ 停止 ZooKeeper 服务:zkServer.sh stop。
④ 重启 ZooKeeper 服务:zkServer.sh restart。

客户端通过 zkCli.sh 连接到 ZooKeeper 服务器,基本操作包括查询、创建节点、修改、删除、quota、ACL 和其他操作。

(1) 服务启动。进行 ZooKeeper 基本 Shell 操作之前,3 台节点都要启动 ZooKeeper 服务,操作如下。

```
[root@master ~]# zkServer.sh start
ZooKeeper JMX enabled by default
Using config: /opt/zookeeper/bin/../conf/zoo.cfg
Starting zookeeper ... STARTED
[root@master ~]# zkServer.sh status
ZooKeeper JMX enabled by default
Using config: /opt/zookeeper/bin/../conf/zoo.cfg
Mode: follower

[root@slave1 ~]# zkServer.sh start
ZooKeeper JMX enabled by default
Using config: /opt/zookeeper/bin/../conf/zoo.cfg
Starting zookeeper ... STARTED
[root@slave1 ~]# zkServer.sh status
```

```
ZooKeeper JMX enabled by default
Using config: /opt/zookeeper/bin/../conf/zoo.cfg
Mode: leader

ZooKeeper JMX enabled by default
Using config: /opt/zookeeper/bin/../conf/zoo.cfg
Starting zookeeper ... STARTED
[root@slave2 ~]# zkServer.sh status
ZooKeeper JMX enabled by default
Using config: /opt/zookeeper/bin/../conf/zoo.cfg
Mode: follower
```

（2）查看命令帮助，命令如下。

```
# zkCli.sh
[zk: localhost:2181(CONNECTED) 0] h
ZooKeeper -server host:port cmd args
    stat path [watch]
    set path data [version]
    ls path [watch]
    delquota [-n|-b] path
    ls2 path [watch]
    setAcl path acl
    setquota -n|-b val path
    history
    redo cmdno
    printwatches on|off
    delete path [version]
    sync path
    listquota path
    rmr path
    get path [watch]
    create [-s] [-e] path data acl
    addauth scheme auth
    quit
    getAcl path
    close
    connect host:port
```

（3）列出指定 ZNode 及其子 ZNode，命令如下。

```
[zk: localhost:2181(CONNECTED) 1] ls /
[zookeeper]
[zk: localhost:2181(CONNECTED) 2] ls /zookeeper
[quota]
```

(4) 创建 ZNode,并指定关联数据,命令如下。

```
[zk: localhost:2181(CONNECTED) 3] create /hello world
Created /hello
```

(5) 获取 ZNode 的数据和状态信息,命令如下。

```
[zk: localhost:2181(CONNECTED) 4] get /hello
world
cZxid = 0x100000036
ctime = Thu Nov 02 09:01:51 CST 2017
mZxid = 0x100000036
mtime = Thu Nov 02 09:01:51 CST 2017
pZxid = 0x100000036
cversion = 0
dataVersion = 0
aclVersion = 0
ephemeralOwner = 0x0
dataLength = 5
numChildren = 0
```

(6) 修改 ZNode 下的数据,命令如下。

```
[zk: localhost:2181(CONNECTED) 5] set /hello swpt
cZxid = 0x100000036
ctime = Thu Nov 02 09:01:51 CST 2017
mZxid = 0x100000037
mtime = Thu Nov 02 09:02:07 CST 2017
pZxid = 0x100000036
cversion = 0
dataVersion = 1
aclVersion = 0
ephemeralOwner = 0x0
dataLength = 4
numChildren = 0
```

(7) 删除 ZNode,命令如下。

```
[zk: localhost:2181(CONNECTED) 6] delete /hello
[zk: localhost:2181(CONNECTED) 7] ls /
[zookeeper]
```

注意,被删除的 ZNode 拥有子 ZNode 时,必须先删除其所有的子 ZNode;也可以用 rmr 命令直接删除 ZNode 及其子 ZNode。

(8) 创建临时节点,并指定关联数据,命令如下。

```
[zk: localhost:2181(CONNECTED) 8] create -e /eph 3456
Created /eph
```

(9) 使用 stat 命令查看临时节点,命令如下。

```
[zk: localhost:2181(CONNECTED) 9] stat /eph
cZxid = 0x10000003a
ctime = Thu Nov 02 09:04:33 CST 2017
mZxid = 0x10000003a
mtime = Thu Nov 02 09:04:33 CST 2017
pZxid = 0x10000003a
cversion = 0
dataVersion = 0
aclVersion = 0
ephemeralOwner = 0x5f76b889e70007
dataLength = 4
numChildren = 0
```

(10) 在节点/sw 下创建一个顺序子节点,命令如下。

```
[zk: localhost:2181(CONNECTED) 10] create /sw father
Created /sw
[zk: localhost:2181(CONNECTED) 11] create -s /sw/item children
Created /sw/item0000000000
[zk: localhost:2181(CONNECTED) 12] create -s /sw/item children
Created /sw/item0000000001
```

(11) 列出指定目录并监听相应事件,命令如下。

```
[zk: localhost:2181(CONNECTED) 13] ls /sw true
[item0000000001, item0000000000]
[zk: localhost:2181(CONNECTED) 14] create -s /sw/item children
WATCHER::

WatchedEvent state:SyncConnected type:NodeChildrenChanged path:/sw
Created /sw/item0000000002
```

(12) 获取 ZNode 的数据和状态信息,并监听该 ZNode 的更新和删除事件,命令如下。

```
[zk: localhost:2181(CONNECTED) 15] get /sw true
father
cZxid = 0x10000003b
ctime = Thu Nov 02 09:05:43 CST 2017
mZxid = 0x10000003b
mtime = Thu Nov 02 09:05:43 CST 2017
pZxid = 0x10000003b
cversion = 0
dataVersion = 0
aclVersion = 0
ephemeralOwner = 0x0
dataLength = 6
```

```
numChildren = 0
[zk: localhost:2181(CONNECTED) 16] create -s /sw/item children
Created /sw/item0000000003
[zk: localhost:2181(CONNECTED) 17] delete /sw/item0000000003
[zk: localhost:2181(CONNECTED) 18] rmr /sw
WATCHER::

WatchedEvent state:SyncConnected type:NodeDeleted path:/sw
```

11.4 ZooKeeper 应用

由于 ZooKeeper 便捷的使用方式、卓越的运行性能及良好的稳定性，它被广泛应用在越来越多的大型系统中，用来解决诸如配置管理、分布式通知/协调、集群管理和 Master 选举等一系列分布式问题，其中最著名的就是在 Hadoop、HBase、Kafka、Hive 和 Spark 等开源系统中的应用。

11.4.1 Hadoop

Hadoop 是一个能够对大量数据进行分布式处理的软件框架，它以一种可靠、高效、可伸缩的方式进行数据处理。Hadoop 由许多元素构成，其最底部是 Hadoop Distributed File System(HDFS)，HDFS 的上一层是 MapReduce 引擎，HDFS 和 MapReduce 分别提供了对海量数据的存储和计算能力。YARN 是 Hadoop 集群的资源管理系统，Hadoop 2.0 对 MapReduce 框架做了彻底的设计重构，人们称 Hadoop 2.0 中的 MapReduce 为 MRv2 或 YARN。

1. YARN 简介

另一种资源协调者(Yet Another Resource Negotiator，Apache Hadoop YARN)是一种新的 Hadoop 资源管理器，它是一个通用资源管理系统，可为上层应用提供统一的资源管理和调度，它的引入为集群在利用率、资源统一管理和数据共享等方面带来了巨大好处。

YARN 的基本思想是将 JobTracker 的两个主要功能(资源管理和作业调度/监控)分离，分别由 ResourceManager(RM)和 ApplicationMaster(AM)管理。其中，ResourceManager 控制整个集群并管理应用程序向基础计算资源的分配，不仅将各个资源精心安排给基础 NodeManager(YARN 的每节点代理)，还与 ApplicationMaster 一起分配资源，与 NodeManager 一起启动和监视它们的基础应用程序。在此上下文中，ApplicationMaster 承担了以前 TaskTracker 的一些角色，ResourceManager 承担了 JobTracker 的角色。整个 YARN 的架构如图 11-4 所示。

2. ResourceManager 单点问题

YARN 架构有一个明显的缺陷——ResourceManager 单点问题。ResourceManager

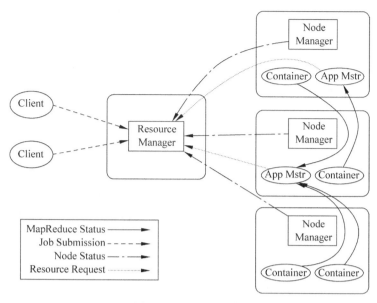

图 11-4　YARN 架构

是 YARN 中非常复杂的一个组件，负责整个系统的资源管理和调度，内部维护了各个应用程序的 ApplictionMaster 信息、NodeManager 信息、资源使用信息等。因此，ResourceManager 的工作状态直接决定了整个 YARN 框架是否可以运转。

3. ResourceManager HA

为了解决 ResourceManager 单点问题，YARN 设计了一套 Active/Standby 模式的 ResourceManager HA 架构，如图 11-5 所示。ResourceManager 由内嵌的一个基于

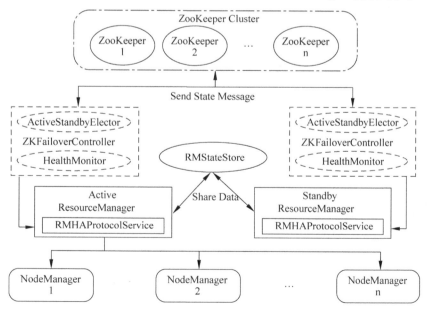

图 11-5　YARN ResourceManager HA 架构

ZooKeeper 的 ActiveStandbyElector 来决定哪个 ResourceManager 应该是 Active，当 Active RM 宕机，另外一个 ResourceManager 将会自动备选为 Active，然后接手工作。

与 HDFS 的高可用不同的是：ZK Failover Controller(ZKFC)属于 RMHAProtocolService 中的一个线程，而 RMHAProtocolService 本身则变成 ResourceManager 内部的一个服务，这意味着无须启动一个单独的 ZKFC。YARN ResourceManager 只负责 ApplicationMaster 的状态维护和容错，ApplicationMaster 内部管理和调度的任务，如 MapTask 和 ReduceTask，则需要由 AppicationMaster 自己容错。

RMStateStore 能够存储一些 ResourceManager 内部状态信息，包括 Application 以及它们的 Atempts 信息、Delegation Token 及 Version Information 等，RMStateStore 中绝大多数状态信息都不需要持久化存储，YARN 的单点故障采用主备切换的方式完成，备节点不会同步主节点的信息，而是在切换之后，从 RMStateStore 中读取所需信息。

4. Fencing

在分布式环境中，由于网络闪断或负载过高会出现诸如单机"假死"的情况。假设 ResourceManager 集群由 RM1 和 RM2 两台机器组成，RM1 为 Active 状态，RM2 为 Standby 状态。在某一时刻 RM1 发生了"假死"，此时 ZooKeeper 认为 RM1 挂掉，RM2 切换为 Active 状态。但在随后，RM1 恢复了正常，依然认为自己还处于 Active 状态，这就是分布式"脑裂(Brain-Split)"现象，这时系统会出现异常。YARN 中引入 Fencing(隔离)机制。借助 ZooKeeper 数据节点的 ACL 权限控制机制来实现不同 RM 之间的隔离。多个 RM 之间通过竞争创建锁节点来实现 Active/Standby 状态的确定，同时创建的锁节点必须携带 ZooKeeper 的 ACL 信息，目的是独占该节点，以防止其他 RM 对其进行更新。RM1"假死"后，ZooKeeper 会将其创建的锁节点移除，RM2 随后创建相应的锁节点并切换为 Active 状态。RM1 恢复正常后发现 ZooKeeper 上的节点不是自己创建的，于是自动切换为 Standby 状态，避免了"脑裂"现象的出现。

5. ResourceManager HA 实现

ResourceManager HA 的实现需要修改 yarn-site.xml 配置文件，内容如下。

```xml
<?xml version = "1.0"?>
<configuration>
  <property>
    <name>yarn.resourcemanager.ha.enabled</name>
    <value>true</value>
  </property>
  <property>
    <name>yarn.resourcemanager.cluster-id</name>
    <value>rm-cluster</value>
  </property>
  <property>
    <name>yarn.resourcemanager.ha.rm-ids</name>
    <value>rm1,rm2</value>
```

```xml
    </property>
    <property>
        <name>yarn.resourcemanager.ha.id</name>
        <value>rm1</value>
    </property>
    <property>
        <name>yarn.resourcemanager.hostname.rm1</name>
        <value>master</value>
    </property>
    <property>
        <name>yarn.resourcemanager.hostname.rm2</name>
        <value>slave1</value>
    </property>
    <property>
        <name>yarn.resourcemanager.recovery.enabled</name>
        <value>true</value>
    </property>
    <property>
        <name>yarn.resourcemanager.store.class</name>
        <value>org.apache.hadoop.yarn.server.resourcemanager.recovery
            .ZKRMStateStore</value>
    </property>
    <property>
        <name>yarn.resourcemanager.zk-address</name>
        <value>master:2181,slave1:2181,slave2:2181</value>
    </property>
    <property>
        <name>yarn.log-aggregation-enable</name>
        <value>true</value>
    </property>
    <property>
        <name>yarn.nodemanager.aux-services</name>
        <value>mapreduce_shuffle</value>
    </property>
</configuration>
```

注意，yarn.resourcemanager.ha.id 不同的节点值不同，在 slave1 节点上，该参数设置为 rm2，其他节点上该配置删除。

mapred-site.xml 配置文件也需要修改，内容如下。

```xml
<?xml version="1.0"?>
<?xml-stylesheet type="text/xsl" href="configuration.xsl"?>
<configuration>
    <property>
        <name>mapreduce.framework.name</name>
        <value>yarn</value>
```

```
    </property>
    <property>
      <name>mapreduce.jobhistory.address</name>
      <value>master:10020,slave1:10020</value>
    </property>
    <property>
      <name>mapreduce.jobhistory.webapp.address</name>
      <value>master:19888,slave1:19888</value>
    </property>
</configuration>
```

6. ResourceManager HA 测试

启动 yarn 服务,执行 start-yarn.sh 命令,操作如下。

```
[root@master ~]# start-yarn.sh
```

slave1 节点启动 ResourceManager 服务,操作如下。

```
[root@slave1 ~]# yarn-daemon.sh start resourcemanager
```

查看进程,操作如下。

```
[root@master ~]# jps
30288 Jps
2164 QuorumPeerMain
29460 Namenode
29689 SecondaryNamenode
29897 ResourceManager

[root@slave1 ~]# jps
2563 Datanode
2692 NodeManager
2886 ResourceManager
1080 QuorumPeerMain
2942 Jps

[root@slave2 ~]# jps
1082 QuorumPeerMain
2683 Datanode
2812 NodeManager
2988 Jps
```

打开浏览器查看 Hadoop YRAN ResourceManager HA 集群信息,Master 节点为 Active 状态,Slave1 为 Standby 状态,如图 11-6 和图 11-7 所示。

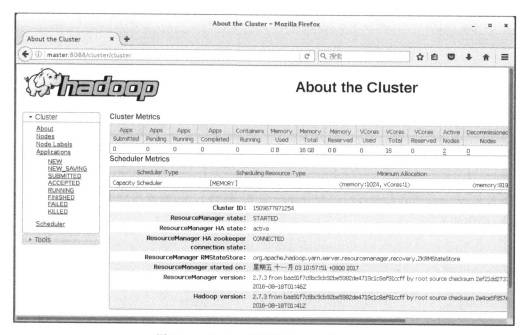

图 11-6　Hadoop 的活动 ResourceManager

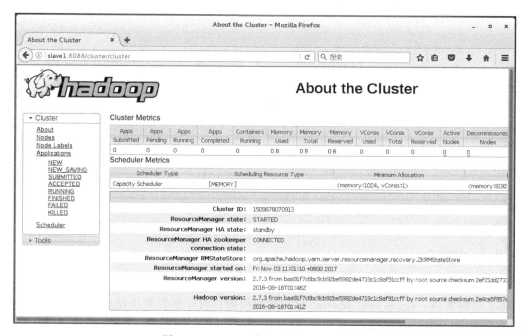

图 11-7　Hadoop 备用 ResourceManager

使用 yarn 命令查看 RM 状态,操作如下。

```
# yarn rmadmin - getServiceState rm1
active
# yarn rmadmin - getServiceState rm2
standby
```

主备切换,操作如下。

```
# yarn rmadmin - transitionToStandby rm1
# yarn rmadmin - transitionToActive rm2
```

关闭 master 节点的 ResourceManager 服务,稍后开启,查看 RM 的 Active/Standby 状态,操作如下。

```
[root@master ~]# yarn - daemon.sh stop resourcemanager
[root@master ~]# yarn - daemon.sh start resourcemanager
[root@master ~]# yarn rmadmin - getServiceState rm1
standby
[root@master ~]# yarn rmadmin - getServiceState rm2
active
```

从上面可以看出,Active/Standy 已经自动切换。

列出 ZNode 信息,YARN 的一些基本信息登记在里面,代码如下。

```
[zk: localhost:2181(CONNECTED) 0] ls /
[zookeeper, yarn - leader - election, rmstore]
[zk: localhost:2181(CONNECTED) 1] ls /rmstore
[ZKRMStateRoot]
[zk: localhost:2181(CONNECTED) 2] ls /rmstore/ZKRMStateRoot
[RMAppRoot, AMRMTokenSecretManagerRoot, EpochNode, RMDTSecretManagerRoot,
RMVersionNode]
[zk: localhost:2181(CONNECTED) 3] ls /yarn - leader - election
[rm - cluster]
[zk: localhost:2181(CONNECTED) 4] ls /yarn - leader - election/rm - cluster
[ActiveBreadCrumb, ActiveStandbyElectorLock]
```

11.4.2 Spark

Spark Standalone 集群是 Master/Slaves 架构的集群模式,与大部分的 Master/Slaves 结构集群一样,存在着 Master 单点故障问题。对于如何解决单点故障问题,Spark 提供了两种方案:基于文件系统的单点恢复(Single-Node Recovery with Local File System)和基于 ZooKeeper 的备用 Master(Standby Masters with ZooKeeper)。

ZooKeeper 提供了一个 Leader Election 机制,利用这个机制可以保证虽然集群存在多个 Master,但是只有一个是 Active 的,其他都是 Standby。当 Active 的 Master 出现故障时，另一个 Standby Master 会被选举出来。由于集群的信息，包括 Worker、Driver 和 Application 的信息都已经持久化到文件系统。因此，在切换的过程中只会影响新 Job 的提交，对于正在进行的 Job 没有任何影响。加入 ZooKeeper 的集群整体架构如图 11-8 所示。

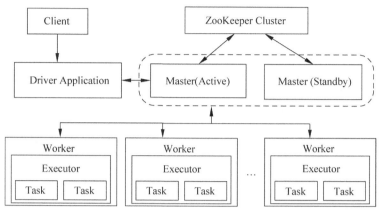

图 11-8　Spark HA 架构

1. 基于 ZooKeeper 的备用 Master 方案

当集群的 Master 的 Active 挂掉后，ZooKeeper 会在 Standby 的 Master 中通过选举机制选择一台 Standby 机器作为 Leader，通过读取 ZooKeeper 中的元数据信息，恢复整个集群的状态，成功恢复后才能作为整个集群的 Master(Active)，开始接受所有的作业提交和资源申请的请求。

Master 分配资源之后，Driver 开始与 Worker 分配的 Excutor 进行通信，这时一般不需要 Master 参与，除非 Excutor 有故障。所以在 Master 出现故障之后，并不影响已有作业的运行。配置 ZooKeeper 的好处是：当集群重启之后系统会将上一次运行的 Driver、Application、Worker 等信息重新恢复。

2. Spark HA 的实现

修改 Spark 配置文件 spark-env.sh，内容如下。

```
export JAVA_HOME = /opt/jdk1.8
export HADOOP_HOME = /opt/hadoop
export HADOOP_CONF_DIR = $HADOOP_HOME/etc/hadoop
export SCALA_HOME = /opt/scala
export SPARK_HOME = /opt/spark
export SPARK_DAEMON_JAVA_OPTS = " - Dspark.deploy.recoveryMode = ZOOKEEPER
    - Dspark.deploy.zookeeper.url = master:2181,slave1:2181,slave2:2181
    - Dspark.deploy.zookeeper.dir = /spark"
export SPARK_WORKER_MEMORY = 2g
```

修改 Spark 配置文件 spark-defaults.conf，内容如下。

```
spark.master spark://master:7077,slave1:7077,slave2:7077
```

启动 Spark 集群，操作如下。

```
[root@master ~]# /opt/spark/sbin/start-all.sh
[root@slave1 ~]# /opt/spark/sbin/start-master.sh
[root@slave2 ~]# /opt/spark/sbin/start-master.sh
```

打开浏览器查看 Spark 集群信息，Master 节点为 Active 状态，Slave1 节点为 Standby 状态，如图 11-9 和图 11-10 所示。

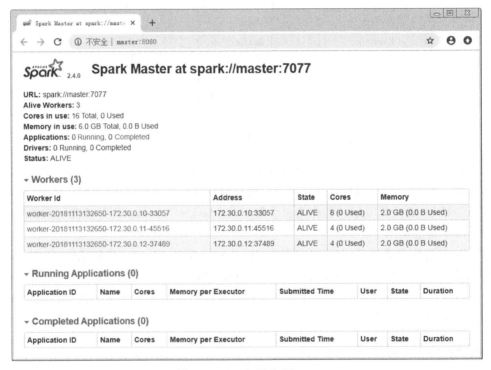

图 11-9 Spark 活动 Master

列出 ZNode 信息，Spark 的一些基本信息登记在里面，如下所示。

```
[zk: localhost:2181(CONNECTED) 1] ls /
[zookeeper, yarn-leader-election, spark, rmstore]
[zk: localhost:2181(CONNECTED) 2] ls /spark
[leader_election, master_status]
[zk: localhost:2181(CONNECTED) 3] ls /spark/master_status
[worker_worker-20171103173110-172.30.0.10-44979,
worker_worker-20171103173115-172.30.0.12-41377,
worker_worker-20171103173116-172.30.0.11-42691]
```

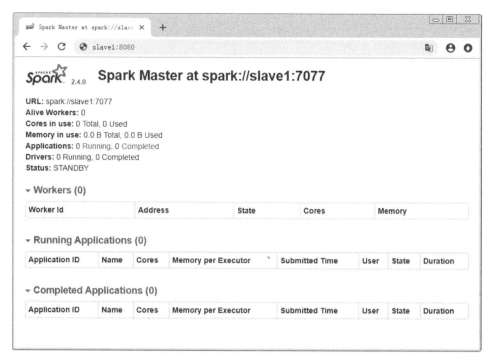

图 11-10　Spark 备用 Master

关闭 Master 节点的 Master 服务,稍后开启,操作如下。

```
[root@master ~]# /opt/spark/sbin/stop-master.sh
[root@master ~]# /opt/spark/sbin/start-master.sh
```

通过浏览器查看 Spark 的 Master 信息,Active/Standy 已经自动切换。
测试词频统计程序,操作如下。

```
# spark-submit -- class org.apache.spark.examples.JavaWordCount \
  examples/jars/spark-examples_2.11-2.1.1.jar  /test
```

11.4.3　Hive

在生产环境中使用 Hive,建议使用 HiveServer2 提供服务。与 hive-cli 方式相比,HiveServer2 不用直接将 HDFS 和 Metastore 暴露给用户;有安全认证机制,并且支持自定义权限校验;有 HA 机制,可以解决应用端的并发和负载均衡问题;使用 JDBC 方式,方便与应用进行数据交互。Hive HA 架构如图 11-11 所示。

视频讲解

1. Hive HA 配置

Hive HA 配置比较简单,修改配置文件 hive-site.xml,添加内容如下。

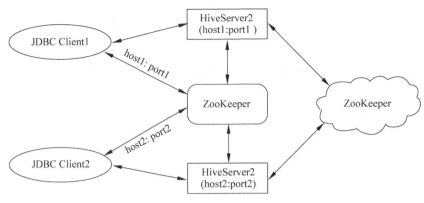

图 11-11 Hive HA 架构

```
< property >
    < name > hive.server2.support.dynamic.service.discovery </name >
    < value > true </value >
</property >
< property >
    < name > hive.server2.zookeeper.namespace </name >
    < value > hiveserver2_zk </value >
</property >
< property >
    < name > hive.zookeeper.quorum </name >
    < value > master:2181,slave1:2181,slave2:2181 </value >
</property >
< property >
    < name > hive.zookeeper.client.port </name >
    < value > 2181 </value >
</property >
< property >
    < name > hive.metastore.uris </name >
    < value > thrift://master:9083,thrift://slave1:9083 </value >
</property >
```

启动服务，操作如下。

```
[root@master ~]# hive -- service metastore &
[root@master ~]# hiveserver2 &
[root@slave1 ~]# hiveserver2 &
```

列出 ZNode 信息，HiveServer2 的一些基本信息登记在里面，代码如下。

```
[zk: localhost:2181(CONNECTED) 1] ls /
[zookeeper, yarn - leader - election, spark, hiveserver2_zk, rmstore]
[zk: localhost:2181(CONNECTED) 2] ls /hiveserver2_zk
[serverUri = master:10000;version = 2.1.1;sequence = 0000000001,
serverUri = slave1:10000;version = 2.1.1;sequence = 0000000000]
```

2. Hive HA 测试

使用 beeline 测试连接，操作如下。

```
# beeline
Beeline version 2.1.1 by Apache Hive
beeline> !connect jdbc:hive2://master:2181,slave1:2181/;
serviceDiscoveryMode=zooKeeper;zooKeeperNamespace=hiveserver2_zk hive ""
Connecting to jdbc:hive2://master:2181,slave1:2181/;
serviceDiscoveryMode=zooKeeper;zooKeeperNamespace=hiveserver2_zk
… INFO jdbc.HiveConnection: Connected to slave1:10000
Connected to: Apache Hive (version 2.1.1)
Driver: Hive JDBC (version 2.1.1)
… WARN jdbc.HiveConnection: Request to set autoCommit to false;
Hive does not support autoCommit=false.
Transaction isolation: TRANSACTION_REPEATABLE_READ
0: jdbc:hive2://master:2181,slave1:2181/> show databases;
+---------------+-+
| database_name |
+---------------+-+
| default       |
+---------------+-+
1 rows selected (3.60 seconds)
```

11.5 ZooKeeper 编程

ZooKeeper 提供丰富的 Java API 接口，方便 Java 调用。

1. 连接 ZooKeeper

```
ZooKeeper zk = new ZooKeeper("ip1:port1,ip2:port2,…", SessionTimeOut,
        new Watcher(){
            //监控所有被触发的事件
            public void process(WatchedEvent event) {
                //do something
            }
        });
```

创建一个 ZooKeeper 实例，第一个参数为目标服务器地址和端口；第二个参数为 Session 超时时间；第三个参数为节点变化时的回调方法。客户端和服务端建立连接后，会话随之建立，生成一个全局唯一的会话 ID(Session ID)。

客户端并不能随意设置 SessionTimeOut，在 ZooKeeper 服务器端对会话超时时间是有限制的，主要由 minSessionTimeout 和 maxSessionTimeout 两个参数设置。如果客户端设置的超时时间不在这个范围，那么会被强制设置为最大或最小时间。默认的 Session 超时时间在 2 * tickTime 和 20 * tickTime 之间。

2. 创建 ZNode

```
String create(final String path, byte data[], List acl, CreateMode createMode)
```

参数：路径、ZNode 内容、ACL（访问控制列表）、ZNode 创建类型。
ZNode 创建类型有以下几种。
① PERSISTENT：持久化节点。
② EPHEMERAL：临时节点，客户端 session 超时类节点就会被自动删除。
③ PERSISTENT_SEQUENTIAL：顺序自动编号持久化节点，根据存在的节点数自动加 1。
④ EPHEMERAL_SEQUENTIAL：临时自动编号节点。
创建成功返回 true，否则返回 false。

3. 删除 ZNode

```
void delete(final String path, int version)
```

参数：路径、版本号。
如果版本号与 ZNode 的版本号不一致，将无法删除，这是一种乐观加锁机制；如果将 version 设置为-1，就不会检测版本，直接删除。
删除成功返回 true，否则返回 false。

4. 判断某个 ZNode 节点是否存在

```
Stat exists(final String path, Watcher watcher)
```

参数：路径、Watcher（监视器）。
当某个 ZNode 节点被改变时，将会触发当前 Watcher。
ZNode 存在返回 true，否则返回 false。

```
Stat exists(String path, boolean watch)
```

参数：路径，并设置是否监控此目录节点。
这里的 Watcher 是在创建 ZooKeeper 实例时指定的 Watcher。
ZNode 存在返回 true，否则返回 false。

5. 设置 ZNode 数据

```
Stat setData(final String path, byte data[], int version)
```

参数：路径、数据、版本号。

如果版本号不一致,就无法进行数据设置;如果 version 为 −1,就跳过版本检查。更新成功返回 true,否则返回 false。

6. 获取 ZNode 数据

```
byte[] getData(final String path, Watcher watcher, Stat stat)
```

参数:路径、监视器、数据版本。

7. 获取 ZNode 的所有子节点

```
List getChildren(final String path, Watcher watcher)
```

参数:路径、监视器。

该方法有多个重载。

11.5.1 ZooKeeper 读/写操作

编写 Java 应用程序,使用 ZooKeeper 提供的 Java API,实现对 ZNode 的增加、删除和设置等操作,监听 ZNode 的变化及处理。

1. 编写代码

运行 Eclipse 开发工具,打开集成环境界面,添加 zookeeper-3.4.10.jar 和 slf4j-api-1.6.1.jar 包。创建 ZooKeeperEx 项目,并在 ZooKeeperEx 项目中新建 ZooKeeperEx1.java 类,代码如下。

```java
import java.io.IOException;
import java.util.List;
import org.apache.zookeeper.CreateMode;
import org.apache.zookeeper.KeeperException;
import org.apache.zookeeper.WatchedEvent;
import org.apache.zookeeper.Watcher;
import org.apache.zookeeper.ZooDefs.Ids;
import org.apache.zookeeper.ZooKeeper;
import org.apache.zookeeper.data.Stat;
import org.slf4j.Logger;
import org.slf4j.LoggerFactory;

public class ZooKeeperEx1{
    private static final int SESSION_TIMEOUT = 30000;
    public static final Logger LOGGER =
        LoggerFactory.getLogger("ZooKeeperEx1Log");
    private static Watcher watcher =   new Watcher() {
        public void process(WatchedEvent event) {
            System.out.println("process : " + event.getType());
            LOGGER.info("process : " + event.getType());
        }
```

```java
    };
    public static void main(String[] args) throws
        IOException,KeeperException,InterruptedException {
      //连接 ZooKeeper
      ZooKeeper zkp = new ZooKeeper("master:2181,slave1:2181,slave2:2181",
          SESSION_TIMEOUT,watcher);
      //创建 ZNode
      zkp.create("/test", "znodedata".getBytes(),
          Ids.OPEN_ACL_UNSAFE,CreateMode.PERSISTENT);
      zkp.create("/test/seq", "childdata".getBytes(),
          Ids.OPEN_ACL_UNSAFE,CreateMode.PERSISTENT_SEQUENTIAL);
      //判断 ZNode 是否存在
      Stat stat = zkp.exists("/test", true);
      if (zkp.exists("/test", false) != null) {
        System.out.println("/test exists now.");
      }
      //更新 ZNode 数据内容
      zkp.setData("/test", "newdata".getBytes(), stat.getVersion());
      //读取 ZNode 数据
      String data = new String(zkp.getData("/test", true, stat));
          System.out.println(data);
      //获取子节点
      List < String > children = zkp.getChildren("/test", true );
      for(String child :children) {
        System.out.println(child);
       }
      //删除 ZNode 子节点
      zkp.delete("/test/" + children.get(0), -1);
      //删除 ZNode 节点
      zkp.delete("/test", -1);
      zkp.close();
    }
}
```

2. 导出 ZooKeeperEx1.jar 运行包

其运行结果如下。

```
# java - jar ZooKeeperEx1.jar
process : None
/test exists now.
process : NodeDataChanged
newdata
seq0000000000
process : NodeChildrenChanged
process : NodeDeleted
```

11.5.2 集群状态监控

为了确保集群能够正常运行,ZooKeeper 可以被用来监视集群状态,这样就可以提高集群可用性。使用 ZooKeeper 的瞬时(Ephemeral)节点概念可以设计一个集群机器状态检测机制。

每一个运行了 ZooKeeper 客户端的生产环境机器都是一个终端进程，它们连接到 ZooKeeper 服务端后创建对应的瞬时节点，可以用/znodename 进行区分。

采用监听(Watcher)方式完成对节点状态的监视，如通过对/znodename 节点的 NodeChildrenChanged 事件进行监听。

由于是一个瞬时节点，所以每次客户端断开时 ZNode 都会立即消失，这样就可以监听到集群节点异常。

NodeChildrenChanged 事件触发后，可以调用 getChildren 方法来查看哪台机器发生了异常。

(1) 编写代码。运行 Eclipse 开发工具，打开集成环境界面，添加 zookeeper-3.4.10.jar 和 slf4j-api-1.6.1.jar 包。在 ZooKeeperEx 项目中新建 ClusterMonitor.java 类，代码如下。

```java
import java.io.IOException;
import java.util.List;
import org.apache.zookeeper.CreateMode;
import org.apache.zookeeper.KeeperException;
import org.apache.zookeeper.WatchedEvent;
import org.apache.zookeeper.Watcher;
import org.apache.zookeeper.ZooDefs.Ids;
import org.apache.zookeeper.ZooKeeper;

public class ClusterMonitor implements Runnable{
    private static String membershipRoot = "/test_zk";
    private final Watcher connectionWatcher;
    private final Watcher childrenWatcher;
    private ZooKeeper zkp;
    boolean alive = true;
    public ClusterMonitor(String HostPort) throws
        IOException,InterruptedException,KeeperException{
      connectionWatcher = new Watcher(){
        @Override
        public void process(WatchedEvent event) {
          if(event.getType() == Watcher.Event.EventType.None &&
             event.getState() == Watcher.Event.KeeperState.SyncConnected){
            System.out.println("\nconnection Watcher Event Received: %s"
                + event.toString());
          }
        }
      };
      childrenWatcher = new Watcher(){
        @Override
        public void process(WatchedEvent event) {
          System.out.println("\nchildren Watcher Event
```

```java
            Received: %s" + event.toString());
        if(event.getType() == Event.EventType.NodeChildrenChanged){
          try{
            List<String> children = zkp.getChildren(membershipRoot,this);
            System.out.println("Cluster Membership change,Members: "
                + children);
          }catch(KeeperException ex){
            throw new RuntimeException(ex);
          }catch(InterruptedException ex){
            Thread.currentThread().interrupt();
            alive = false;
            throw new RuntimeException(ex);
          }
        }
      }
    };
zkp = new ZooKeeper(HostPort,3000,connectionWatcher);
//Ensure the parent znode exists
if(zkp.exists(membershipRoot, false) == null){
  zkp.create(membershipRoot, "ClusterMonitor".getBytes(),
      Ids.OPEN_ACL_UNSAFE, CreateMode.PERSISTENT);
}
//Set a watch on the parent znode
List<String> children = zkp.getChildren(membershipRoot, childrenWatcher);
  System.err.println("Members:" + children);
}
public synchronized void close(){
  try{
    zkp.close();
  }catch(InterruptedException ex){
    ex.printStackTrace();
  }
}
@Override
public void run() {
  try{
    synchronized(this){
      while(alive){
        wait();
      }
    }
  }catch(InterruptedException ex){
    ex.printStackTrace();
    Thread.currentThread().interrupt();
  }finally{
    this.close();
  }
}
```

```java
    public static void main(String[] args) throws
        IOException,InterruptedException,KeeperException{
      String hostPort = "master:2181,slave1:2181,slave2:2181";
      new ClusterMonitor(hostPort).run();
    }
}
```

在 ZooKeeperEx 项目中新建 ClusterClient.java 类,代码如下。

```java
import java.io.IOException;
import java.lang.management.ManagementFactory;
import org.apache.zookeeper.CreateMode;
import org.apache.zookeeper.KeeperException;
import org.apache.zookeeper.WatchedEvent;
import org.apache.zookeeper.Watcher;
import org.apache.zookeeper.ZooDefs.Ids;
import org.apache.zookeeper.ZooKeeper;

public class ClusterClient implements Watcher,Runnable{
    private static String membershipRoot = "/test_zk";
    ZooKeeper zkp;
    public ClusterClient(String hostPort,Long pid){
      String processId = pid.toString();
      try{
        zkp = new ZooKeeper(hostPort,3000,this);
      }catch(IOException ex){
        ex.printStackTrace();
      }
      if(zkp!= null){
        try{
          zkp.create(membershipRoot + '/' + processId, processId.getBytes(),
              Ids.OPEN_ACL_UNSAFE,CreateMode.EPHEMERAL);
        }catch(KeeperException ex){
          ex.printStackTrace();
        } catch(InterruptedException ex){
          ex.printStackTrace();
        }
      }
    }
    public synchronized void close(){
      try{
        zkp.close();
      }catch(InterruptedException ex){
        ex.printStackTrace();
      }
    }
    @Override
    public void process(WatchedEvent event) {
```

```java
      System.out.println("\nEvent Received: % s" + event.toString());
    }
    @Override
    public void run() {
      try{
        synchronized(this){
          while(true){
            wait();
          }
        }
      }catch(InterruptedException ex){
        ex.printStackTrace();
        Thread.currentThread().interrupt();
      }finally{
        this.close();
      }
    }
    public static void main(String[] args){
      String hostPort = "master:2181,slave1:2181,slave2:2181";
      //Get the process id
      String name = ManagementFactory.getRuntimeMXBean().getName();
      int index = name.indexOf('@');
      Long processId = Long.parseLong(name.substring(0,index));
      new ClusterClient(hostPort,processId).run();
    }
}
```

(2) 导出 ClusterMonitor.jar 和 ClusterClient.jar 运行包,其运行结果如下。

```
[root@master ~]# java - jar ClusterMonitor.jar
connection Watcher Event Received: % sWatchedEvent state:SyncConnected
    type:None path:null
Members:[]
[root@slave1 ~]# java - jar ClusterClient.jar
Event Received: % sWatchedEvent state:SyncConnected type:None path:null

[root@master ~]# …
children Watcher Event Received: % sWatchedEvent state:SyncConnected
    type:NodeChildrenChanged path:/test_zk
Cluster Membership change,Members: [2302]
```

(3) 运行 zkCli 客户端工具,查看节点信息如下。

```
[root@slave2 ~]# zkCli.sh
[zk: localhost:2181(CONNECTED) 0] ls /test_zk
[2302]
[zk: localhost:2181(CONNECTED) 1] get /test_zk/2302
2302
```

```
cZxid = 0x2000000a1
ctime = Sun Nov 05 22:11:09 CST 2017
mZxid = 0x2000000a1
mtime = Sun Nov 05 22:11:09 CST 2017
pZxid = 0x2000000a1
cversion = 0
dataVersion = 0
aclVersion = 0
ephemeralOwner = 0x25f86de49bc0009
dataLength = 4
numChildren = 0
```

小结

ZooKeeper 是一个分布式服务协调框架,实现同步服务、配置维护和命名服务等分布式应用,是一个高性能的分布式数据一致性解决方案。本章详细讲解了 ZooKeeper 的重要协议 ZAB,介绍了 ZooKeeper 数据模型和 Watcher,给出了集群部署的过程和配置,详细介绍了使用 zkCli.sh 客户端工具对 ZNode 进行创建、设置、删除等操作,然后详细讲解了 ZooKeeper 在 Hadoop YARN ResourceManager HA、Spark HA 和 Hive HA 的应用及配置方法,重点讲解了 ZooKeeper 在单节点故障的解决办法,最后介绍了 ZooKeeper 基本的读/写操作编程案例,以及集群监控和集群客户端操作案例。

习题

1. 简述 ZooKeeper 协调服务。
2. 简述 ZAB 协议。
3. 简述 ZooKeeper 事件监听器工作流程。
4. 简述 ZooKeeper 如何解决 YARN 中的 ResourceManager 单点故障问题。
5. 编写程序实现 ZNode 的增加、删除等。

应用篇

第 12 章

医药大数据案例分析

本章通过医药电商大数据平台的分析和开发,介绍如何开发一个基于 Hadoop 平台的大数据系统以及大数据的可视化问题,让读者在实践中学习大数据相关技术,掌握大数据相关技术和大数据系统的开发原理,从而利用现有大数据技术解决现实中相关的问题。

12.1 项目概述

近几年,电子商务的崛起给零售行业造成的竞争压力越来越大,一方面是由于移动信息技术日渐成熟、物流快递行业迅速发展以及政策的支持;另一方面在于"互联网+"存在的客观竞争优势,低成本运营、信息化应用带来的较高运营效率和创新模式带来的短期利好等因素;最后,长久以来,传统零售业习惯了粗犷经营,加上医药零售业半封闭的政策环境,导致了传统的销售模式和电商企业竞争时劣势更加明显。

医药电子商务是以医药企业、医疗机构、支付机构、医药信息服务提供商等为网络成员,通过互联网技术,为用户提供安全、可靠、开放并易于维护的医药电子商务平台。随着大数据技术的发展,硬件设备不断升级,计算能力不断增强,大数据技术逐步被引入各个行业。在医药产品销售过程中,通过利用大数据技术在海量数据计算、统计、分析等方面的优势,开展精准化营销和实现线上商品优化,例如基于顾客行为和消费习惯,开展个性化营销,提高成交率;根据医药产品销售数据和库存数据来优化医药商品,提高销售率,降低过期损耗,优化商品组合,充分挖掘数据的价值,提高企业的综合竞争力。因此,通过建立医药电商大数据分析平台采集医药电商平台数据、分析电商平台数据、可视化电商平台数据很有必要。

12.2 功能需求

为了让读者了解该医药电商大数据分析平台,下面介绍该医药电商大数据分析平台的功能需求,后面章节将针对部分功能设计、开发进行详细介绍。

(1) 流量分析。按照每日、月度、年度分析用户的行为数据,如浏览量、访客数、访问次数、平均访问深度等。

(2) 经营状况分析。按照月度或年度对销售状况进行统计,统计指标包括下单金额、下单客户数、下单量、下单商品件数、客单价。

(3) 大数据可视化系统。所有的分析结果最终通过大数据可视化系统进行展示,整个大数据分析系统建立在 Hadoop 之上,用户可以直接通过可视化的界面查询分布式数据库 HBase 中的数据,并进行展示。

12.3 软件关键技术

医药电商大数据分析平台的关键技术有以下几种。
(1) Hadoop 作为分布式计算平台。
(2) HBase 作为分布式数据存储数据库。
(3) Bootstrap 作为页面搭建框架。
(4) jQuery 进行后台交互操作。
(5) EChart 实现数据可视化。

12.4 效果展示

前端系统设计完成后,系统的效果展示如图 12-1 和图 12-2 所示。

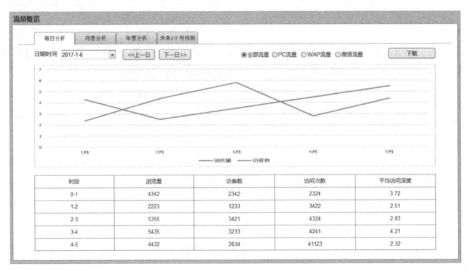

图 12-1 流量概览每日分析结果

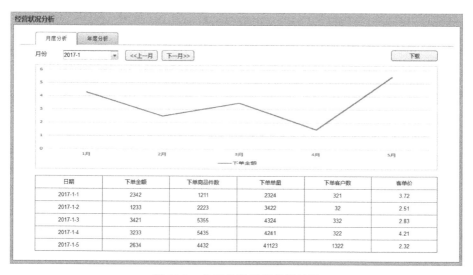

图 12-2　经营状况月度分析结果

12.5　系统构架设计

12.5.1　系统组成

医药大数据分析平台方案主要分为 3 个部分：大数据采集子系统、大数据统计分析子系统和大数据报表呈现子系统，如表 12-1 所示。

表 12-1　系统组成

子　系　统	系　统　定　义	交　互　接　口
大数据采集子系统	系统以离线批处理方式，推送采集结果数据给大数据分析平台	(1) 采集大数据接收的格式 (2) 大数据接口定义
大数据统计分析子系统	具有接收采集系统的数据、客户行为分析、不同药品的精准预测算法、药品推荐算法等特色功能。生成分析结果数据	(1) 大数据的存储 (2) 客户行为模型 (3) 流量分析模型 (4) 统计分析模型
大数据报表呈现子系统	采用 Web 的方案，进行大数据分析，结果以报表、图表的方式呈现给医药电子商务商家	以交互接口、调用报表数据的方式获取需要的结果

12.5.2　系统协作方式

医药大数据系统的子系统间的协作方式如图 12-3 所示。

12.5.3　系统网络拓扑

大数据分析系统的网络拓扑图如图 12-4 所示。

医药电商系统以批处理方式，推送采集数据给大数据分析平台，存储到 Hadoop 服务器集群，大数据报表服务器通过交换机和集群相连。

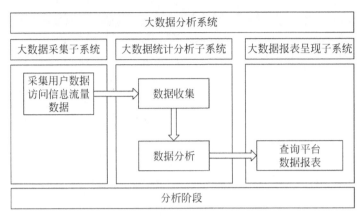

图 12-3　系统协作图

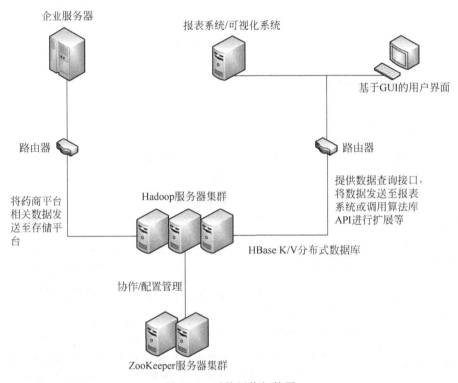

图 12-4　系统网络拓扑图

12.5.4　系统建设方案

1. 数据采集子系统建设方案

1）流量数据建设方案

医药电商平台采用在页脚添加跟踪脚本的方案,能够获取所有访问用户的访问URL、时间、USER AGENT 串,并可根据 cookie 获得用户是否曾经访问过站点,以及记录用户的 ID 来获取是访客还是会员在浏览站点。当用户访问时,PC 平台通过页面跟踪脚本将用户的信息异步发送至大数据平台,数据结构如表12-2所示。

表12-2 用户信息数据结构表

字 段 名	含 义	描 述
sessionId	会话ID	一次连续的访问为一个会话,若用户关闭了浏览器重新打开,则为一个新的会话
userId	登录用户的ID	登录后的用户将具备该信息
trackUid	用户标识	用户标识记录,在机器上永久地记录该标识,只要机器不清理缓存,该标记就永远存在,并在下次访问时发送给服务器。若没有该标识,则表示新用户
userAgent	用户UA	判断用户属于哪个终端,用于区分PC/WAP/微信
referer	要访问的页面地址	当访问商品数据时,给出商品页面的访问URL规则,包含伪静态的和直接访问的地址。若判断请求在这些地址,则表示在访问商品详细页,并从地址获得商品的ID

2)订单数据建设方案

医药电商平台在用户生成订单,以及订单进行支付或货到付款订单确认时将对应的数据发送给大数据分析子系统,并在订单实际支付时通过接口通知大数据分析子系统该订单已完成支付,用于分析成交订单和流量之间的关系。订单数据建设方案包含的数据结构如表12-3和表12-4所示。

表12-3 订单信息表

字 段 名	含 义	描 述
orderId	订单ID	订单的ID编号,数据库序号
userId	下单的用户ID	
orderNum	订单编号	冗余项,用于核对数据
payment	支付金额	订单打折促销后实际收取的支付金额
productTotalAmount	订单商品金额	订单商品的销售金额
isCod	是否货到付款	T为货到付款,否则为先款订单
orderStatus	订单状态	0=刚生成订单;1=用户已支付订单;2=用户已确认订单。款到发货订单推送两次的状态序列是[0,1],货到付款订单推送两次的状态序列是[0,2]

表12-4 订单项信息表

字 段 名	含 义	描 述
orderItemId	订单项ID	订单项的流水序号
orderId	订单ID	
productId	商品ID	在进行URL分析时,提取商品ID后可以和商品关联
productUnitPrice	商品单价	订单打折促销后实际收取的支付金额
num	商品数量	订单商品的销售金额

2. 大数据统计分析子系统建设方案

集群搭建过程请参考前面章节,此处不再赘述,由于本书应用于教学,因此选用3台PC

服务器,其中包括一台 Namenode+ResourseManager、一台 Datanode+SecondNamenode、一台 Datanode,并可以根据需求动态扩展。

(1) Hadoop 选择。由于可靠性需求和容错性需求,本书选择 Hadoop-2.7.3、zookeeper-3.4.9 和 HBase-1.3.1。

(2) JDK。系统自带的 Java 不需要卸载,使用 jdk-8u131-linux-x64.tar.gz 即可。

3. 大数据报表呈现子系统建设方案

报表数据查询接口使用 webservice 进行数据查询。

12.6 数据存储设计

结合医药电商数据的具体特点和上述的设计及优化策略,为了满足用户进行流量分析、销售分析、药品推荐等需求,从而设计流量数据表、订单数据表、会员评价表,具体内容如表 12-5~表 12-7 所示。

1. 流量数据表

表名: tb_data;

行键:由数据来源的平台类型标识、用户访问时间、用户 ID(注册用户的电话号码,具有很好的离散性)后 4 位组合而成。

列族:名称为 cf,使用单列族设计。

列所包含的信息:会话 ID(sessionId)、用户登录 ID(userId)、用户标识(trackUid)、用户 UA(userAgent)、访问的页面地址(referer)等,还可以根据业务的需要动态扩展。

表 12-5 流量数据表

RowKey	列族 cf				
\<platformtype\>\<clicktime\>\<userId\>	sessionId	userId	trackUid	userAgent	referer

2. 订单数据表

表名: tb_order;

行键:由订单 ID 和用户 ID 后 4 位组合而成。

列族:名称为 cf,使用单列族设计。

列所包含的信息:订单 ID(orderId)、下单用户 ID(userId)、订单编号(orderNum)、支付总金额(Payment)、订单商品总金额(totalAmount)、支付方式(isCod)和订单状态(orderStatus)等。

表 12-6 订单数据表

RowKey	列族 cf						
<orderId><userId>	orderId	userId	orderNum	Payment	totalAmount	isCod	orderStatus

3. 订单项数据表

表名：tb_orderItem;

行键：由订单 ID 和用户 ID 后 4 位随机数组成。
列族：名称为 cf，使用单列族设计。
列所包含的信息：商品 ID($productId0$)、商品单价($unitPrice0$)、商品数据($num0$)等。若订单项中包含多个商品，则对应增加相应的列项即可。

表 12-7 订单项数据表

RowKey	列族 cf						
<orderId><userId>	productId0	unitPrice0	num0	productId1	unitPrice1	num1	…

12.7 数据分析

本节描述了利用 Eclipse 开发工具，分析用户流量数据、用户订单数据以及数据存储接口实现的详细过程；基于 Hbase 数据库，对存储流量数据、存储订单数据进行了测试。具体实施步骤如下：

（1）打开 Eclipse，右击 Project Explorer 项目列表空间，选择 New→Dynamic Web Project 选项，这里命名为 test01，如图 12-5 所示。

（2）导入 HBase 全部的 jar 包。

（3）添加本地访问集群所需要的配置文件 core-site.xml、hdfs-site.xml 及 log4j.properties。

图 12-5 项目 test01

① 添加 core-site.xml，配置文件，代码如下。

```
<?xml version = "1.0" encoding = "UTF-8"?>
<?xml-stylesheet type = "text/xsl" href = "configuration.xsl"?>
<!--
  Licensed under the Apache License, Version 2.0 (the "License");
  you may not use this file except in compliance with the License.
  You may obtain a copy of the License at
    http://www.apache.org/licenses/LICENSE-2.0
  Unless required by applicable law or agreed to in writing, software
  distributed under the License is distributed on an "AS IS" BASIS,
  WITHOUT WARRANTIES OR CONDITIONS OF ANY KIND, either express or implied.
  See the License for the specific language governing permissions and
```

```
  limitations under the License. See accompanying LICENSE file.
-->
<!-- Put site-specific property overrides in this file. -->
<configuration>
    <property>
        <name>fs.defaultFS</name>
        <value>hdfs://master:9000</value>
    </property>
    <property>
        <name>hadoop.tmp.dir</name>
        <value>/home/hadoop/hadoopdata</value>
    </property>
</configuration>
```

② 添加 hdfs-site.xml 配置文件,代码如下。

```
<?xml version="1.0" encoding="UTF-8"?>
<?xml-stylesheet type="text/xsl" href="configuration.xsl"?>
<!--
  Licensed under the Apache License, Version 2.0 (the "License");
  you may not use this file except in compliance with the License.
  You may obtain a copy of the License at
    http://www.apache.org/licenses/LICENSE-2.0
  Unless required by applicable law or agreed to in writing, software
  distributed under the License is distributed on an "AS IS" BASIS,
  WITHOUT WARRANTIES OR CONDITIONS OF ANY KIND, either express or implied.
  See the License for the specific language governing permissions and
  limitations under the License. See accompanying LICENSE file.
-->
<!-- Put site-specific property overrides in this file. -->
<configuration>
    <property>
        <name>dfs.replication</name>
        <value>1</value>
    </property>
    <property>
        <name>dfs.permissions</name>
        <value>false</value>
    </property>
</configuration>
```

③ 添加 log4j.properties 配置文件,代码如下。

```
log4j.rootLogger = debug, stdout, R

log4j.appender.stdout = org.apache.log4j.ConsoleAppender
log4j.appender.stdout.layout = org.apache.log4j.PatternLayout

log4j.appender.stdout.layout.ConversionPattern = %5p - %m%n

log4j.appender.R = org.apache.log4j.RollingFileAppender
```

```
log4j.appender.R.File = firestorm.log

log4j.appender.R.MaxFileSize = 100KB
log4j.appender.R.MaxBackupIndex = 1

log4j.appender.R.layout = org.apache.log4j.PatternLayout
log4j.appender.R.layout.ConversionPattern = %p %t %c - %m%n

log4j.logger.com.codefutures = DEBUG
```

（4）根据数据采集子系统建设方案，实现采集数据 WsSysAgentAccessVo 类、WsOrderVo 类及 WsOrderItemVo 类。

① 采集用户流量数据，在 WsSysAgentAccessVo 类中添加如下代码。

```java
package receiveVo;
import java.io.Serializable;
public class WsSysAgentAccessVo implements Serializable {
    private static final long serialVersionUID = 1L;
    private String sessionId;        //会话 ID
    private Integer userId;          //登录用户的 ID
    private String trackUid;         //用户标识
    private String userAgent;        //用户 UA
    private String referer;          //访问的页面地址
    public String getSessionId() {
        return sessionId;
    }
    public void setSessionId(String sessionId) {
        this.sessionId = sessionId;
    }
    public Integer getUserId() {
        return userId;
    }
    public void setUserId(Integer userId) {
        this.userId = userId;
    }
    public String getTrackUid() {
        return trackUid;
    }
    public void setTrackUid(String trackUid) {
        this.trackUid = trackUid;
    }
    public String getUserAgent() {
        return userAgent;
    }
    public void setUserAgent(String userAgent) {
        this.userAgent = userAgent;
    }
```

```java
    public String getReferer() {
        return referer;
    }
    @Override
    public String toString() {
        return "WsSysAgentAccessVo [sessionId = " + sessionId + ", userId = " + userId + ", trackUid = " + trackUid
                + ", userAgent = " + userAgent + ", referer = " + referer + "]";
    }
    public void setReferer(String referer) {
        this.referer = referer;
    }
}
```

② 采集用户订单数据,在 WsOrderVo 类中添加如下代码。

```java
package receiveVo;
import java.io.Serializable;
public class WsOrderVo implements Serializable{
    private static final long serialVersionUID = 1L;
    public WsOrderVo()
    {
    }
    public int getOrderId()
    {
        return orderId;
    }
    public void setOrderId(int orderId)
    {
        this.orderId = orderId;
    }
    public int getUserId()
    {
        return userId;
    }
    public void setUserId(int userId)
    {
        this.userId = userId;
    }
    public String getOrderNum()
    {
        return orderNum;
    }
    public void setOrderNum(String orderNum)
    {
        this.orderNum = orderNum;
    }
    public Double getPayment()
```

```java
{
    return payment;
}
public void setPayment(Double payment)
{
    this.payment = payment;
}
public Double getProductTotalAmount()
{
    return productTotalAmount;
}
public void setProductTotalAmount(Double productTotalAmount)
{
    this.productTotalAmount = productTotalAmount;
}
public String getIsCod()
{
    return isCod;
}
public void setIsCod(String isCod)
{
    this.isCod = isCod;
}
public String getOrderStatus()
{
    return orderStatus;
}
public void setOrderStatus(String orderStatus)
{
    this.orderStatus = orderStatus;
}
public String toString()
{
    return (new StringBuilder()).append("WsOrderVo{orderId = ").append(orderId).append(", userId = ").append(userId).append(", orderNum = '").append(orderNum).append('\'').append(", payment = ").append(payment).append(", productTotalAmount = ").append(productTotalAmount).append(", isCod = '").append(isCod).append('\'').append(", orderStatus = '").append(orderStatus).append('\'').append('}').toString();
}
private int orderId;                        //订单 ID
private int userId;                         //下单的用户 ID
private String orderNum;                    //订单编号
private Double payment;                     //支付总金额
private Double productTotalAmount;          //订单商品总金额
private String isCod;                       //是否货到付款
private String orderStatus;                 //订单状态
}
```

③ 采集用户订单项数据，在 WsOrderItemVo 类中添加如下代码。

```java
package receiveVo;
import java.io.Serializable;
public class WsOrderItemVo implements Serializable{
    private static final long serialVersionUID = 1L;
    public WsOrderItemVo()
    {
    }
    public int getOrderItemId()
    {
        return orderItemId;
    }
    public void setOrderItemId(int orderItemId)
    {
        this.orderItemId = orderItemId;
    }
    public int getOrderId()
    {
        return orderId;
    }
    public void setOrderId(int orderId)
    {
        this.orderId = orderId;
    }
    public int getProductId()
    {
        return productId;
    }
    public void setProductId(int productId)
    {
        this.productId = productId;
    }
    public Double getProductUnitPrice()
    {
        return productUnitPrice;
    }
    public void setProductUnitPrice(Double productUnitPrice)
    {
        this.productUnitPrice = productUnitPrice;
    }
    public int getNum()
    {
        return num;
    }
    public void setNum(int num)
    {
        this.num = num;
    }
    public String toString()
```

```
        {
            return (new StringBuilder()).append("WsOrderItemVo{orderItemId = ").append
(orderItemId).append(", orderId = ").append(orderId).append(", productId = ").append
(productId).append(", productUnitPrice = ").append(productUnitPrice).append(", num = ").
append(num).append('}').toString();
        }
    private int orderItemId;              //订单项 ID
    private int orderId;                  //订单 ID
    private int productId;                //商品 ID
    private Double productUnitPrice;      //商品单价
    private int num;                      //商品数量
}
```

（5）根据功能需求流量分析及经营状况分析的设计，实现需要展示数据的 WsSysDataResultVo 类和 OrderVo 类。

① 流量分析数据，在 WsSysDataResultVo 类中添加如下代码。

```
package hbase.resultVo;
import java.io.Serializable;
public class WsSysDataResultVo implements Serializable{
    private static final long serialVersionUID = 1L;
    //时间维度
    private String reportTime;
    //浏览量
    private int pageView;
    //访客数
    private int userView;
    //访问次数
    private int visitView;
    //平均访问深度
    private float visitDepth;
    public String getReportTime() {
        return reportTime;
    }
    public void setReportTime(String reportTime) {
        this.reportTime = reportTime;
    }
    public int getPageView() {
        return pageView;
    }
    public void setPageView(int pageView) {
        this.pageView = pageView;
    }
    public int getUserView() {
        return userView;
    }
    public void setUserView(int userView) {
```

```
            this.userView = userView;
    }
    public int getVisitView() {
        return visitView;
    }
    public void setVisitView(int visitView) {
        this.visitView = visitView;
    }
    public float getVisitDepth() {
        return visitDepth;
    }
    public void setVisitDepth(float visitDepth) {
        this.visitDepth = visitDepth;
    }
}
```

② 经营状况分析数据，在 OrderVo 类中添加如下代码。

```
package hbase.resultVo;
import java.io.Serializable;
public class OrderVo implements Serializable{
        private static final long serialVersionUID = 1L;
        private String reportTime;          //时间维度
    private Double orderPriceTotal;         //下单金额
    private int orderProductCount;          //下单商品件数
    private int orderCount;                 //下单数量
    private int orderUserCount;             //下单客户数
    private Double orderSalePrice;          //客单价
    public String getReportTime() {
        return reportTime;
    }
    public void setReportTime(String reportTime) {
        this.reportTime = reportTime;
    }
    public Double getOrderPriceTotal() {
        return orderPriceTotal;
    }
    public void setOrderPriceTotal(Double orderPriceTotal) {
        this.orderPriceTotal = orderPriceTotal;
    }
    public int getOrderProductCount() {
        return orderProductCount;
    }
    public void setOrderProductCount(int orderProductCount) {
        this.orderProductCount = orderProductCount;
    }
    public int getOrderCount() {
        return orderCount;
```

```
    }
    public void setOrderCount(int orderCount) {
        this.orderCount = orderCount;
    }
    public int getOrderUserCount() {
        return orderUserCount;
    }
    public void setOrderUserCount(int orderUserCount) {
        this.orderUserCount = orderUserCount;
    }
    public Double getOrderSalePrice() {
        return orderSalePrice;
    }
    public void setOrderSalePrice(Double orderSalePrice) {
        this.orderSalePrice = orderSalePrice;
    }
}
```

（6）实现数据存储接口，创建 HBaseDAO 类和 HBaseDAOImp 类，项目结构如图 12-6 所示。

图 12-6　项目结构

① 创建 HBaseDAO 类，代码如下。

```
package hbase.dao;
import java.util.List;
import receiveVo.WsOrderItemVo;
import receiveVo.WsOrderVo;
import receiveVo.WsSysAgentAccessVo;

public interface HBaseDAO {
    //存储流量数据
    public void saveAgentAccessData(List<WsSysAgentAccessVo> list) ;
    //存储订单数据及订单项数据
    public void saveOrderVo(WsOrderVo orderVo, List<WsOrderItemVo> list);
}
```

② 创建 HBaseDAOImp 类,代码如下。

```java
package hbase.dao.imp;

import hbase.dao.HBaseDAO;
import java.io.IOException;
import java.text.SimpleDateFormat;
import java.util.ArrayList;
import java.util.Date;
import java.util.List;
import org.apache.hadoop.conf.Configuration;
import org.apache.hadoop.hbase.client.HConnection;
import org.apache.hadoop.hbase.client.HConnectionManager;
import org.apache.hadoop.hbase.client.HTableInterface;
import org.apache.hadoop.hbase.client.Put;
import receiveVo.WsOrderItemVo;
import receiveVo.WsOrderVo;
import receiveVo.WsSysAgentAccessVo;
import utils.ObjectAndByte;

public class HBaseDAOImp implements HBaseDAO{
    HConnection hTbalePool = null;
    public HBaseDAOImp()
    {
        Configuration conf = new Configuration();
        String zk_list = "39.108.130.215,39.108.130.65,120.77.181.4";
        conf.set("hbase.zookeeper.quorum",zk_list);
        try {
            hTbalePool = HConnectionManager.createConnection(conf);
        } catch (IOException e) {
            //TODO Auto-generated catch block
            e.printStackTrace();
        }
    }

    @Override
    public void saveAgentAccessData(List<WsSysAgentAccessVo> list) {
        //TODO Auto-generated method stub
        HTableInterface table = null;
        try {
            table = hTbalePool.getTable("tb_data");
            for (WsSysAgentAccessVo vo : list) {
                //行键:用户类型 + 当前时间 + 用户标识
                //用户类型,目前只分析 PC 端
                String userTypeString = "1";    //1 = PC 流量,2 = WAP 流量,3 = 微信流量
                //当前时间
                Date now = new Date();
                SimpleDateFormat dateFormat = new SimpleDateFormat("yyyyMMddHHmmss");
                                                //可以方便地修改日期格式
```

```java
                String timeStr = dateFormat.format( now ) ;
                String rowkey = userTypeString+"-"+timeStr+"-"
                                              +System.currentTimeMillis() ;
                System.out.println("rowkey:"+rowkey) ;
                Put p = new Put(rowkey.getBytes()) ;
                ObjectAndByte objToByte = new ObjectAndByte() ;
                System.out.println(objToByte.toByteArray(vo)) ;
                p.add("cf".getBytes(), "wsSysAgentAccessVo".getBytes(),
                                              objToByte.toByteArray(vo)) ;
                table.put(p) ;
            }
        }catch (Exception e){
            e.printStackTrace() ;
        }finally{
            try {
                table.close() ;
            } catch (IOException e) {
                e.printStackTrace();
            }
        }
    }

    @Override
    public void saveOrderVo(WsOrderVo orderVo, List<WsOrderItemVo> list) {
        //TODO Auto-generated method stub
        HTableInterface table = null ;
        try {
            int orderProductCount = 0;            //订单商品件数
            for (WsOrderItemVo vo : list) {
                orderProductCount = orderProductCount + vo.getNum();
            }
            table = hTbalePool.getTable("tb_order") ;
            //orderStatus 为 1 ,货到付款; 0,在线支付
            //当前时间
            Date now = new Date() ;
            SimpleDateFormat dateFormat = new SimpleDateFormat("yyyyMMddHHmmss") ;
                                          //可以方便地修改日期格式
            String timeStr = dateFormat.format( now ) ;
            String rowkey = timeStr+"-"+System.currentTimeMillis() ;
            System.out.println("rowkey:"+rowkey) ;
            Put p = new Put(rowkey.getBytes()) ;
            p.add("cf".getBytes(), "orderProductCount".getBytes(),
                                  String.valueOf(orderProductCount).getBytes());
            ObjectAndByte objToByte = new ObjectAndByte() ;
            p.add("cf".getBytes(), "wsOrderVo".getBytes(), objToByte.toByteArray(orderVo)) ;
            table.put(p);
        }catch (Exception e){
            e.printStackTrace() ;
```

```java
        }finally{
            try {
                table.close();
            } catch (IOException e) {
                e.printStackTrace();
            }
        }

        try {
            table = hTbalePool.getTable("tb_orderItem");
            int i = list.size();
            for (WsOrderItemVo vo : list) {
                String rowkey = String.valueOf(vo.getOrderId()) + "_" + i;
                System.out.println("rowkey:" + rowkey);
                Put p = new Put(rowkey.getBytes());
                ObjectAndByte objToByte = new ObjectAndByte();
                p.add("cf".getBytes(), "wsOrderItemVo".getBytes(), objToByte.toByteArray(vo));
                i++;
                table.put(p);
            }
        }catch (Exception e){
            e.printStackTrace();
        }finally{
            try {
                table.close();
            } catch (IOException e) {
                e.printStackTrace();
            }
        }
    }

    public static void main(String[] args) {
        HBaseDAO dao = new HBaseDAOImp();
        //测试存储流量数据
        List < WsSysAgentAccessVo > list = new ArrayList < WsSysAgentAccessVo >();
        int i = 20;
        int type = 1;
        while (i > 0) {
            WsSysAgentAccessVo vo = new WsSysAgentAccessVo();
            vo.setSessionId("000000" + i);
            vo.setUserId(1867500 + i);
            vo.setTrackUid("111100" + i);
            vo.setUserAgent(String.valueOf(type));
            vo.setReferer("www.baidu.com" + i);
```

```java
            i--;
            list.add(vo);
        }
        dao.saveAgentAccessData(list);

        //测试存储订单数据
        List<WsOrderVo> list1 = new ArrayList<WsOrderVo>();
        int i = 20;
        while (i > 0) {
            WsOrderVo vo = new WsOrderVo();
            vo.setOrderId(20170829);
            vo.setUserId(1867500);
            vo.setOrderNum("111100" + i);
            vo.setPayment(18.50);
            vo.setProductTotalAmount(128.6);
            vo.setIsCod("1");
            vo.setOrderStatus(String.valueOf(i % 3));
            i--;
            list.add(vo);
        }
        dao.saveOrderData(list);

        //测试存储订单项数据
        List<WsOrderItemVo> list = new ArrayList<WsOrderItemVo>();
        int i = 10;
        while(i > 0){
            WsOrderItemVo vo = new WsOrderItemVo();
            vo.setOrderId(i);
            vo.setNum(10 + i);
            vo.setOrderItemId(100 + i);
            vo.setProductId(100000 + i);
            vo.setProductUnitPrice(99.0);
            i--;
            list.add(vo);
        }
        dao.saveOrderItemData(list);

        System.err.println("save successful !");
    }
}
```

（7）由于医药电商平台数据采集及离线传输所使用的技术与 Hadoop 无关，因此本书没有对其进行介绍，此处以本地生成数据对存储接口进行测试。运行 HBaseDAOImp 类中 main 函数，运行结果如图 12-7 所示。

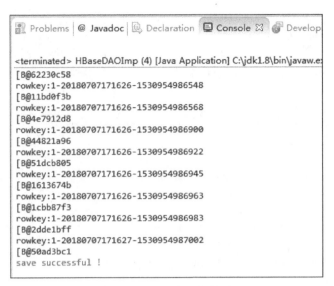

图 12-7　控制台运行结果

用 HBase shell 命令查看数据是否存入数据库,结果显示数据已成功写入,如图 12-8 所示。

图 12-8　数据查询结果

(8) 新建 Servlet。

在 src 文件下,新建一个 controller 包,右击该包,选择 NEW→Other 选项,选中 Web →Servlet 文件,如图 12-9 所示。

然后单击 Next 按钮,在 Class name 输入框中输入"OrderController",如图 12-10 所示。

第12章 医药大数据案例分析

图 12-9 新建 Servlet 1

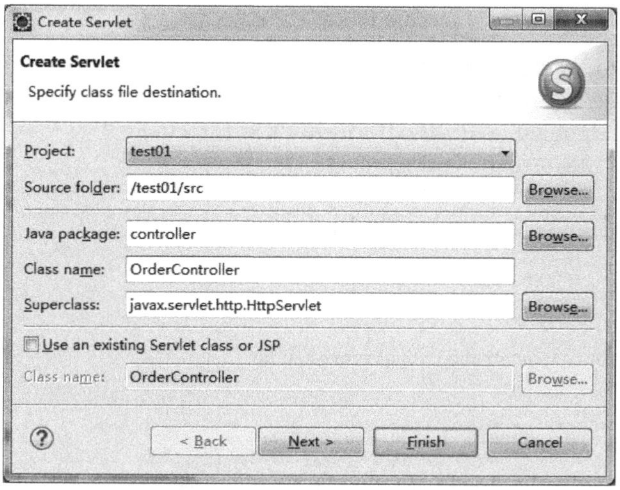

图 12-10 新建 Servlet 2

① 在 OrderController 类中加入以下代码。

```
package controller;

import resultVo.OrderVo;
import java.io.IOException;
import java.text.SimpleDateFormat;
import java.util.ArrayList;
import java.util.Calendar;
```

```java
import java.util.List;
import java.util.Map;
import javax.servlet.ServletException;
import javax.servlet.annotation.WebServlet;
import javax.servlet.http.HttpServlet;
import javax.servlet.http.HttpServletRequest;
import javax.servlet.http.HttpServletResponse;
import net.sf.json.JSONArray;
import org.apache.hadoop.conf.Configuration;
import org.apache.hadoop.hbase.client.HConnection;
import org.apache.hadoop.hbase.client.HConnectionManager;
import org.apache.hadoop.hbase.client.HTableInterface;
import org.apache.hadoop.hbase.client.Result;
import org.apache.hadoop.hbase.client.ResultScanner;
import org.apache.hadoop.hbase.client.Scan;
import org.apache.hadoop.hbase.util.Bytes;
import utils.Jsonp;
import utils.ObjectAndByte;
import receiveVo.WsOrderVo;

/**
 * Servlet implementation class OrderController
 */
@WebServlet("/OrderController")
public class OrderController extends HttpServlet {
    private static final long serialVersionUID = 1L;
    HConnection hTbalePool = null ;
    /**
     * @see HttpServlet#HttpServlet()
     */
    public OrderController() {
        super();
        //TODO Auto-generated constructor stub
    }

    /**
     * @see HttpServlet#doGet(HttpServletRequest request, HttpServletResponse response)
     */
    protected void doGet(HttpServletRequest request, HttpServletResponse response) throws ServletException, IOException {

        //远程访问集群
        Configuration conf = new Configuration();
        String zk_list = "39.108.130.215,39.108.130.65,120.77.181.4" ;
        conf.set("hbase.zookeeper.quorum",zk_list) ;
        try {
            hTbalePool = HConnectionManager.createConnection(conf) ;
        } catch (IOException e) {
```

```java
            //TODO Auto-generated catch block
            e.printStackTrace();
}

//reportType 0=月度分析,1=年度分析
//dataStr 月度分析 格式为 yyyy-MM
//dataStr 年度分析 格式为 yyyy

String reportType = request.getParameter("reportType");
String dateStr = request.getParameter("dateStr");

if(reportType == null || dateStr == null){
    System.out.println("请求参数不能为空");
}

HTableInterface table = null;
ArrayList<OrderVo> list = null;
String startRow = "";
String stopRow = "";

try {
    list = new ArrayList<OrderVo>();
    table = hTbalePool.getTable("tb_order");

    if (reportType.equals("0")) {
        //月度分析
        dateStr = dateStr + "-01";
        SimpleDateFormat sdf = new SimpleDateFormat("yyyy-MM-dd");
        Calendar calendar = Calendar.getInstance();
        calendar.setTime(sdf.parse(dateStr));

        int days = calendar.getActualMaximum(Calendar.DAY_OF_MONTH);
        String startDay;
        String stopDay;
        for (int i=1;i<=days;i++) {
            OrderVo orderVo = new OrderVo();

            orderVo.setReportTime(sdf.format(calendar.getTime()).substring(5))
;  //MM-dd

            int orderCount = 0;
            Double orderPriceTotal = 0.0;
            int orderProductCount = 0;
            Double orderSalePrice = 0.0;
            int orderUserCount = 0;
            List<String> userIdList = new ArrayList<String>();

            startDay = sdf.format(calendar.getTime()).replaceAll("-", "");
```

```java
            System.out.println("start:" + startDay);
            calendar.add(Calendar.DATE, 1);
            stopDay = sdf.format(calendar.getTime()).replaceAll("-", "");
            System.out.println("stop:" + stopDay);

            //20171130 151024-1512025824579
            startRow = startDay + "000000-0000000000000";
            stopRow = stopDay + "000000-0000000000000";
            Scan scan = new Scan();
            scan.setStartRow(startRow.getBytes());
            scan.setStopRow(stopRow.getBytes());
            ResultScanner scanner = table.getScanner(scan);
            byte[] famA = Bytes.toBytes("cf"); //列族

            for (Result result : scanner) {
                System.out.println("scan:" + result + "\n");
                if (result == null) {
                    orderCount = 0;
                    orderPriceTotal = 0.0;
                    orderProductCount = 0;
                    orderSalePrice = 0.0;
                    orderUserCount = 0;
                }else {
                    orderCount++;
                    byte[] bytes = result.getValue(famA, "wsOrderVo".getBytes());
                    ObjectAndByte objToByte = new ObjectAndByte();
                    WsOrderVo vo = (WsOrderVo) objToByte.toObject(bytes);

                    Double payment = vo.getPayment();
                    orderPriceTotal = orderPriceTotal + payment;    //支付金额
                    int oneOrderProductCount = Integer.valueOf(Bytes.toString
(result.getValue(famA, "orderProductCount".getBytes()))).intValue();
                    orderProductCount = orderProductCount + oneOrderProductCount;
                    String userId = Integer.toString(vo.getUserId());
                    if (!userIdList.contains(userId)) {
                        userIdList.add(userId);
                    }
                }
            }

            orderVo.setOrderCount(orderCount);
            orderVo.setOrderPriceTotal(orderPriceTotal);
            orderVo.setOrderProductCount(orderProductCount);
            orderSalePrice = orderCount == 0 ? 0.0 : orderPriceTotal/orderCount;
            orderVo.setOrderSalePrice(orderSalePrice);
            orderUserCount = userIdList.size();
            orderVo.setOrderUserCount(orderUserCount);
            list.add(orderVo);
```

```java
        }
    }else {
        //年度分析
        //dataStr 格式为 yyyy 2017      2017-01 2017-02
        dateStr = dateStr + "-01";
        SimpleDateFormat sdf = new SimpleDateFormat("yyyy-MM");
        Calendar calendar = Calendar.getInstance();
        calendar.setTime(sdf.parse(dateStr));

        String startDay;
        String stopDay;
        calendar.add(Calendar.MONTH, -1);
        for (int i = 1;i <= 12;i++) {
            OrderVo orderVo = new OrderVo();
            int orderCount = 0;
            Double orderPriceTotal = 0.0;
            int orderProductCount = 0;
            Double orderSalePrice = 0.0;
            int orderUserCount = 0;
            List<String> userIdList = new ArrayList<String>();

            startDay = sdf.format(calendar.getTime()).replaceAll("-", "");
            System.out.println("start:" + startDay);
            startRow = startDay + "00000000-0000000000000";

            calendar.add(Calendar.MONTH, 1);
            stopDay = sdf.format(calendar.getTime()).replaceAll("-", "");

            orderVo.setReportTime(sdf.format(calendar.getTime()).substring(0, 7));
            stopRow = stopDay + "00000000-0000000000000";

            Scan scan = new Scan();
            scan.setStartRow(Bytes.toBytes(startRow));
            scan.setStopRow(Bytes.toBytes(stopRow));
            byte[] famA = Bytes.toBytes("cf"); //列族
            ResultScanner scanner = table.getScanner(scan);
            for (Result result : scanner) {
                System.out.println("scan:" + result + "\n");
                if (result == null) {
                    orderCount = 0;
                    orderPriceTotal = 0.0;
                    orderProductCount = 0;
                    orderSalePrice = 0.0;
                    orderUserCount = 0;
                }else {

                    orderCount++;
                    byte[] bytes = result.getValue(famA, "wsOrderVo".getBytes());
                    ObjectAndByte objToByte = new ObjectAndByte();
```

```java
                                WsOrderVo vo   = (WsOrderVo) objToByte.toObject(bytes);

                                Double payment = vo.getPayment();
                                orderPriceTotal = orderPriceTotal + payment;    //支付金额
                                int oneOrderProductCount = Integer.valueOf(Bytes.toString
(result.getValue(famA, "orderProductCount".getBytes()))).intValue();
                                orderProductCount = orderProductCount + oneOrderProductCount;
                                String userId = Integer.toString(vo.getUserId());
                                if (!userIdList.contains(userId)) {
                                    userIdList.add(userId);
                                }
                            }
                        }
                        orderVo.setOrderCount(orderCount);
                        orderVo.setOrderPriceTotal(orderPriceTotal);
                        orderVo.setOrderProductCount(orderProductCount);
                        orderSalePrice = orderCount == 0 ? 0 : orderPriceTotal/orderCount;
                        orderVo.setOrderSalePrice(orderSalePrice);
                        orderUserCount = userIdList.size();
                        orderVo.setOrderUserCount(orderUserCount);
                        list.add(orderVo);
                    }
                }
            }catch (Exception e) {
                e.printStackTrace();
            }finally{
                try {
                    table.close();
                } catch (IOException e) {
                    e.printStackTrace();
                }
            }
            try {
                JSONArray result = JSONArray.fromObject(list);
                response.getWriter().println(new Jsonp(request, result.toString()));
            } catch (Exception e) {
                System.out.println("数组转json失败");
            }
        }
        /**
         * @see HttpServlet#doPost(HttpServletRequest request, HttpServletResponse response)
         */
        protected void doPost(HttpServletRequest request, HttpServletResponse response) throws
ServletException, IOException {
            doGet(request, response);
        }
    }
```

② 使用同样的方法创建 WsSysController 类，并加入以下代码。

```java
package controller;

import resultVo.WsSysDataResultVo;

import java.io.IOException;
import java.text.SimpleDateFormat;
import java.util.ArrayList;
import java.util.Calendar;
import java.util.List;

import javax.servlet.ServletException;
import javax.servlet.annotation.WebServlet;
import javax.servlet.http.HttpServlet;
import javax.servlet.http.HttpServletRequest;
import javax.servlet.http.HttpServletResponse;
import net.sf.json.JSONArray;
import org.apache.hadoop.conf.Configuration;
import org.apache.hadoop.hbase.client.HConnection;
import org.apache.hadoop.hbase.client.HConnectionManager;
import org.apache.hadoop.hbase.client.HTableInterface;
import org.apache.hadoop.hbase.client.Result;
import org.apache.hadoop.hbase.client.ResultScanner;
import org.apache.hadoop.hbase.client.Scan;
import org.apache.hadoop.hbase.util.Bytes;
import utils.Jsonp;
import utils.ObjectAndByte;

import receiveVo.WsSysAgentAccessVo;

/**
 * Servlet implementation class WsSysController
 */
@WebServlet("/WsSysController")
public class WsSysController extends HttpServlet {
    private static final long serialVersionUID = 1L;
    HConnection hTbalePool = null ;

    /**
     * @see HttpServlet#HttpServlet()
     */
    public WsSysController() {
        super();
        //TODO Auto-generated constructor stub
    }

    /**
```

```java
 * @see HttpServlet#doGet(HttpServletRequest request, HttpServletResponse response)
 */
protected void doGet(HttpServletRequest request, HttpServletResponse response) throws ServletException, IOException {

    //远程访问集群
    Configuration conf = new Configuration();
    String zk_list = "39.108.130.215,39.108.130.65,120.77.181.4";
    conf.set("hbase.zookeeper.quorum",zk_list);
    try {
        hTbalePool = HConnectionManager.createConnection(conf);
    } catch (IOException e) {
        //TODO Auto-generated catch block
        e.printStackTrace();
    }


    //sourceType 终端类型   0=全部流量,1=PC流量,2=WAP流量,3=微信流量
    //reportType 0=日分析,1=月度分析,3=年度分析,4=未来3个月预测
    //dateStr    未来3个月分析该参数传空字符串,其他传时间戳
    //日分析              dataStr 格式为 yyyy-MM-dd
    //月分析              dataStr 格式为 yyyy-MM
    //年度分析            dataStr 格式为 yyyy
    String sourceType = request.getParameter("sourceType");
    String reportType = request.getParameter("reportType");
    String dateStr = request.getParameter("dateStr");
    if(sourceType == null || reportType == null || dateStr == null){
        System.out.println("请求参数不能为空");
    }

    HTableInterface table = null;
    ArrayList<WsSysDataResultVo> list = null;
    String tableName = "tb_data";

    try {
        table = hTbalePool.getTable(tableName);
        list = new ArrayList<WsSysDataResultVo>();

        String startRow;
        String stopRow;
        //用户类型
        String userTypeString = "1";   这里只分析PC流量

        String timeStr;
        if (reportType.equals("0")) {
            //日分析          00:00:00 - 01:00:00 - 24:00:00
            //dataStr 格式为 yyyy-MM-dd

            timeStr = dateStr + "-01";
```

```java
SimpleDateFormat sdf = new SimpleDateFormat("yyyy-MM-dd-HH");
Calendar calendar = Calendar.getInstance();
calendar.setTime(sdf.parse(timeStr));

calendar.add(Calendar.HOUR, -1);
for (int i = 1; i <= 24; i++) {
    String startHours;
    String stopHours;
    int pv = 0;
    float visitD = 0.0f;
    List<String> trackList = new ArrayList<String>();
    List<String> sessionList = new ArrayList<String>();
    WsSysDataResultVo sysDataResultVo = new WsSysDataResultVo();
    sysDataResultVo.setReportTime(i-1 + "-" + i);
    System.out.println(i-1 + "-" + i);

    startHours = sdf.format(calendar.getTime()).replaceAll("-", "");
    System.out.println("start:" + startHours);

    calendar.add(Calendar.HOUR, 1);
    stopHours = sdf.format(calendar.getTime()).replaceAll("-", "");
    System.out.println("stop: " + stopHours);

    startRow = userTypeString + "-" + startHours + "0000";
    stopRow = userTypeString + "-" + stopHours + "0000";

    Scan scan = new Scan();
    scan.setStartRow(Bytes.toBytes(startRow));
    scan.setStopRow(Bytes.toBytes(stopRow));
    byte[] famA = Bytes.toBytes("cf"); //列族
    ResultScanner rs = table.getScanner(scan);
    for (Result result : rs) {
        pv++;
        System.out.println(startRow + ":" + stopRow);
        byte[] bytes = result.getValue(famA,"wsSysAgentAccessVo".getBytes());
        ObjectAndByte objToByte = new ObjectAndByte();
        WsSysAgentAccessVo vo = (WsSysAgentAccessVo)objToByte.toObject(bytes);

        String sessionId = vo.getSessionId();
        String trackUid = vo.getTrackUid();
        if (!trackList.contains(trackUid)) {
            trackList.add(trackUid);
        }
        if (!sessionList.contains(sessionId)) {
            sessionList.add(sessionId);
        }
    }
    if (sessionList.size()!= 0) {
```

```java
                    visitD = pv * 1.0f/sessionList.size();
                }
                sysDataResultVo.setVisitDepth(visitD);
                                                //平均访问深度,浏览量/访问次数
                sysDataResultVo.setVisitView(sessionList.size());
                                                //访问次数,统计所有的 sessionId
                sysDaLaResultVo.setUserView(trackList.size());
                                                //访客数,统计所有的 trackUid
                sysDataResultVo.setPageView(pv);    //浏览量
                list.add(sysDataResultVo);
            }
        }else if (reportType.equals("1")) {
            //月分析 2017-05-01 : 2017-06-01
            //dataStr 格式为 yyyy-MM
            timeStr = dateStr + "-01";

            SimpleDateFormat sdf = new SimpleDateFormat("yyyy-MM-dd");
            Calendar calendar = Calendar.getInstance();
            calendar.setTime(sdf.parse(timeStr));

            int days = calendar.getActualMaximum(Calendar.DAY_OF_MONTH);

            for (int i = 1; i <= days; i++) {
                String startDay;
                String stopDay;
                int pv = 0;
                float visitD = 0.0f;
                List<String> trackList = new ArrayList<String>();
                List<String> sessionList = new ArrayList<String>();

                WsSysDataResultVo sysDataResultVo = new WsSysDataResultVo();

                sysDataResultVo.setReportTime( sdf. format ( calendar. getTime ( )).
substring(5));   //MM-dd

                System.out.println(sdf.format(calendar.getTime()).substring(5));

                startDay = sdf.format(calendar.getTime()).replaceAll("-", "");
                System.out.println("start:" + startDay);
                calendar.add(Calendar.DATE, 1);
                stopDay = sdf.format(calendar.getTime()).replaceAll("-", "");
                System.out.println("stop:" + stopDay);

                //2017:05:04:00:00 - 20170505:00:00
                startRow = userTypeString + "-" + startDay + "0000";
                stopRow = userTypeString + "-" + stopDay + "0000";

                Scan scan = new Scan();
                scan.setStartRow(Bytes.toBytes(startRow));
                scan.setStopRow(Bytes.toBytes(stopRow));
```

```java
            byte[] famA = Bytes.toBytes("cf");            //列族
            ResultScanner rs = table.getScanner(scan);
            for (Result result : rs) {
                pv++;
                System.out.println("scan:" + result +"\n");
                byte[] bytes = result.getValue(famA,"wsSysAgentAccessVo".getBytes());
                ObjectAndByte objToByte = new ObjectAndByte();
                WsSysAgentAccessVo vo = (WsSysAgentAccessVo)objToByte.toObject(bytes);

                String sessionId = vo.getSessionId();
                String trackUid = vo.getTrackUid();
                if (!trackList.contains(trackUid)) {
                    trackList.add(trackUid);
                }
                if (!sessionList.contains(sessionId)) {
                    sessionList.add(sessionId);
                }
            }
            if (sessionList.size()!= 0) {
                visitD = pv * 1.0f/sessionList.size();
            }
            sysDataResultVo.setVisitDepth(visitD);
            sysDataResultVo.setVisitView(sessionList.size());
            sysDataResultVo.setUserView(trackList.size());
            sysDataResultVo.setPageView(pv);
            list.add(sysDataResultVo);
    }
}else if (reportType.equals("3")) {
    //年度分析
    //dataStr 格式为 yyyy 2017:01 - 2017:02 - 2018
    timeStr = dateStr + " - 01";
    SimpleDateFormat sdf = new SimpleDateFormat("yyyy-MM");
    Calendar calendar = Calendar.getInstance();
    calendar.setTime(sdf.parse(timeStr));

    calendar.add(Calendar.MONTH, -1);
    for (int i = 1;i <= 12;i++) {
        String startMonth;
        String stopMonth;
        int pv = 0;
        float visitD = 0.0f;
        List<String> trackList = new ArrayList<String>();
        List<String> sessionList = new ArrayList<String>();
        WsSysDataResultVo sysDataResultVo = new WsSysDataResultVo();

        startMonth = sdf.format(calendar.getTime()).replaceAll("-", "");
        System.out.println("start:" + startMonth);
```

```java
                        calendar.add(Calendar.MONTH, 1);
                        stopMonth = sdf.format(calendar.getTime()).replaceAll("-", "");
                        System.out.println("stop:" + stopMonth);

                        sysDataResultVo.setReportTime(sdf.format(calendar.getTime()).
    substring(0, 7));     //yyyy-MM
                        startRow = userTypeString + "-" + startMonth + "000000";
                        stopRow = userTypeString + "-" + stopMonth + "000000";
                        Scan scan = new Scan();
                        scan.setStartRow(Bytes.toBytes(startRow));
                        scan.setStopRow(Bytes.toBytes(stopRow));
                        byte[] famA = Bytes.toBytes("cf");      //列族
                        ResultScanner rs = table.getScanner(scan);
                        for (Result result : rs) {
                            pv++;
                            System.out.println("scan:" + result + "\n");
                            byte[] bytes = result.getValue(famA,"wsSysAgentAccessVo".getBytes());
                            ObjectAndByte objToByte = new ObjectAndByte();
                            WsSysAgentAccessVo vo = (WsSysAgentAccessVo)objToByte.toObject(bytes);

                            String sessionId = vo.getSessionId();
                            String trackUid = vo.getTrackUid();
                            if (!trackList.contains(trackUid)) {
                                trackList.add(trackUid);
                            }
                            if (!sessionList.contains(sessionId)) {
                                sessionList.add(sessionId);
                            }
                        }
                        if (sessionList.size()!= 0) {
                            visitD = pv * 1.0f/sessionList.size();
                        }
                        sysDataResultVo.setVisitDepth(visitD);
                        sysDataResultVo.setVisitView(sessionList.size());
                        sysDataResultVo.setUserView(trackList.size());
                        sysDataResultVo.setPageView(pv);
                        list.add(sysDataResultVo);
                    }
                }
            }catch (Exception e) {
                e.printStackTrace();
            }finally{
                try {
                    table.close();
                } catch (IOException e) {
                    e.printStackTrace();
                }
            }
            try {
                //数据对象格式
```

```
                JSONArray result = JSONArray.fromObject(list);
                response.getWriter().println(new Jsonp(request, result.toString()));
            } catch (Exception e) {
                System.out.println("数组转 json 失败");
            }
        }

        /**
         * @see HttpServlet#doPost(HttpServletRequest request, HttpServletResponse response)
         */
        protected void doPost(HttpServletRequest request, HttpServletResponse response) throws ServletException, IOException {
            doGet(request, response);
        }
    }
```

(9) 启动 Tomcat 服务。

Tomcat 服务启动后,在浏览器输入以下网址。

```
http://localhost:8080/test01/WsSysController?sourceType = 0&reportType = 0&dateStr = 2018
    - 01 - 18
http://localhost:8080/test01/WsSysController?sourceType = 0&reportType = 1&dateStr = 2018
    - 03
http://localhost:8080/test01/WsSysController?sourceType = 0&reportType = 3&dateStr = 2018
http://localhost:8080/test01/OrderController?reportType = 1&dateStr = 2018
http://localhost:8080/test01/OrderController?reportType = 0&dateStr = 2018 - 05
```

12.8 数据展示

本节基于 Bootstrap 页面搭建框架、jQuery 交互操作框架和 EChart 数据可视化组件,详细分析了 Web 大数据展示项目的实现过程,具体步骤如下:

(1) 创建 Web 项目。选择"文件"→"新建"→"Web 项目"选项,在打开的"创建 Web 项目"对话框中选中"默认项目"复选框,输入项目名称,选择项目创建位置,创建项目,如图 12-11 和图 12-12 所示。

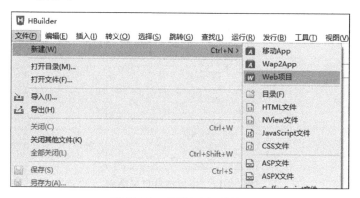

图 12-11 创建 Web 项目 1

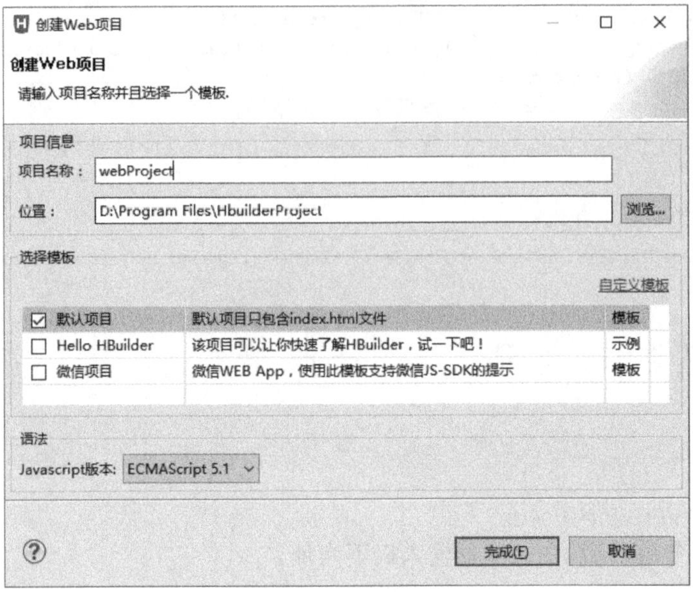

图 12-12　创建 Web 项目 2

右击新创建的项目,在打开的快捷菜单中选择"新建"→"目录"选项,创建项目需要的文件目录,如图 12-13 所示。生成项目目录结构,如图 12-14 所示。

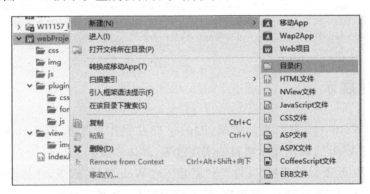

图 12-13　创建 Web 项目目录

图 12-14　生成项目目录结构

plugin 文件夹内存放本项目需要用到的插件、框架等公共的 CSS 文件和 JS 文件等。本项目需要用到 jquery3.0.0.js、bootstrap.js、echarts.js 和 moment.js 文件,结构目录如图 12-15 所示。

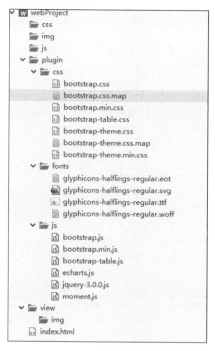

图 12-15 本项目目录结构

(2) 新建 view/flowSurvey.html,加入以下代码。

```
orderVo.setOrderCount(orderCount);
orderVo.setOrderPriceTotal(orderPriceTotal);

<!DOCTYPE html>
<html>

  <head>
    <meta charset = "UTF - 8">
    <title></title>
    <script src = "../plugin/js/jquery - 3.0.0.js" type = "text/javascript" charset = "utf - 8"></script>
    <script src = "../plugin/js/echarts.js" type = "text/javascript" charset = "utf - 8"></script>
    <script src = "../plugin/js/bootstrap.js" type = "text/javascript" charset = "utf - 8"></script>
    <script src = "../plugin/js/bootstrap - table.js"></script>
    <script src = "../plugin/js/moment.js" type = "text/javascript" charset = "utf - 8"></script>
    <script src = "../js/config.js" type = "text/javascript" charset = "utf - 8"></script>
    <script src = "../js/flowSurvey.js" type = "text/javascript" charset = "utf - 8"></script>
    <link rel = "stylesheet" type = "text/css" href = "../plugin/css/bootstrap.css" />
```

```html
        < link rel = "stylesheet" type = "text/css" href = "../plugin/css/bootstrap - table.css">
        < link rel = "stylesheet" type = "text/css" href = "../css/flowSurvey.css" />
    </head>

    < body >
        < div class = "content">
            < h4 class = "container">流量概览</h4 >
            < div class = "container chart">
                <!-- 选项卡按钮 -->
                < ul class = "nav nav - tabs" role = "tablist" id = "myTabs">
                    < li role = "presentation" reportType = "0" class = "active">
                        < a href = "#home" aria - controls = "home" role = "tab" data - toggle = "tab">每日分析</a >
                    </li >
                    < li role = "presentation" reportType = "1">
                        < a href = "#profile" aria - controls = "profile" role = "tab" data - toggle = "tab">月度分析</a >
                    </li >
                    < li role = "presentation" reportType = "3">
                        < a href = "#messages" aria - controls = "messages" role = "tab" data - toggle = "tab">年度分析</a >
                    </li >
                    < li role = "presentation" reportType = "4">
                        < a href = "#settings" aria - controls = "settings" role = "tab" data - toggle = "tab">未来3个月预测</a >
                    </li >
                </ul >
                < form class = "option">
                    < label for = "dateStr">日期时间</label >
                    < input type = "date" name = "dateStr" id = "dateStr" value = "2018 - 02 - 18" />
                    < button id = "preDay" class = "btn btn - default timePre"><<上一日</button >
                    < button id = "nextDay" class = "btn btn - default timeNext">下一日>></button >
                    < button id = "preMonth" class = "btn btn - default timePre"><<上一月</button >
                    < button id = "nextMonth" class = "btn btn - default timeNext">下一月>></button >
                    < button id = "preYear" class = "btn btn - default timePre"><<上一年</button >
                    < button id = "nextYear" class = "btn btn - default timeNext">下一年>></button >
                    < input type = "radio" name = "sourceType" id = "allFlow" value = "0" checked = "checked" />< label for = "allFlow">全部流量</label >
                    < input type = "radio" name = "sourceType" id = "pcFlow" value = "1" />< label for = "pcFlow">PC流量</label >
                    < input type = "radio" name = "sourceType" id = "wapFlow" value = "2" />< label for = "wapFlow">WAP流量</label >
                    < input type = "radio" name = "sourceType" id = "weChatFlow" value = "3" />< label for = "weChatFlow">微信流量</label >
                    < button id = "download" class = "btn btn - default">下载</button >
                </form >
                <!-- 选项卡内容 -->
                < div class = "tab - content">
```

```html
            <!-- 折线图 -->
            <div id = "main">
            </div>
            <!-- 初始化表格 -->
            <table id = "table">
            </table>
        </div>
      </div>
    </div>
  </body>
</html>
```

(3) 新建 css/flowSurvey.css,加入以下代码。

```css
body {
  background-color: rgb(204, 204, 204);
}

.content {
  background-color: rgb(204, 204, 204);
  padding-top: 10px;
}

.chart {
  background-color: rgb(255, 255, 255);
  border: 1px solid #000;
  padding: 40px;
}

.option {
  margin-top: 10px;
  height: 32px;
  border: 1px solid #fff;
}

.option input#dateStr {
  margin-left: 10px;
  height: 24px;
}

.option input#allFlow {
  margin-left: 25%;
}

.option button {
  margin-left: 10px;
  height: 24px;
```

```css
  line-height: 0;
  vertical-align: middle;
}
.option button#download {
  float: right;
  width: 80px;
}

#main {
  height: 300px;
}
```

(4) 新建 css/flowSurvey.js,加入以下代码。

```javascript
$(function() {
  $("#preMonth, #nextMonth").hide();
  $("#preYear, #nextYear").hide();
  var myChart = echarts.init( $("#main").get(0));

  function createChart() {
    //指定图表的配置项和数据
    var option = {
      tooltip: {
        trigger: 'axis'
      },
      legend: {
        data: ['访客数', '浏览量']
      },
      xAxis: {
        type: 'category',
        gridIndex: 0,
        boundaryGap: false,
        data: xAxis,

      },
      yAxis: {
        type: 'value',
        splitNumber: 8,
        axisLabel: {
          formatter: '{value}'
        }
      },
      series: [{
        name: '访客数',
        type: 'line',
        data: userView
      },
      {
```

```
              name: '浏览量',
              type: 'line',
              data: pageView
          }
        ]
    };
    //使用刚指定的配置项和数据显示图表
    myChart.setOption(option);
    //图表自适应 --- https://zhidao.baidu.com/question/2011060705121251708.html
    window.onresize = myChart.resize;
}

$('#table').bootstrapTable({
    columns: [{
        field: 'reportTime',
        title: '时段'
    }, {
        field: 'pageView',
        title: '浏览量'
    }, {
        field: 'userView',
        title: '访客数'
    }, {
        field: 'visitView',
        title: '访问次数'
    }, {
        field: 'visitDepth',
        title: '平均访问深度'
    }],
    pagination: true,              //是否显示分页(*)
    pageNumber: 1,                 //初始化加载第一页,默认第一页
    pageSize: 10,                  //每页的记录行数(*)
    data: data
});

var urlData = {
    sourceType: 0,
    reportType: 0,
    dateStr: getNow()
}

var chart = $("#main").clone();
var table = $("#table").clone();

var data;
var xAxis = [];
var userView = [];
var pageView = [];
```

```javascript
getData();

//单击按钮改变日期
var dateNow = new Date();
$("#dateStr").val(moment(dateNow).format("YYYY-MM-DD"));
$(".timeNext").attr("disabled", "disabled");
$(".timePre").click(function() {
  $(".timeNext").attr("disabled", false);
  var date;
  var newDate;
  if( $(this).attr("id") == "preDay") {
    date = moment( $("#dateStr").val()).format("YYYY-MM-DD");
    newDate = moment(date).subtract(1, "days").format("YYYY-MM-DD");
    urlData.dateStr = newDate;
  } else if( $(this).attr("id") == "preMonth") {
    date = moment( $("#dateStr").val()).format("YYYY-MM-DD");
    newDate = moment(date).subtract(1, "months").format("YYYY-MM-DD");
    urlData.dateStr = moment(newDate).format("YYYY-MM");
  } else if( $(this).attr("id") == "preYear") {
    date = moment( $("#dateStr").val()).format("YYYY-MM-DD");
    newDate = moment(date).subtract(1, "years").format("YYYY-MM-DD");
    urlData.dateStr = moment(newDate).format("YYYY");
  }
  if(moment(newDate).isSame(getMinData()) || moment(newDate).isBefore(getMinData())) {
    $(".timePre").attr("disabled", "disabled");
  }
  $("#dateStr").val(newDate);
  getData();
  return false;
});
$(".timeNext").click(function() {
  $(".timePre").attr("disabled", false);
  var date;
  var newDate;
  if( $(this).attr("id") == "nextDay") {
    date = moment( $("#dateStr").val()).format("YYYY-MM-DD");
    newDate = moment(date).add(1, "days").format("YYYY-MM-DD");
    urlData.dateStr = newDate;
  } else if( $(this).attr("id") == "nextMonth") {
    date = moment( $("#dateStr").val()).format("YYYY-MM-DD");
    newDate = moment(date).add(1, "months").format("YYYY-MM-DD");
    urlData.dateStr = moment(newDate).format("YYYY-MM");
  } else if( $(this).attr("id") == "nextYear") {
    date = moment( $("#dateStr").val()).format("YYYY-MM-DD");
    newDate = moment(date).add(1, "years").format("YYYY-MM-DD");
    urlData.dateStr = moment(newDate).format("YYYY");
  }
  if(moment(newDate).isSame(getNow()) || moment(newDate).isAfter(getNow())) {
```

```javascript
        $(".timeNext").attr("disabled", "disabled");
    }
    $("#dateStr").val(newDate);
    getData();
    return false;
});
$("#dateStr").change(function() {
    var now = moment($("#dateStr").val()).format("YYYY-MM-DD");
    if(moment(now).isSame(getNow()) || moment(now).isAfter(getNow())) {
        $(".timeNext").attr("disabled", "disabled");
    } else if(moment(now).isSame(getMinData()) || moment(now).isBefore(getMinData())) {
        $(".timePre").attr("disabled", "disabled");
    } else {
        $(".timeNext").attr("disabled", false);
        $(".timePre").attr("disabled", false);
    }
})

//流量分类按钮
$(".option input[type='radio']").click(function() {
    urlData.sourceType = parseInt($(this).val());
    console.log(urlData);
    getData();
});

//每日分析,月度分析的 Tab 栏切换
$("#myTabs li").click(function() {
    urlData.reportType = parseInt($(this).attr("reportType"));
    if($(this).attr("reportType") == 0) {
        $("#preDay, #nextDay").show();
        $("#preMonth, #nextMonth").hide();
        $("#preYear, #nextYear").hide();
        $("#dateStr").show();
        urlData.dateStr = moment($("#dateStr").val()).format("YYYY-MM-DD");
    } else if($(this).attr("reportType") == 1) {
        $("#preDay, #nextDay").hide();
        $("#preMonth, #nextMonth").show();
        $("#preYear, #nextYear").hide();
        $("#dateStr").show();
        urlData.dateStr = moment($("#dateStr").val()).format("YYYY-MM");
    } else if($(this).attr("reportType") == 3) {
        $("#preDay, #nextDay").hide();
        $("#preMonth, #nextMonth").hide();
        $("#preYear, #nextYear").show();
        $("#dateStr").show();
        urlData.dateStr = moment($("#dateStr").val()).format("YYYY");
    } else if($(this).attr("reportType") == 4) {
        $("#preDay, #nextDay").hide();
```

```javascript
            $("#preMonth,#nextMonth").hide();
            $("#preYear,#nextYear").hide();
            $("#dateStr").hide();
        }
        console.log(urlData);
        getData();
    });

    //获取数据的封装
    function getData() {
        //          $(".tab-content").empty();
        //          $(".tab-content").append(chart,table);
        xAxis = [];
        userView = [];
        pageView = [];
        $.get("http://39.108.130.215:8080/lecshop/WsSysController", urlData, function(res) {
            if(res.length > 0) {
                data = res;
                console.log(data);
                for(var i = 0; i < data.length; i++) {
                    xAxis.push(i + 1);
                    userView.push(data[i].userView);
                    pageView.push(data[i].pageView);
                }
                createChart();
                //                  createTable();
                $('#table').bootstrapTable("load", data);
            }
        }, "jsonp");
    }
})
```

(5) 新建 view/operatingConditions.html，加入以下代码。

```html
<!DOCTYPE html>
<html>

<head>
    <meta charset="UTF-8">
    <title></title>
    <script src="../plugin/js/jquery-3.0.0.js" type="text/javascript" charset="utf-8"></script>
    <script src="../plugin/js/echarts.js" type="text/javascript" charset="utf-8"></script>
    <script src="../plugin/js/bootstrap.js" type="text/javascript" charset="utf-8"></script>
    <script src="../plugin/js/bootstrap-table.js"></script>
```

```html
<script src="../plugin/js/moment.js" type="text/javascript" charset="utf-8"></script>
<script src="../js/config.js" type="text/javascript" charset="utf-8"></script>
<script src="../js/operatingConditions.js" type="text/javascript" charset="utf-8"></script>
<link rel="stylesheet" type="text/css" href="../plugin/css/bootstrap.css" />
<link rel="stylesheet" type="text/css" href="../plugin/css/bootstrap-table.css">
<link rel="stylesheet" type="text/css" href="../css/operatingConditions.css" />
</head>

<body>
    <div class="content">
        <h4 class="container">经营状况分析</h4>

        <div class="container chart">

            <!-- 选项卡按钮 -->
            <ul class="nav nav-tabs" role="tablist" id="myTabs">
                <li role="presentation" reportType="0" class="active">
                    <a href="#home" aria-controls="home" role="tab" data-toggle="tab">月度分析</a>
                </li>
                <li role="presentation" reportType="1">
                    <a href="#profile" aria-controls="profile" role="tab" data-toggle="tab">年度分析</a>
                </li>
            </ul>
            <form class="option">
                日期时间
                <input type="date" name="dateStr" id="dateStr" value="2018-01-18" />
                <button id="preMonth" class="btn btn-default timePre"><<上一月</button>
                <button id="nextMonth" class="btn btn-default timeNext">下一月>></button>
                <button id="preYear" class="btn btn-default timePre"><<上一年</button>
                <button id="nextYear" class="btn btn-default timeNext">下一年>></button>
                <button id="download">下载</button>
            </form>
            <!-- 选项卡内容 -->
            <div class="tab-content">
                <!-- 折线图 -->
                <div id="main">
                </div>
                <!-- 表格 -->
                <table id="table">
                </table>
            </div>

        </div>
    </div>
```

```
    </body>
</html>
```

(6) 新建 css/operatingConditions.css,加入以下代码。

```css
body {
  background-color: rgb(204, 204, 204);
}

.content {
  background-color: rgb(204, 204, 204);
  padding-top: 10px;
}

.chart {
  background-color: rgb(255, 255, 255);
  border: 1px solid #000;
  padding: 40px;
}

.option {
  margin-top: 10px;
  height: 32px;
  border: 1px solid #fff;
}

.option input#dateStr {
  margin-left: 10px;
  height: 24px;
}

.option button {
  margin-left: 10px;
  height: 24px;
  line-height: 0;
  vertical-align: middle;
}

.option button#download {
  float: right;
  width: 80px;
}

#main {
  height: 300px;
}
```

(7) 新建 js/operatingConditions.js，加入以下代码。

```javascript
$(function() {
  $("#preYear,#nextYear").hide();
  var myChart = echarts.init($("#main").get(0));

  function createChart() {
    //指定图表的配置项和数据
    var option = {
      tooltip: {
        trigger: 'axis'
      },
      legend: {
        data: ['下单全部']
      },
      xAxis: {
        type: 'category',
        gridIndex: 0,
        boundaryGap: false,
        data: xAxis
      },
      yAxis: {
        type: 'value',
        splitNumber: 8,
        axisLabel: {
          formatter: '{value}'
        }
      },
      series: [{
        name: '下单全部',
        type: 'line',
        data: orderCount
      }]
    };

    //使用刚指定的配置项和数据显示图表
    myChart.setOption(option);
    window.onresize = myChart.resize;
  }

  $('#table').bootstrapTable({
    columns: [{
      field: 'reportTime',
      title: '时间维度'
    }, {
      field: 'orderPriceTotal',
      title: '下单金额'
    }, {
```

```
        field: 'orderProductCount',
        title: '下单商品件数'
    }, {
        field: 'orderCount',
        title: '下单量'
    }, {
        field: 'orderUserCount',
        title: '下单客户数'
    }, {
        field: 'orderSalePrice',
        title: '客单价'
    }],
    pagination: true,                    //是否显示分页(*)
    pageNumber: 1,                       //初始化加载第一页,默认第一页
    pageSize: 10,                        //每页的记录行数(*)
    data: data
});

var urlData = {
    reportType: 0,
    dateStr: getNow()
}
var data;
var xAxis = [];
var orderCount = [];

var chart = $("#main").clone();
var table = $("#table").clone();
getData();

//单击按钮改变日期
var dateNow = new Date();
$("#dateStr").val(moment(dateNow).format("YYYY-MM-DD"));
$(".timeNext").attr("disabled", "disabled");
$(".timePre").click(function() {
    $(".timeNext").attr("disabled", false);
    var date;
    var newDate;
    if( $(this).attr("id") == "preMonth") {
        date = moment( $("#dateStr").val()).format("YYYY-MM-DD");
        newDate = moment(date).subtract(1, "months").format("YYYY-MM-DD");
        urlData.dateStr = moment(newDate).format("YYYY-MM");
    } else if( $(this).attr("id") == "preYear") {
        date = moment( $("#dateStr").val()).format("YYYY-MM-DD");
        newDate = moment(date).subtract(1, "years").format("YYYY-MM-DD");
        urlData.dateStr = moment(newDate).format("YYYY");
    }
    if(moment(newDate).isSame(getMinData()) || moment(newDate).isBefore(getMinData())) {
```

```javascript
        $(".timePre").attr("disabled", "disabled");
      }
      $("#dateStr").val(newDate);
      getData();
      return false;
    });
    $(".timeNext").click(function() {
      $(".timePre").attr("disabled", false);
      var date;
      var newDate;
      if($(this).attr("id") == "nextMonth") {
        date = moment($("#dateStr").val()).format("YYYY-MM-DD");
        newDate = moment(date).add(1, "months").format("YYYY-MM-DD");
        urlData.dateStr = moment(newDate).format("YYYY-MM");
      } else if($(this).attr("id") == "nextYear") {
        date = moment($("#dateStr").val()).format("YYYY-MM-DD");
        newDate = moment(date).add(1, "years").format("YYYY-MM-DD");
        urlData.dateStr = moment(newDate).format("YYYY");
      }
      if(moment(newDate).isSame(getNow()) || moment(newDate).isAfter(getNow())) {
        $(".timeNext").attr("disabled", "disabled");
      }
      $("#dateStr").val(newDate);
      getData();
      return false;
    });
    $("#dateStr").change(function() {
      var now = moment($("#dateStr").val()).format("YYYY-MM-DD");
      if(moment(now).isSame(getNow()) || moment(now).isAfter(getNow())) {
        $(".timeNext").attr("disabled", "disabled");
      } else if(moment(now).isSame(getMinData()) || moment(now).isBefore(getMinData())) {
        $(".timePre").attr("disabled", "disabled");
      } else {
        $(".timeNext").attr("disabled", false);
        $(".timePre").attr("disabled", false);
      }
    })

    $("#myTabs li").click(function() {
      urlData.reportType = parseInt($(this).attr("reportType"));
      if($(this).attr("reportType") == 0) {
        $("#preMonth, #nextMonth").show();
        $("#preYear, #nextYear").hide();
        urlData.dateStr = moment($("#dateStr").val()).format("YYYY-MM");
      } else if($(this).attr("reportType") == 1) {
        $("#preMonth, #nextMonth").hide();
        $("#preYear, #nextYear").show();
        urlData.dateStr = moment($("#dateStr").val()).format("YYYY");
```

```javascript
      }
      getData();
    });

    function getData() {
      xAxis = [];
      orderCount = [];
      console.log(urlData);
      $.get("http://39.108.130.215:8080/lecshop/OrderController", urlData, function(res) {
        if(res.length > 0) {
          data = res;
          console.log(data);
          for(var i = 0; i < data.length; i++) {
            xAxis.push(i + 1);
            orderCount.push(data[i].orderCount);
          }
          createChart();
          $('#table').bootstrapTable("load", data);
        }
      }, "jsonp");
    }
})
```

(8) 新建 js/config.js，加入以下代码。

```javascript
//获取最小日期
function getMinData() {
  var data = new Date("2014 - 01 - 01");
  return moment(data).format("YYYY - MM - DD");
}
//获取当前日期(最大日期)
function getNow() {
  var data = new Date();
  return moment(data).format("YYYY - MM - DD");
}
```

小结

本章通过医药大数据案例介绍如何开发一个基于 Hadoop 实现的数据分析及可视化平台，讲解大数据实时分析平台的开发原理及相关的技术知识。本章涉及的关键技术包括 Hadoop 分布式计算平台、HBase 分布式存储数据库、Bootstrap 页面搭建框架、jQuery 交互操作框架和 EChart 数据可视化组件。案例首先介绍了大数据平台的功能需求，然后介绍了系统的组成和协作方式，接着介绍了流量子系统、订单子系统、分析子系统及数据的存储方式，最后给出了系统的程序实现。

习题

1. 简述大数据分析平台的原理及功能。
2. 简述大数据分析平台的功能。
3. 简述大数据分析平台的系统组成。
4. 简述大数据分析平台涉及的关键技术。
5. 简述大数据分析平台的存储方式。

参考文献

[1] 任磊,杜一,马帅.大数据可视分析综述[J].软件学报,2014,25(9):1909-1936.
[2] 高志鹏,牛琨,刘杰.面向大数据的分析技术[J].北京邮电大学学报,2015,38(3):1-12.
[3] 李国杰.从复杂性角度看大数据面临的挑战[N].中国信息化周报,2016-04-25(22).
[4] 迈尔-舍恩伯格,库克耶.大数据时代[M].盛杨燕,周涛,译.杭州:浙江人民出版社,2013.
[5] 2011年5月全球知名咨询机构麦肯锡全球研究所发布[OL]. https://www.mckinsey.com/business-functions/mckinsey-digital/our-insights/big-data-the-next-frontier-for-innovation.
[6] Lipcon T. Design Patterns for Distributed Non-relational Databases[R]. 2009.
[7] Ghemawat S, Gobioff H, Leung S T. The Google File System[J]. 19th ACM Symposium on Operating Systems Principles, 2003. 37(5): 29-43.
[8] Dean J, Ghemawat S. MapReduce: Simplified Data Processing on Large Clusters[J]. Sixth Symposium on Operating System Design and Implementation, 2004. 12.
[9] Zhou M. Taobao Distributed Data Processing Practice[R]. Hadoop in China 2010, 2010.
[10] 张浩然.农业大数据综述[J].计算机科学,2014,41(11):387-392.
[11] 胡叠泉,邢启顺.大数据背景下的旅游产业发展研究综述[J].贵州师范大学学报,2016,32(10):42-46.
[12] 张良均.Hadoop大数据分析与挖掘实战[M].北京:机械工业出版社,2015.
[13] George L. HBase权威指南[M].代志远,刘佳,蒋杰,译.北京:人民邮电出版社,2013.
[14] 李俊杰,石慧,谢志明,等.云计算和大数据技术实战[M].北京:人民邮电出版社,2015.
[15] 王晓华.Spark MLlib机器学习实践[M].2版.北京:清华大学出版社,2013.
[16] 王万良.人工智能及其应用[M].北京:高等教育出版社,2016.
[17] Edward Capriolo,Dean Wampler,Jason Rutberglen,著.Hive编程指南[M].曹坤,译.北京:人民邮电出版社,2013.
[18] 王鹏,李俊杰,谢志明,等.云计算和大数据技术:概念、应用与实战[M].北京:人民邮电出版社,2016.
[19] 李俊杰,谢志明,石慧.大数据技术与应用基础项目教程[M].北京:人民邮电出版社,2017.
[20] 倪超.从Paxos到ZooKeeper:分布式一致性原理与实践[M].北京:电子工业出版社,2015.

图书资源支持

感谢您一直以来对清华版图书的支持和爱护。为了配合本书的使用,本书提供配套的资源,有需求的读者请扫描下方的"书圈"微信公众号二维码,在图书专区下载,也可以拨打电话或发送电子邮件咨询。

如果您在使用本书的过程中遇到了什么问题,或者有相关图书出版计划,也请您发邮件告诉我们,以便我们更好地为您服务。

我们的联系方式:

地　　址: 北京市海淀区双清路学研大厦A座701

邮　　编: 100084

电　　话: 010-83470236　010-83470237

资源下载: http://www.tup.com.cn

客服邮箱: 2301891038@qq.com

QQ: 2301891038(请写明您的单位和姓名)

书圈

扫一扫,获取最新目录

课程直播

用微信扫一扫右边的二维码,即可关注清华大学出版社公众号"书圈"。